Thermal Spray Technology for Advanced Composite Materials

Thermal Spray Technology for Advanced Composite Materials presents advancements in multifunctional, sustainable, and smart materials towards enhancing performance and improving processes across a wide range of engineering fields, including renewable energy, automotive, and aerospace.

From nano-composite feedstock materials to CNT reinforcement and high entropy alloy composites, the book provides an authoritative examination of cutting-edge developments that address the limitations of traditional single-component materials. It guides readers through the complete lifecycle of composite coatings, from material selection and fabrication techniques to performance optimization and parameter control using machine learning approaches.

This book will interest researchers and graduate students studying composite materials processing and fabrication.

Advances in Surface Engineering and Tribology

Series Editors: Hitesh Vasudev, Jashanpreet Singh, and Jeyaprakash Natarajan

Advances in Surface Engineering and Tribology explores the interdisciplinary aspects of surface science and engineering, as well as tribology discussions pertaining to the examination of friction, wear, and lubrication. It covers topics ranging from surface modification techniques, nanomaterials and their applications in nanotechnology, catalytic and photocatalytic processes, synthesis of materials, colloids and surfaces, thin film deposition processes, methods for characterizing surfaces, and surface engineering for diverse industrial purposes. The volumes comprehensively explore the complex mechanisms that regulate surface phenomena, presenting both established principles and emerging developments in the field of surface science and engineering. The series offers a forum for professionals, engineers, and researchers in the energy, automotive, aerospace, and biomedical sectors to delve into innovative developments in these interrelated disciplines.

Should you have an idea for this series, please get in contact with Kyra Lindholm (kyra.lindholm@taylorandfrancis.com).

Fabrication of Wear and Corrosion Resistant Materials
Surface Processing Techniques
Edited by Gurbhej Singh and Hitesh Vasudev

Metal Oxides
Advances in Sustainability and Nanotechnology
Edited by Virat Khanna, Rohit Jasrotia, Tejeet Singh, and Gulab Singh

Multifunctional and Smart Materials
Advances in Sustainable Engineering Applications
Edited by Hitesh Vasudev and Gurbhej Singh

Microwave Processing of Metallic Materials
Revolutionizing Surface Engineering for Green Manufacturing
Edited by Hitesh Vasudev, Gurbhej Singh, and Amit Bansal

Slurry Erosion
Flow Phenomenon, Complexities, and Protection Methods
Edited by Jashanpreet Singh and Satish Kumar

Artificial Intelligence in Tribology
Edited by Jashanpreet Singh and Hitesh Vasudev

Thermal Spray Technology for Advanced Composite Materials
Edited by Hitesh Vasudev, Gaurav Prashar, and Lalit Thakur

Thermal Spray Technology for Advanced Composite Materials

Edited by

Hitesh Vasudev, Gaurav Prashar, and Lalit Thakur

CRC Press
Taylor & Francis Group
Boca Raton London New York

CRC Press is an imprint of the
Taylor & Francis Group, an informa business

Designed cover image: Shutterstock

First edition published 2027
by CRC Press
2385 NW Executive Center Drive, Suite 320, Boca Raton FL 33431

and by CRC Press
4 Park Square, Milton Park, Abingdon, Oxon, OX14 4RN

CRC Press is an imprint of Taylor & Francis Group, LLC

ISBN: 978-1-041-22098-5 (hbk)
ISBN: 978-1-041-22099-2 (pbk)
ISBN: 978-1-003-73063-7 (ebk)

DOI: 10.1201/9781003730637

Typeset in Times
by Newgen Publishing UK

Contents

About the Editors

Hitesh Vasudev is a Professor in the School of Mechanical Engineering and Head of the Department of Research Impact and Outcome, Research and Development Cell at Lovely Professional University, India. He obtained his PhD from Guru Nanak Dev Engineering College, India. His primary research interests are surface engineering, with particular emphasis on thermal spray technologies; development of nanostructured, multimodal, and high-entropy alloy coatings for high-temperature oxidation and corrosion resistance; thermal barrier coatings (TBCs); and microwave processing of materials. Dr. Vasudev has guided numerous undergraduate and postgraduate students, with 13 PhD scholars being successfully awarded degrees under his supervision. He has authored over 250 publications across journals, conferences, and book chapters, co-edited more than 10 books, and holds one granted patent. Dr. Vasudev is a member of the American Society of Mechanical Engineers (ASME).

Gaurav Prashar is a Professor and Head of the Mechanical Engineering Department at the Rayat Bahra Professional University, India with over 16 years of teaching experience. He earned his PhD degree from Lovely Professional University, India. He has published 20 papers, 13 book chapters, and 7 international conference papers. Dr. Prashar is the co-editor of two book publications with CRC Press, including *Thermal Spray Coatings for High-Temperature Conditions* (2025) and *Artificial Intelligence and Machine Learning in Thermal Spray Industry: Practices, Implementation, and Challenges* (2023). He works on the post-processing of thermal spray coatings and has contributed extensively to journal articles on the subject. In addition, Dr. Prashar serves as a reviewer in numerous journals, and he is currently working on the development of nano-structured bimodal alloy coatings for high-temperature oxidation and erosion resistance.

Lalit Thakur is an Assistant Professor in the Department of Mechanical Engineering at the National Institute of Technology, India. He obtained his M.Tech and PhD degrees from the Indian Institute of Technology (IIT), India. For the past 13 years, Dr. Thakur has been continuously exploring new possibilities in the fields of welding engineering and thermal spray technology. His current research areas include thermal spray coatings and tungsten inert-gas (TIG) weld claddings for wear, corrosion, biomedical applications, ferromagnetic lubricants, advanced composite development by powder metallurgy, stir casting, and friction stir processing. Dr. Thakur has authored more than 60 research articles in various journals, conferences, and edited books and serves as a reviewer for several journals. He is the co-editor of four book publications with CRC Press, including *Thermal Spray Coatings for High-Temperature Conditions* (2025), *Thermal Claddings for Engineering Applications* (2024), *Artificial Intelligence and Machine Learning in Thermal Spray Industry: Practices, Implementation, and Challenges* (2023), and *Thermal Spray Coatings* (2021). He has published two patents in the field of thermal spray coatings and friction stir processing.

Contributors

Satyajeet Gaikwad
CitiBank,
United Kingdom

Umesh Ghorpade
Department of Mechanical Engineering,
Dnyanshree Institute of Engineering and Technology,
India

Suresh Gudala
Department of Mechanics and Mechanical Engineering,
Siverian State Industrial University,
Russia

Abdul Wahab Hashmi
School of Mechanical Engineering,
Shandong University of Technology,
China

Shivam Kumar
WA School of Mines,
Curtin University,
Australia

Virinder Kumar
University Centre for Research and Development,
Chandigarh University,
India

Amrinder Mehta
Research and Development Cell,
Lovely Professional University,
India

Jeyaprakash Natarajan
School of Mechanical and Electrical Engineering,
China University of Mining and Technology,
China

Sule Emmanuel Oshioriameh
Industrial Engineering and Management,
American University of Beirut,
Lebanon

Brijesh Prasad
Department of Mechanical Engineering,
Graphic Era (Deemed to be University),
India

Durga Prasad
Department of Mechanical Engineering,
RV Institute of Technology and Management,
India

Gaurav Prashar
University School of Engineering,
Rayat Bahra Professional University,
Punjab, Hoshiarpur,
India

Harminder Singh Saggu
School of Mechanical Engineering,
Lovely Professional University,
India

Jashanpreet Singh
Department of Mechanical Engineering,
Chandigarh University,
India

Rajwinder Singh
Department of Materials Science and Engineering,
Uppsala University,
Sweden

Shubhangi Suryawanshi
Department of Information Technology,
Dr. D.Y. Patil Institute of Technology,
India

Lalit Thakur
Department of Mechanical Engineering,
National Institute of Technology, Kurukshetra,
India

Hitesh Vasudev
Research and Development Cell,
Lovely Professional University,
India

Praveen Kumar Verma
Research and Development Cell,
Lovely Professional University,
India

1 Introduction to Thermal Spray Composite Coatings

Gaurav Prashar, Hitesh Vasudev, and Lalit Thakur

1.1 INTRODUCTION

Among the various coating deposition methods, thermal spray processes are extensively utilized to apply thick coatings aimed at improving the surface characteristics of components composed of diverse material classes [1, 2]. Thermal spray techniques encompass a series of coating deposition processes that utilize energy generated from a heat source to melt thermal spray feedstock, which may be powder feedstock, rod, or in the form of wire. The melted or semi-melted particles then accelerate toward a prepared substrate at designated velocity (m/s), resulting in the formation of a coating build-up [3]. The range of materials suitable for thermal spraying is extensive, rendering the thermal spray coating industry highly effective in delivering engineering and design ideas aimed at protecting components and improving their performance.

Modern advancements in surface engineering are recognized for their significant popularity with researchers. This interest stems from the numerous benefits associated with the optimization of parameters with the help of artificial intelligence during thermal spraying to obtain the desired microstructures and finally the coating's enhanced resistance to the environmental factors in which it is to be used [4, 5]. The fabrication of composite coatings exhibiting distinctive mechanical, chemical, and physical properties is an emerging subject in this field. Materials that exhibit the ability to maintain strength and stability under conditions of extreme temperatures, substantial mechanical loads, and harsh environments are of particular interest [6]. The enhancement of the performance characteristics of materials can be achieved through the application of coatings, which are produced via thermal spraying technologies. High velocity oxy-fuel (HVOF) spraying, High velocity air-fuel (HVAF), plasma spraying (PS), and cold spray (CS) technology are some of the most widely utilized methods in the industry [7, 8]. The methods described above facilitate the creation of high-density, composite coatings characterized by strong adhesion, thereby providing effective protection for surfaces against harsh conditions. This technology is sought after across multiple sectors, such as aviation, energy, and metallurgy, due to its substantial capability to enhance the service life of crucial components [9, 10]. The next section discus the different thermal spray approaches used to fabricate composite coatings.

DOI: 10.1201/9781003730637-1

1.2 COMPOSITE COATINGS AN OVERVIEW

Various combinations have been suggested in composite coatings, as summarized in Table 1.1. Composite coatings can be considered advantageous for enhancing the tribological properties of materials, thereby facilitating extensive applications across significant areas [11–14]. Figure 1.1 illustrates the different thermal spray coating strategies employed in the deposition of composites. Table 1.1 summarizes the various categories of composite coatings developed through thermal spray technologies, classified based on their structural characteristics or functional applications.

1.3 RELIABLE COATING METHODS FOR DEPOSITION OF COMPOSITES

HVOF is extensively utilized for applications requiring hardness, wear resistance, adhesion, and density. Wear-resistant coatings are produced by mixing oxygen with a fuel gas, which is then burned inside a combustion chamber. This process generates high-velocity flames that melt and propel the powder materials onto the substrate to be coated [15]. Figure 1.2 illustrates the diagrammatical representation of the operational concept behind the HVOF spray operation. The primary effectiveness of HVOF developed coatings exceeds that of other thermal spraying approaches, making them suitable and appropriate for applications involving coatings such as cemented carbides. The interaction of high-speed particles with the substrate surface results in a high density of uniform coatings, creating a series of condensed

TABLE 1.1
Various categories of composite coatings developed through thermal spray technologies

Type	Primary matrix	Reinforcement added	Important property
Metal matrix composite (MMC)	Metal	Ceramic (e.g. YSZ, WC)	Corrosion and wear resistance
Ceramic matrix composite (CMC)	Ceramic for (e.g. YSZ)	Ceramic (e.g. Al_2O_3)	Chemical or thermal resistance
Cermet composites	Ceramic + metal		Microhardness and fracture toughness
Polymer matrix composite (PMC)	Polymer	Metal or ceramic	Corrosion and lightweight
Functionally graded coatings (FGC)	Gradient		Resistance to thermal shock (e.g. in thermal barrier coatings)
Self-lubricating composites	Metal/ ceramic	Solid lubricants like graphite	Low friction in bearings and aerospace actuators
Nanocomposite coatings	Any	Nanoparticles	Improved mechanical and thermal properties
Oxide composite	Ceramic	Ceramic oxides	Thermal and electrical properties
Hybrid composites	Mixed	Mixed	Multi-functionality

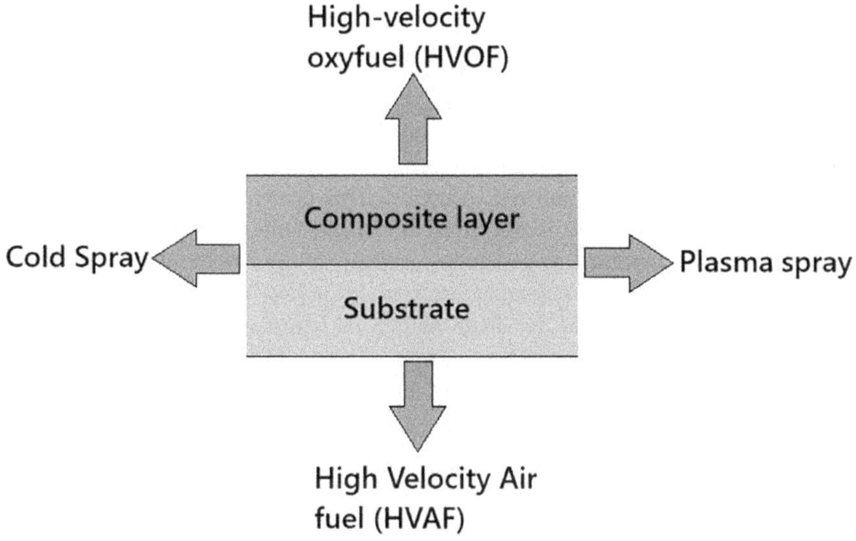

FIGURE 1.1 Composite coatings by thermal spray techniques.

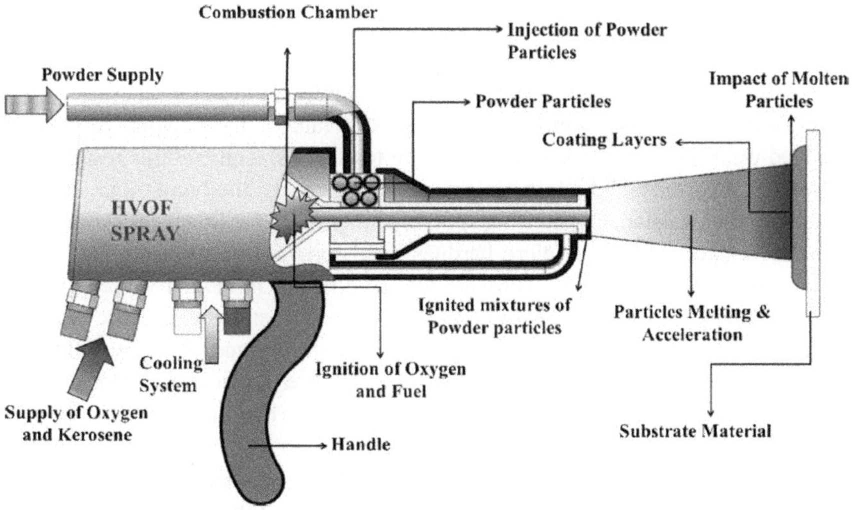

FIGURE 1.2 HVOF working principle [18].

splats [16]. The primary advantages of HVOF compared to other thermal spraying processes include increased particle velocity and a relatively lower development of thermal energy during operation. The elevated velocities decrease the fuelling duration for the particle feedstock, resulting in denser coating buildup and minimizing oxidation [17, 18].

Current research in surface engineering emphasizes the creation of coatings that resist wear and corrosion, aimed at improving material performance in demanding environments such as those found in the aviation, energy, and nuclear sectors. Metal–ceramic and composite coatings, including Cr_3C_2-NiCr, receive special attention due to their ability to offer protection against oxidation, and wear under high-temperature conditions. In comparison to PS, HVOF technology results in coatings characterized by reduced porosity and minimized oxidation. This is attributed to the elevated jet velocity and decreased flame temperature associated with the HVOF process [19]. HVOF spraying is favoured due to its comparatively lower cost in relation to vacuum plasma spraying. Cr_3C_2-NiCr coatings derived from HVOF have been extensively researched and implemented for their ability to withstand oxidation, corrosion, and abrasion at elevated temperatures, reaching up to 850°C [20–22]. Cr_3C_2-NiCr coatings have demonstrated effectiveness in minimizing fretting damage to parts within elevated-temperature gas-cooled reactors in the domain of corrosion-resistant coatings [23–25].

At present, HVOF spraying and, in recent years, HVAF spraying are the predominant methods employed for the application of Cr_3C_2-NiCr coatings [26]. The differences noticed between Cr_3C_2-NiCr coatings produced by the HVOF and HVAF approaches arise from deviations in cooling rates and process temperatures, which have a significant effect on the phase composition of the developed coatings and final coating microstructures. Elevated spraying temperatures facilitate the partial decarburization of Cr_3C_2, resulting in the growth of Cr_7C_3 and $Cr_{23}C_6$, which affects mechanical properties. The fast-cooling process in HVOF leads to the development of amorphous phases, which enhances hardness while diminishing ductility. In contrast, HVAF maintains the initial crystalline structure of the powder, resulting in improved wear resistance. Extended exposure to elevated temperatures results in the coarsening of carbides, which diminishes the plasticity of the binder and adversely impacts erosion resistance. Adjusting spraying parameters facilitates the optimization of fracture toughness values, micro-hardness, and corrosion resistance.

Polymer-based composites (PMCs) are significant in numerous engineering applications owing to their extremely low density, superior specific strength, corrosion resistance, low manufacturing costs, superior fatigue, and ease of forming [27]. Consequently, these components find extensive application across various domains such as cars, airplanes, spacecraft, ships, and turbines [28]. The aviation and automotive sectors are transitioning to a higher utilization of plastics and composite components, motivated by the growing need for lightweight assemblies and enhanced fuel efficiency [29]. Nonetheless, their application is constrained by factors including poor electrical and thermal conductivity, vulnerability to erosion, and restricted operating temperature zones [30]. The lightweight and high-strength characteristics of PMCs facilitate their application in the aerospace and automobile sectors; however, their constraints in electrical and thermal conductivity require the implementation of advanced metallization techniques. The CS technique has emerged as a viable solution for fixing these challenges. The working principle of the process is shown in Figure 1.3. This process is also known as supersonic deposition process [31–33].

CS is an advanced surface engineering approach that facilitates the spraying of metallic coatings onto polymer substrates at relatively low processing temperature

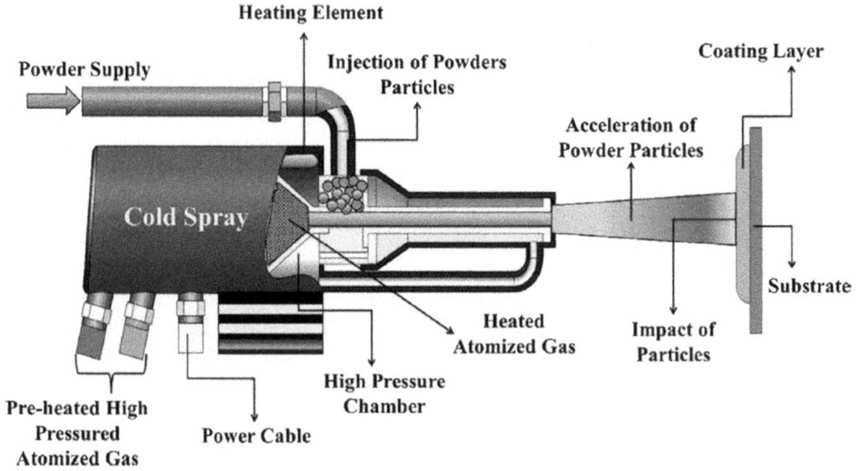

FIGURE 1.3 CS working principle [18].

ranges. The process of cold spraying metallic feedstock powders onto polymer substrates frequently leads to intricate results, primarily due to the erosion of the base material [34]. In recent years, scientists/researchers have made notable advancements in the metallization of polymeric substrates through CS techniques, utilizing a variety of metals including tin, iron, 316L stainless steel, AlSi10Mg, copper, aluminium, and titanium. Frequently identified problems include substrate damage, relatively low deposition efficiencies, and layer delamination. Therefore, metallic feedstock materials must possess ductility, as the presence of brittle materials can result in suboptimal coating formation and insufficient bonding. For the initial application onto the PMC substrate, it is essential to achieve a specific surface roughness to facilitate mechanical interlocking. To address these issues, it is necessary to optimize process parameters and feedstock materials further.

Research in the CS metallization of polymers has predominantly concentrated on the direct application of metal particles onto fibre-reinforced polymer substrates [35]. The study carried by Schwenger et al. [36] examined a metal–polymer composite feedstock, comprising PEEK powder and different ratios of Cu flakes applied to a PEEK substrate. Figure 1.4 illustrates a schematic illustration of the CS process utilizing a mixture of composite powder. The Cu-PEEK powder mixtures are made using two methods: manual mixing and cryo-milling. The results indicated that prepared mixtures of Cu-PEEK can be effectively deposited onto PEEK using low-pressure CS conditions, achieving a significantly higher deposition efficiency. Among the evaluated powder mixture compositions, a volume fraction of 30% Cu resulted in the highest deposition efficiency. The effective CS and application of Cu-PEEK composite powders facilitated the metallization of polymer-rich substrates and structures through the formation of bonding layers within PEEK matrix composites. This finding indicates the potential for the production of stronger, thicker, and more adherent metallic coatings [36].

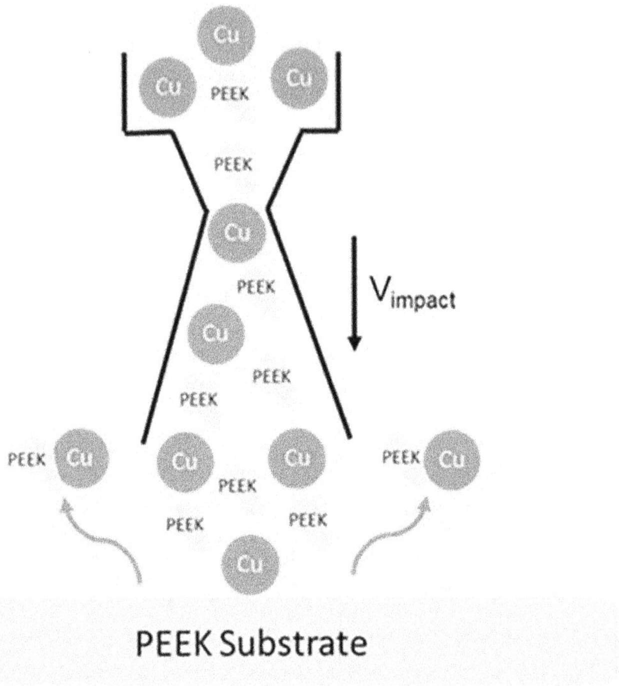

FIGURE 1.4 Schematic illustration of the CS technique utilizing a composite mixture [37].

The selection of the metal–polymer material combination significantly affects the deposition process and the resultant properties of the fabricated coating. Bortolussi et al. [38] found that the combination of Cu and polymer particles results in the formation of composite coatings. This study examined the influence of coating microstructure on the electrical conductivity of coatings. Different types of copper powders, characterized by varying morphologies, sizes, and O_2 contents, were combined with PEEK powder. The CS of these powders produced composite coatings, and an analysis of their microstructures and electrical properties was conducted based on the characteristics of the powders and the parameters of the spraying process. The selection of copper and PEEK powder mixtures, along with optimized spraying parameters, facilitated the production of metal–polymer coatings exhibiting high electrical conductivity. This approach resulted in coatings that are adherent, conductive, and heterogeneous.

WC-based coatings generated via CS technology are being utilized more frequently to improve the reliability and efficiency of diverse industrial components, offering an efficient approach to prolonging the service life of machines exposed to rigorous operational conditions. Alidokht et al. [39] utilized the CS technique for the fabrication of Ni-WC composite structures. The results of reciprocating sliding wear experiments demonstrated that the addition of WC particles to the Ni matrix improved the sliding wear performance. Recent research emphasizes the utilization of CS technology in the restoration of aerospace elements, the development of wear-resistant coatings for automobile parts and surface modification of biomedical implants. The applications illustrate the versatility of CS technology in the area of MMC coatings, as well as its potential to improve performance and prolong the lifespan of industrial goods.

Despite notable advancements in the preparation of MMC coatings, challenges persist regarding the coating's uniformity, adhesion strength, and microstructure control. Current research emphasizes the optimization of coating properties through the manipulation of process parameters, the knowledge of particle behaviour in the CS spray method and how it interacts with the substrate, and establishing a scientific foundation for the fabrication of coatings that meet specified performance criteria. This technology offers several key advantages, such as minimal thermal input, elevated deposition rates, and extensive compatibility with various materials. CS technology encounters specific technical challenges, including inadequate bond strength and difficulties in porosity control. Furthermore, achieving uniform coating with intricate geometries continues to present a technical challenge. Subsequent advancements involve the application of machine learning algorithms to forecast and enhance the spraying process. By utilizing comprehensive experimental data during training, these algorithms are capable of predicting how various spraying parameters affect coating performance, thus facilitating process optimization. Ongoing developments in the field of materials science, technology, and the manufacturing of equipment position CS technology to expand its applications across diverse fields, facilitating the discovery of new composite materials and processes [40].

1.4 HYBRID APPROACHES

Hybrid thermal spray represents a novel approach that utilizes two separate feedstocks within the thermal spray process, resulting in coatings that exhibit enhanced properties. This approach facilitates the integration of multiple feedstocks in diverse configurations for the production of coatings. Independent control of the feedstocks enables the achievement of various coating architectures, including composites [41–44]. Investigation of hybrid PS [45], hybrid HVOF [46], and hybrid HVAF [47] is consequently attracting growing attention [48].

The concept of introducing nano powder particles into the plasma plume to produce nanostructured composite coatings has attracted growing attention. The refined microstructures demonstrate superior performance in contrast to coatings produced from conventional-sized powders across multiple parameters, including wear resistance, erosion resistance, mechanical strength, and thermal diffusivity [49–52]. However, this has necessitated that researchers modify the conventional PS technology appropriately, as the processing of nanoparticles using the conventional PS method, especially when the feedstock is injected perpendicular to the plasma plume direction, poses significant challenges when employing the standard spraying procedure. The challenges are mainly related to the insufficient momentum of fine powder particles, which results in reduced flowability and potential blockage of the injector nozzle [53].

Multiple efforts have been undertaken to tackle the aforementioned challenges through the pre-processing of nano powder particles utilizing agglomeration and sintering methods [53–55]. However, this approach necessitates stringent monitoring and process control throughout the sintering and agglomeration stages, which are time-intensive as well as resource-consuming. Additionally, it may result in the inclusion of non-molten particles within the deposited coating, adversely affecting coating performance [55, 56].

Another approach that is gaining traction involves the use of suspensions, wherein nano powder particles are suspended in H_2O or ethanol before the spraying process. The method of processing suspensions through plasma spraying is referred to as suspension PS. In this process, the suspension is introduced into the plasma plume either in a radial or axial manner. Radial injection has been predominantly employed in PS [49, 57]. However, axial injection is receiving increased focus due to its capability to enhance thermal exchange among the plasma plume and the injected feedstock material [58, 59]. The use of axial injection is particularly advantageous when employing a suspension feedstock, as a considerable amount of heat is required to evaporate the solvent. Enhanced thermal exchange and elevated plasma power levels facilitate the complete melting of fine solute powder particle [60]. Axial suspension PS refers to the method of processing suspension through axial injection utilizing plasma spraying techniques.

The challenges related to the spraying of nano powder particles through PS and recent advancements in axial injection systems for feedstocks have prompted the investigation of a hybrid spraying approach. This method involves the axial injection of both a micron-sized powder and a suspension containing nano powder, aimed at producing various function-specific coating architectures and microstructures (inset Figure 1.5.) [61]. The hybrid processing idea utilizing the axial injection path with suspension and powder was initially introduced by Björklund et al. [61] for the production of various function-specific coating architectures. The idea was further investigated by [62] and [63] to show the coating application of MMC coatings intended for wear applications. The analogous concept was employed by [64] and [65] to illustrate the application of ceramic matrix Al composite coatings for sliding wear applications.

The processing of (CMCs) that include constituents with varying melting characteristics, for instance Al_2O_3 and ZrO_2, presents significant challenges when using conventional deposition methods. These challenges are exacerbated when there is a substantial difference in size by an order of magnitude between the two feedstocks employed in the fabrication of CMCs. The hybrid processing method has shown potential in addressing these challenges, as evidenced in the literature [61, 64–66]. The research carried out by Ganvir et al. [45] focused on illustrating the hybrid processing method utilizing axial injection to apply an Al_2O_3-YSZ CMC coating for tribological issues. The addition of a fine second phase (YSZ) demonstrated an improvement in wear resistance when compared to the traditional PS deposited Al_2O_3 coating. The enhanced wear resistance noticed in tribological experiments for the CMC coating can be ascribed to the microstructural characteristics of the coatings, along with the associated hardness and fracture toughness readings. The primary causes may include the presence of increased porosity and the softer second phase (YSZ) within the CMC coating. Although the CMC coating exhibited increased porosity and retained alpha-alumina phase, its toughness was approximately four times greater than that of the

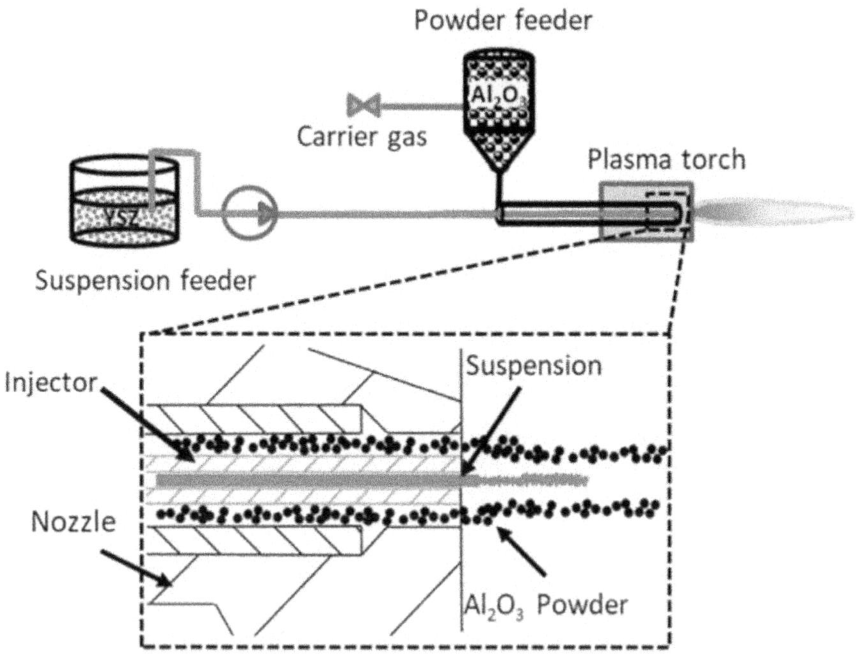

FIGURE 1.5 Schematic representation of the hybrid powder-suspension spraying system employed for the deposition of CMC coatings [45].

monolithic counterpart, attributed to the reinforcement provided by the fine-sized YSZ. The existence of 2-D distinct types of "building blocks" (splats) represents a fundamental and potentially beneficial characteristic of coatings produced from a hybrid feedstock. SEM images at extremely high magnification were taken to analyze the interface between the Al_2O_3 and YSZ splats. The interface among the Al_2O_3 and YSZ splats, as depicted in Figure 1.6, demonstrates a smooth transition, indicating strong cohesion within the CMC coating.

Exploration of additional thermal spray versions has been conducted, highlighting the evident flexibility of the hybrid thermal spray approach. The HVOF method is a thermal spray technique wherein feedstock materials undergo heating and directed toward a substrate by a high-temperature, high-velocity flame generated in a combustion chamber [14]. Suspension HVOF (SHVOF) and solution precursor HVOF (SPHVOF) spray are versions of the HVOF technique that utilize a liquid feedstock [67].

Memon et al. [68] proposed a hybrid feedstock design that incorporates both suspension and solution precursor approaches for the application of micrometric SiC particles into composite coatings via spray techniques. The design's innovation is attributed to the thermal spraying of silicon carbide (SiC) suspended in a yttrium aluminium garnet (YAG) solution precursor. This method effectively limits the decomposition of SiC during the thermal spray process, providing a streamlined coating deposition technique for the production of dense SiC composite coatings.

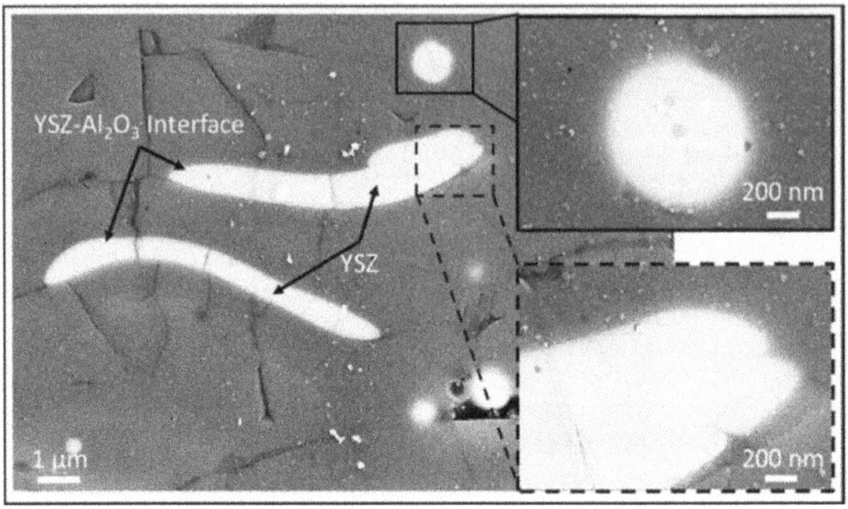

FIGURE 1.6 Scanning electron microscopy micrographs of the 20-weight percent yttria-stabilized zirconia (YSZ) reinforced Al_2O_3 matrix (CMC) coating [45].

The suggested feedstock design for coatings was applied using suspension and solution precursor HVOF (SSP-HVOF) thermal spray techniques. A comprehensive analysis of material composition was performed to elucidate the final microstructure of the fabricated coatings. Additionally, durability assessments, specifically ambient-temperature sliding wear tests, were executed to assess their mechanical performance.

The YAG matrix covering around SiC particles was assessed by collecting splats from a single spray pass. The microstructural study demonstrates that the SiC/YAG feedstock was coated in the form of molten YAG splats (inset Figure 1.7). The generation of single-pass splats suggests that the SiC particles were sufficiently covered throughout the coating deposition process. The formation and coverage of YAG surrounding the SiC particles provided protection against both oxidation and sublimation processes. The as-sprayed coatings did not show any evidence of oxide formation, and the SiC particles were effectively protected throughout the spraying process. The proposed route provides a single-step coating application technique for SiC/YAG coatings utilizing the thermal spray technique. Additionally, the present investigation focuses on micro-sized particles, characterized by an average diameter (d50) of 3 µm. These particles previously presented challenges in the spraying process due to their micrometric dimensions, which result in an increased specific surface area and reduced in-flight momentum. Figure 1.8 illustrates the mechanism involved in the hybrid SSP-HVOF. Consequently, the properties obtained through a hybrid approach have been proposed and demonstrated to produce innovative coating architectures.

Arc spray is a thermal spray method that utilizes an electric arc for melting metal wires. The resulting molten material is then propelled onto surfaces using a high-velocity gas stream. This approach is efficient for big areas and is commonly

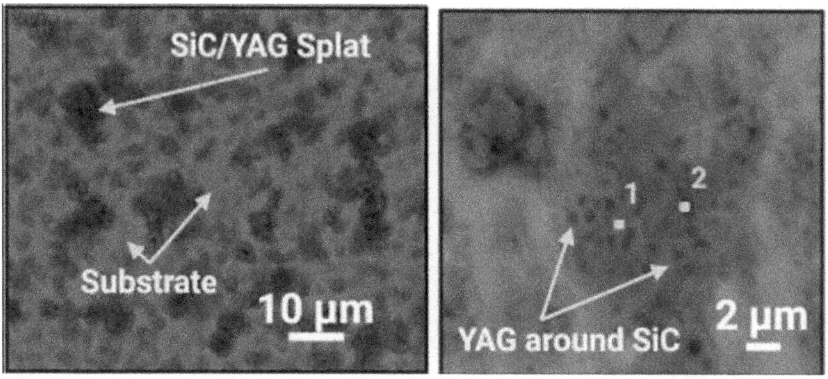

FIGURE 1.7 The microstructural SEM micrographs of single-pass splats formation [68].

implemented to safeguard against corrosive wear in industrial environments [69]. The application of cored wires, which contain ceramic or hard particles, serves as a technique for depositing composite coatings via the arc spray technique. However,

cored wires may incur high costs, and the melting of the wires can lead to the melting of hard particles, potentially resulting in undesirable phase transformation [70].

The hybrid arc spray (HAS) approach is an innovative process that involves simultaneous application of wire and powder, representing an additional method for achieving composite coatings. The research done by Wilden et al. [70] utilized the HAS technique for manufacturing composite coatings. The research concentrated on the incorporation of corundum and steel particles within the Zn and Al matrix. Tillmann et al. [71] investigated the application of the HAS method for the embedding of WC within a low-alloy steel matrix. Figure 1.9 illustrates the schematic representation of the HAS procedure, indicating that the powder can be injected into the arc zone either axially or radially.

The research conducted by Eskandari et al. [48] effectively illustrated the application of the HAS process for the integration of secondary particles, including Cu, Mo, and WC, into Babbitt coatings. The composite coatings exhibited enhanced wear resistance relative to the Babbitt coatings alone during sliding wear tests, positioning them as a viable option for applications that necessitate improved durability.

A recent development in thermal spray technology is HVAF spraying. During this procedure, feedstock materials undergo heating and are propelled toward a surface utilizing a high-velocity flame generated within a combustion chamber. This approach utilizes air, leading to reduced temperatures and increased kinetic energy in comparison to HVOF approach. The capability of HVAF to spray at lower temperatures and higher velocities makes it a viable option for depositing materials susceptible to decarburization and oxidation during the spraying process, including WC-Co, and Cr_3C_2–NiCr. This innovation presents new opportunities for improving coating performance and prolonging the life expectancy of components exposed to extreme service conditions [50].

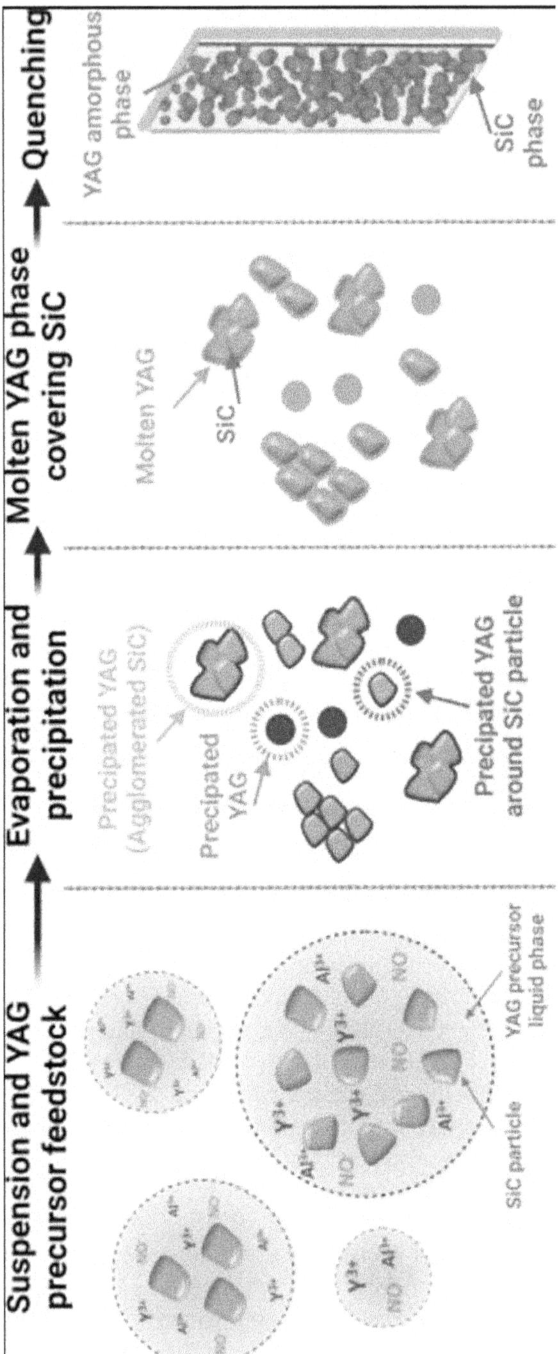

FIGURE 1.8 Mechanism involved in the hybrid solution precursor and suspension approach [68].

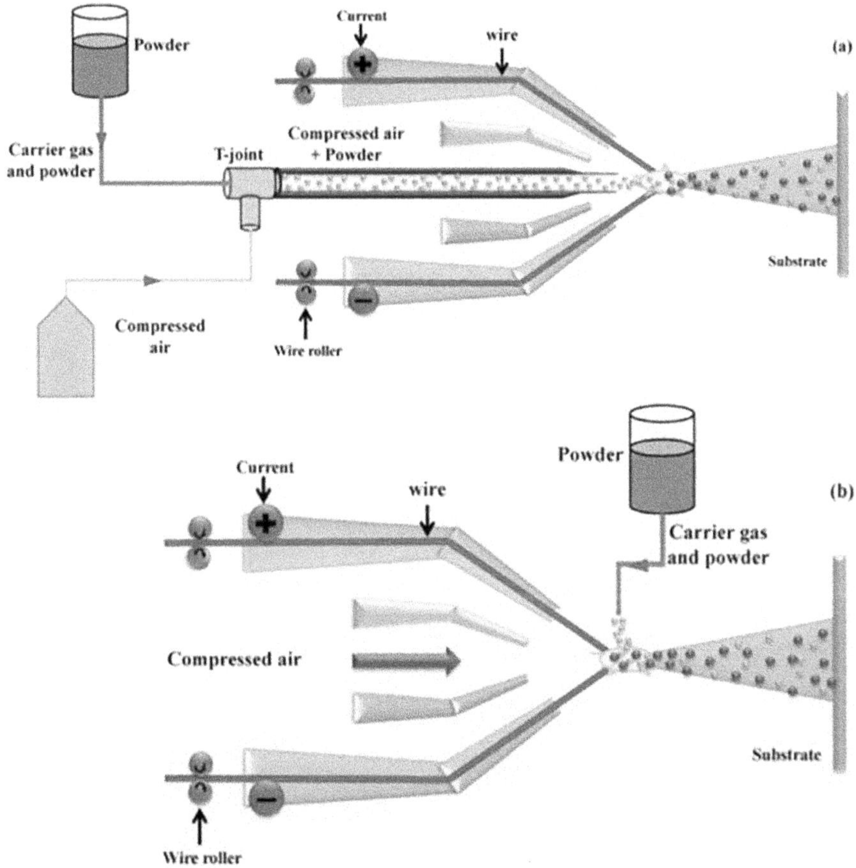

FIGURE 1.9 Schematic of HAS (a) axial feeding and (b) radial feeding [44].

The hybrid HVAF technique is a thermal spray method that integrates the advantages of suspension/solution precursor-based methodologies with those of traditional powder-based processes. Figure 1.10 illustrates the schematic representation of the hybrid HVAF methodology. The powder in suspension/solution can be introduced axially into the gun, contingent upon the feedstocks and application requirements. The study conducted by Mahade et al. [47] presents chromium carbide suspension that has been co-sprayed with superalloy-625 powder using the hybrid HVAF. A high-powered SEM/EDS examination of the dark phase in the cross-section images of hybrid coating revealed the existence of (chromium) and (carbon) as indicated by the elemental maps in Figure 1.11, suggesting their origin from the suspension feedstock. Additionally, the elemental analysis of the dark sections (Spectrum 1, 2) indicated a chromium content of around 88 wt% and a carbon content of 12 wt%. The application temperature of hybrid coatings using HVAF is less than the melting temperature of Cr_3C_2. Additionally, the brittle characteristics of Cr_3C_2 inhibit plastic deformation of the carbide, unlike IN-625. It is anticipated that Cr_3C_2 will maintain its feedstock size

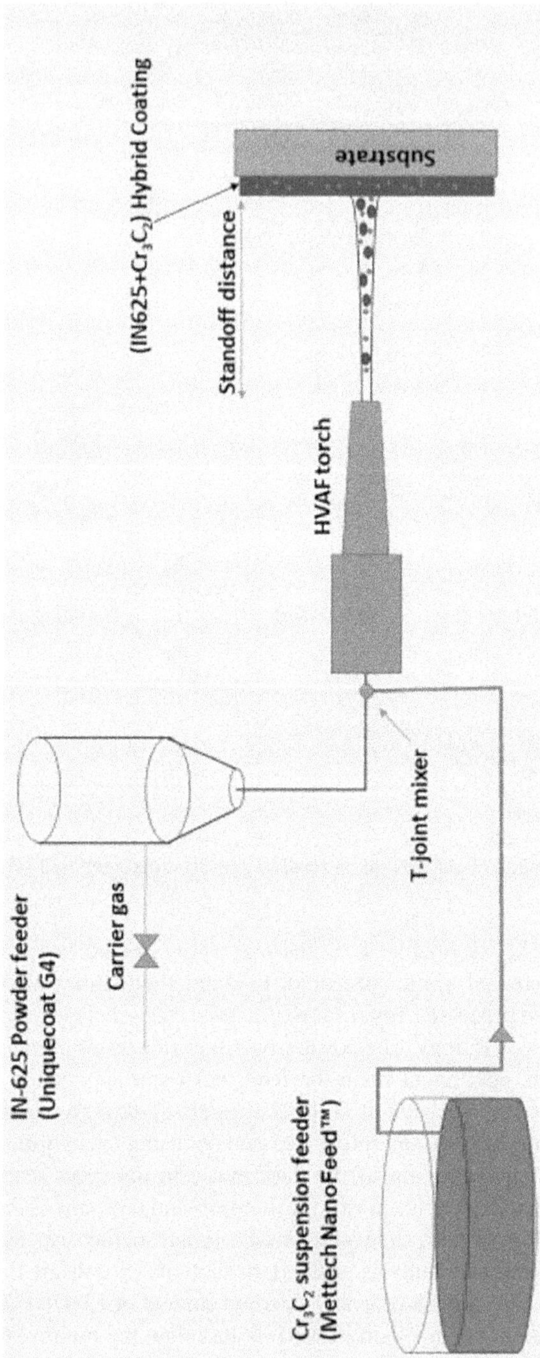

FIGURE 1.10 Depiction of the powder-suspension injection process utilizing HVAF for the deposition of hybrid coating microstructure [47].

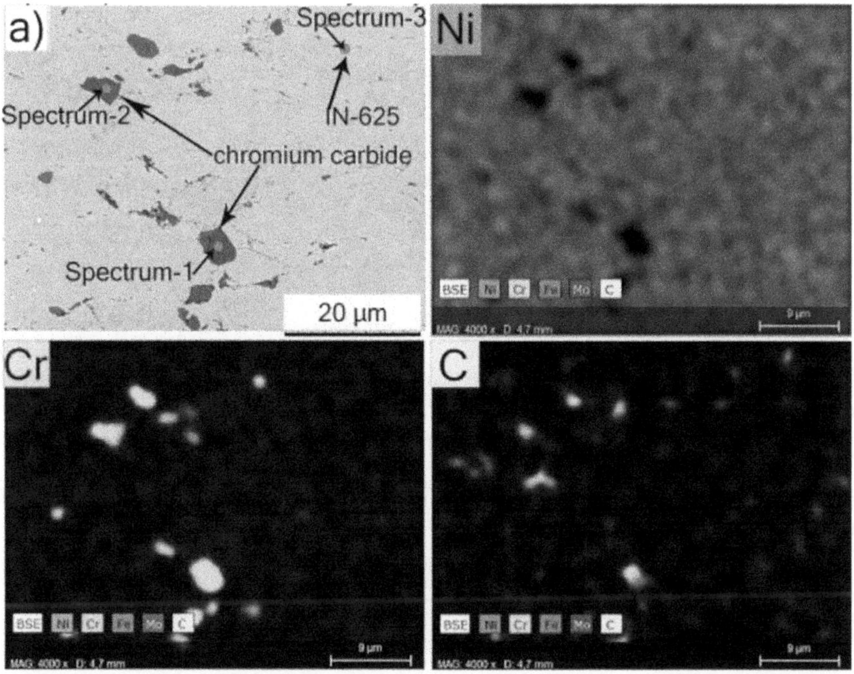

FIGURE 1.11 Analysis of the cross section of HVAF hybrid coating using SEM/EDS, along with the associated elemental maps [47].

of less than 5 µm in the hybrid coatings fabricated via HVAF. The potential mechanism for the development of the hybrid coating microstructure of HVAF-treated IN-625 + Cr_3C_2, is illustrated in Figure 1.12.

Microstructural analysis revealed the presence of carbides distributed throughout the bulk of the coating. However, the engineering challenges associated with addressing the high back pressure in a HVAF system have yet to be resolved.

The combination of hybrid CS and electroless/electroplating methods for metallizing polymeric composite materials, including CFRPs, offers significant advancements in improving their surface properties and effectiveness. More studies and developments are necessary to fully leverage the prospective benefits of these methods and to address the current challenges. The combination of hybrid metallization methods with advanced technologies like AM and smart materials has the potential to provide innovative solutions for intricate engineering challenges. The integration of CS and electroplating into 3D printing procedures enables the development of metal–polymer composite structures with customized attributes for particular purposes. The versatility of layer-by-layer deposition in three-dimensional printing, when combined with the benefits of high deposition rates and low temperatures in CS, facilitates the manufacturing of multi-material components that exhibit graded or site-specific characteristics. Future research should investigate these hybrid processes for the fabrication of components featuring complex geometries and multifunctional

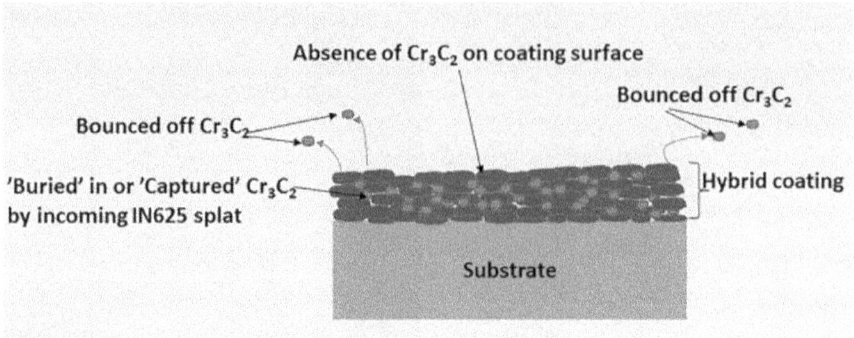

FIGURE 1.12 Schematic representation of the microstructure of the IN-625 + Cr3C2 hybrid coating as deposited through (HVAF) process [47].

attributes, including thermal management, integrated electrical conductivity, and structural reinforcement [31].

1.5 APPLICATIONS OF THERMALLY SPRAYED COMPOSITE COATINGS

Recent advancements in thermal spray technologies enable them to effectively address a diverse array of industrial requirements. Significant advancements have been made in the thermal spray coating system, which now encompasses a wide range of feeder materials in addition to process control. Composite coatings can be employed to enhance the tribological properties of various engineered materials, resulting in cost savings related to service lifespan extension, repair, and replacement.

Alumina ceramics are extensively employed across various technical sectors due to their exceptional durability, high mechanical toughness, and robust corrosion resistance. Mechanical components such as pistons, bearings, cylinders, and valves can enhance their wear and corrosion resistance through the application of a plasma-sprayed (Al_2O_3–Y_2O_3-TiO_2) composite coating. This type of coating exhibits enhanced thermal shock resistance attributed to the incorporation of TiO_2 particles [72].

Transition metal-based ceramic coatings, including TiC, ZrC, Ti_5Si_3, and $MoSi_2$, are extensively utilized as protective coatings in aerospace applications. This widespread use is attributed to their different characteristics, including high hardness, higher melting point, materials strength at high temperatures, thermal stability, and excellent wear and corrosion resistance. An in-situ Cr_7C_3–$CrSi_2$ composite coating was developed, resulting in a finer, denser, and smoother microstructure [73]. WC-based coatings are widely utilized in aerospace and various other industries. The lifespan of the components can be significantly improved due to their high micro-hardness, fracture toughness, and thermal stability.

Recent years have witnessed significant increases in the temperature applications, speeds, and workloads of the different manufacturing equipment used across industries. High-performance components and parts require the application of wear resistant coatings. YSZ thermal barrier coating is applied to Ni-based superalloy sections to protect the underlying engine parts during operation at temperatures approaching their melting point [74, 75].

The research and development of novel biomaterials stands as a top priority among the various international efforts dedicated to biomedical engineering. Biomaterials utilized in orthopaedic surgery are often subjected to demanding service conditions, requiring materials that possess a diverse array of properties. Hydroxyapatite (HA) was developed to address the inadequate biocompatibility of alloys including SS, cobalt-based alloys, titanium, and Ti–6Al–4V. These materials often experience a decline in functionality over time due to factors such as wear, disease, or injury, as well as a propensity to release metallic ions, which increases the likelihood of corrosion in biological environments. The HA/graphene-nanosheet composite coatings generated through vacuum cold spraying offer a comprehensive insight into the potential biological applications of these composites [76]. The Ti-HA composite coatings exhibit high porosity along with robust adhesion and bonding characteristics.

1.6 CONCLUSION

Thermal spray composite coating has become a key technology to help meet the performance, durability, and efficiency requirements throughout the biomedical, gas turbine, aerospace, and automotive industries. The composite coatings provide the desired material versatility, affordability, and flexibility, allowing custom designed solutions to extreme operating environments like high temperatures, severe corrosion, erosion, and wear. Through the addition of ceramics, metals, and polymers to engineered composite systems, composite coatings allow massive improvements in surface properties without damaging the substrate upon which they are applied. Another application of thermal spray coating is in the energy sector where it improves the quality and life of a component that is in a high-stress environment, such as boiler tubes, valves, and pipelines. The advantages of these types of coatings on the gas turbine is that they provide more thermal barrier protection and oxidative resistance which directly translate into a higher fuel efficiency and longer maintenance cycles. High thermal and mechanical loads on critical engine and structural components in aerospace are reduced by the use of thermal spray composites, which ensures flight safety and low operation costs. Similarly, in the motor industry, these surface coatings have been applied to improve engine efficiency, reduce frictional losses, and to increase the service life of the parts to align with the increasing sustainability and fuel economy demands. In general, the thermal spray composite coating is one of the major enabling technologies of advanced engineering. Hybrid thermal spray represents a novel approach that utilizes two separate feedstocks within the thermal spray process, resulting in coatings that exhibit enhanced properties. This approach facilitates the integration of multiple feedstocks in diverse configurations for the production of coatings. Independent control of the feedstocks enables the achievement of various coating architectures, including composites. Investigation on hybrid PS, hybrid HVOF, hybrid HVAF, and hybrid CS is consequently attracting growing attention.

REFERENCES

1. Rezzoug, A., Abdi, S., Kaci, A. and Yandouzi, M., 2018. Thermal spray metallisation of carbon fibre reinforced polymer composites: Effect of top surface modification on coating adhesion and mechanical properties. *Surface and Coatings Technology*, *333*, pp.13–23.
2. Smith, R.W. and Knight, R., 1995. Thermal spraying I: Powder consolidation—From coating to forming. *Jom*, *47*(8), pp.32–39.
3. Davis, J.R. ed., 2004. *Handbook of Thermal Spray Technology*. ASM international.
4. Chong, K., Zou, Y., Wu, D., Tang, Y. and Zhang, Y., 2021. Pulsed laser remelting supersonic plasma sprayed Cr3C2-NiCr coatings for regulating microstructure, hardness and corrosion properties. *Surface and Coatings Technology*, *418*, p.127258.
5. Ping, X.L., Fu, H.G., Sun, S.T., Lin, J., Guo, X.Y. and Lei, Y.P., 2020. Microstructure and performance of Nb-bearing Ni60A-Cr3C2 coatings manufactured by laser cladding. *Surface Engineering*, *36*(12), pp.1294–1306.
6. Sadeghimeresht, E., Markocsan, N. and Nylen, P., 2017. Microstructural characteristics and corrosion behavior of HVAF-and HVOF-sprayed Fe-based coatings. *Surface and Coatings Technology*, *318*, pp.365–373.
7. Zhu, L., Wang, S., Pan, H., Yuan, C. and Chen, X., 2020. Research on remanufacturing strategy for 45 steel gear using H13 steel powder based on laser cladding technology. *Journal of Manufacturing Processes*, *49*, pp.344–354.
8. Mehta, J., Grewal, J.S. and Gupta, P., 2020. Analysis of D-gun sprayed coating on medium carbon steel. *Materials Today: Proceedings*, *21*, pp.1403–1406.
9. Romanchenko, O., Porkuian, O., Kharlamov, Y., Sokolov, V. and Krol, O., 2022. Research of features of oxide coatings deposition by D–gun spraying. In *International Conference Innovation in Engineering* (pp. 305–316). Springer International Publishing.
10. Li, X., Zhai, H., Li, W., Cui, S., Ning, W. and Qiu, X., 2020. Dry sliding wear behaviors of Fe-based amorphous metallic coating synthesized by D-gun spray. *Journal of Non-Crystalline Solids*, *537*, p.120018.
11. Prashar, G. and Vasudev, H., 2022. Structure-property correlation and high-temperature erosion performance of Inconel625-Al2O3 plasma-sprayed bimodal composite coatings. *Surface and Coatings Technology*, *439*, p.128450.
12. Saxena, A., Mehta, A., Vasudev, H., Prashar, G. and Jasim, N.Y., 2023. A short review on machine learning for the purpose of optimizing and predicting the properties of polymeric nanocomposites. *Materials Today: Proceedings*. https://doi.org/10.1016/j.matpr.2023.07.347
13. Prashar, G., Vasudev, H. and Mehta, A., 2024. Application of Bimodal stuructured thermal sprayed coatings in power plant components. In Ram K. Gupta, Amir Motallebzadeh Koc, Saeid Kakooei, and Ajit Behera (Eds.) *Advanced Ceramic Coatings for Energy Applications* (pp. 273–287). Elsevier.
14. Prashar, G. and Vasudev, H., 2020. Nanocomposite coatings: HVOF spray processing, microstructural evolution and performance. In Cherry Bhargava, Amit Sachdeva, Pradeep Kumar Sharma (Eds.) *Smart Nanotechnology with Applications* (pp. 115–130). CRC Press.
15. Asgari, H., Saha, G. and Mohammadi, M., 2017. Tribological behavior of nanostructured high velocity oxy-fuel (HVOF) thermal sprayed WC-17NiCr coatings. *Ceramics International*, *43*(2), pp.2123–2135.
16. Ghadami, F. and Aghdam, A.S.R., 2019. Improvement of high velocity oxy-fuel spray coatings by thermal post-treatments: A critical review. *Thin Solid Films*, *678*, pp.42–52.

17. Yuan, J., Ma, C., Yang, S., Yu, Z. and Li, H., 2016. Improving the wear resistance of HVOF sprayed WC-Co coatings by adding submicron-sized WC particles at the splats' interfaces. *Surface and Coatings Technology*, *285*, pp.17–23.
18. Sathish, M., Radhika, N. and Saleh, B., 2023. Duplex and composite coatings: A thematic review on thermal spray techniques and applications. *Metals and Materials International*, *29*(5), pp.1229–1297.
19. Matthews, S. and Berger, L.M., 2016. Long-term compositional/microstructural development of Cr3C2-NiCr coatings at 500 C, 700 C and 900 C. *International Journal of Refractory Metals and Hard Materials*, *59*, pp.1–18.
20. Kamal, S., Jayaganthan, R. and Prakash, S., 2009. Evaluation of cyclic hot corrosion behaviour of detonation gun sprayed Cr3C2–25% NiCr coatings on nickel-and iron-based superalloys. *Surface and Coatings Technology*, *203*(8), pp.1004–1013.
21. Vats, A., Patnaik, A., Meena, M.L. and Shringi, D., 2021. Role of micro-factors on microstructure and on the tribological performance of HVOF coatings: A review. *IOP Conference Series: Materials Science and Engineering*, *1017*(1), p.012010.
22. James, B., Matthews, S. and Hyland, M., 2005. Microstructural influence on erosion behaviour of thermal-spray coatings. *Microscopy and Microanalysis*, *11*(S02), pp.1710–1711.
23. Lindgren, J.R. and Johnson, W.R., 1987. Friction and wear behavior of chromium carbide coatings. *Surface and Coatings Technology*, *32*(1–4), pp.249–260.
24. Kumar, R., Bhandari, D. and Goyal, K., 2021. Slurry erosion behaviour of HVOF sprayed coatings on hydro turbine steel: A review. *IOP Conference Series: Materials Science and Engineering*, *1033*(1), p.012064.
25. Goyal, D.K., Singh, H., Kumar, H. and Sahni, V., 2012. Slurry erosive wear evaluation of HVOF-spray Cr2O3 coating on some turbine steels. *Journal of Thermal Spray Technology*, *21*(5), pp.838–851.
26. Berger, L.M., 2014. Coatings by thermal spray. In Sarin, V.K., Mari, D., Llanes, L., and Nebel, C., Eds., *Comprehensive Hard Materials* (pp. 471–506). Elsevier.
27. Liu, H., Ren, Y., Zhao, L., Yang, Y., Cui, X., Wang, J. and Xiong, T., 2022. Influence of Al2O3 content on microstructure, electrical conductivity and adhesion strength of cold sprayed Al-Al2O3 coatings on PEEK substrate. *Surface and Coatings Technology*, *446*, p.128752.
28. Monteforte, M., 2017. Polymer-matrix composites, technology map business opportunities in technology commercialization, strategic business insights, explorer. *Explorer*, 33, pp.882–894.
29. Pervaiz, M., Panthapulakkal, S., Sain, M. and Tjong, J., 2016. Emerging trends in automotive lightweighting through novel composite materials. *Materials Sciences and Applications*, *7*(1), p.26.
30. Chen, M., Che, H. and Yue, S., 2022. Exploring surface preparation for cold spraying on polymers. *Surface and Coatings Technology*, *450*, p.128993.
31. Prashar, G. and Vasudev, H., 2021. A comprehensive review on sustainable cold spray additive manufacturing: State of the art, challenges and future challenges. *Journal of Cleaner Production*, *310*, p.127606.
32. Prashar, G. and Vasudev, H., 2023. Deposition and fabrication of biomaterials using cold spray technique: A review on the application of biomedical-implants. *Applied Surface Science Advances*, *18*, p.100533.
33. Prashar, G. and Vasudev, H., 2024. A comprehensive review on the analysis of adhesion strength of cold spray deposits. *Results in Surfaces and Interfaces*, *16*, p.100263.
34. Zhang, D., Shipway, P.H. and McCartney, D.G., 2005. Cold gas dynamic spraying of aluminum: The role of substrate characteristics in deposit formation. *Journal of Thermal Spray Technology*, *14*(1), pp.109–116.

35. Della Gatta, R., Perna, A.S., Viscusi, A., Pasquino, G. and Astarita, A., 2022. Cold spray deposition of metallic coatings on polymers: A review. *Journal of Materials Science*, *57*(1), pp.27–57.
36. Schwenger, M.S., Kaminskyj, M.S., Haas, F.M. and Stanzione III, J.F., 2024. Mixed-material feedstocks for cold spray additive manufacturing of metal–polymer composites. *Journal of Thermal Spray Technology*, *33*(2), pp.619–628.
37. Karahan, B., Ozdemir, I., Grund, T., Hanisch, N. and Lampke, T., 2025. Advancements in cold spraying for polymer matrix composites: Enhanced LSP and EMI shielding performance—Review and future directions. *Journal of Materials Science: Materials in Engineering*, *20*(1), p.13.
38. Bortolussi, V., Figliuzzi, B., Willot, F., Faessel, M. and Jeandin, M., 2020. Electrical conductivity of metal–polymer cold spray composite coatings onto carbon fiber-reinforced polymer. *Journal of Thermal Spray Technology*, *29*(4), pp.642–656.
39. Alidokht, S.A. and Chromik, R.R., 2021. Sliding wear behavior of cold-sprayed Ni-WC composite coatings: Influence OF WC content. *Wear*, *477*, p.203792.
40. Dai, S., Cui, M., Li, J. and Zhang, M., 2024. Cold spray technology and its application in the manufacturing of metal matrix composite materials with carbon-based reinforcements. *Coatings*, *14*(7), p.822.
41. Björklund, S., Goel, S. and Joshi, S., 2018. Function-dependent coating architectures by hybrid powder-suspension plasma spraying: Injector design, processing and concept validation. *Materials & Design*, *142*, pp.56–65.
42. Joshi, S.V., Sivakumar, G., Raghuveer, T. and Dusane, R.O., 2014. Hybrid plasma-sprayed thermal barrier coatings using powder and solution precursor feedstock. *Journal of thermal spray technology*, *23*(4), pp.616–624.
43. Joshi, S.V. and Sivakumar, G., 2015. Hybrid processing with powders and solutions: A novel approach to deposit composite coatings. *Journal of Thermal Spray Technology*, *24*(7), pp.1166–1186.
44. Eskandari, A., Gupta, M. and Joshi, S., 2025. Hybrid thermal spray: A pathway to realize novel coating microstructures and properties. *Journal of Thermal Spray Technology*, *34*, pp.1517–1544.
45. Ganvir, A., Goel, S., Govindarajan, S., Jahagirdar, A.R., Björklund, S., Klement, U. and Joshi, S., 2021. Tribological performance assessment of Al2O3-YSZ composite coatings deposited by hybrid powder-suspension plasma spraying. *Surface and Coatings Technology*, *409*, p.126907.
46. Derelizade, K., Venturi, F., Wellman, R.G., Kholobysov, A. and Hussain, T., 2021. Wear performance of graphene nano platelets incorporated WC-Co coatings deposited by hybrid high velocity oxy fuel thermal spray. *Wear*, *482*, p.203974.
47. Mahade, S., Baiamonte, L., Sadeghimeresht, E., Björklund, S., Marra, F. and Joshi, S., 2021. Novel utilization of powder-suspension hybrid feedstock in HVAF spraying to deposit improved wear and corrosion resistant coatings. *Surface and Coatings Technology*, *412*, p.127015.
48. Eskandari, A., Björklund, S., Gupta, M., Fadaei, A.H., Leiro, A. and Joshi, S., 2025. Microstructure and tribological properties of hybrid feedstock derived arc sprayed composite coatings. *Surface and Coatings Technology*, *515*, p.132598.
49. Zois, D., Lekatou, A., Vardavoulias, M., Panagiotopoulos, I. and Vazdirvanidis, A., 2008. A comparative microstructural investigation of nanostructured and conventional Al_2O_3 coatings deposited by plasma spraying. *Journal of Thermal Spray Technology*, *17*(5), pp.887–894.
50. Sathish, S., Geetha, M., Aruna, S.T., Balaji, N., Rajam, K.S. and Asokamani, R., 2011. Sliding wear behavior of plasma sprayed nanoceramic coatings for biomedical applications. *Wear*, *271*(5–6), pp.934–941.

51. Wang, Y., Darut, G., Poirier, T., Stella, J., Liao, H. and Planche, M.P., 2017. Cavitation erosion of plasma sprayed YSZ coatings produced by feedstocks with different initial sizes. *Tribology International*, *111*, pp.226–233.
52. Wang, Q., Ramachandran, C.S., Smith, G.M. and Sampath, S., 2017. Sliding wear behavior of air plasma sprayed Al_2O_3 coatings sealed with aluminum phosphate. *Tribology International*, *116*, pp.431–439.
53. Singh, V.P., Sil, A. and Jayaganthan, R., 2011. A study on sliding and erosive wear behaviour of atmospheric plasma sprayed conventional and nanostructured alumina coatings. *materials & Design*, *32*(2), pp.584–591.
54. Bandyopadhyay, P.P., Chicot, D., Venkateshwarlu, B., Racherla, V., Decoopman, X. and Lesage, J., 2012. Mechanical properties of conventional and nanostructured plasma sprayed alumina coatings. *Mechanics of Materials*, *53*, pp.61–71.
55. Lima, R.S., Kucuk, A. and Berndt, C.C., 2002. Bimodal distribution of mechanical properties on plasma sprayed nanostructured partially stabilized zirconia. *Materials Science and Engineering: A*, *327*(2), pp.224–232.
56. Lamuta, C., Di Girolamo, G. and Pagnotta, L., 2015. Microstructural, mechanical and tribological properties of nanostructured YSZ coatings produced with different APS process parameters. *Ceramics International*, *41*(7), pp.8904–8914.
57. Song, X., Suhonen, T., Varis, T., Huang, L., Zheng, X. and Zeng, Y., 2014. Fabrication and characterization of amorphous alumina-yttria-stabilized zirconia coatings by air plasma spraying. *Journal of Thermal Spray Technology*, *23*(8), pp.1302–1311.
58. Goel, S., Björklund, S., Curry, N., Wiklund, U. and Joshi, S.V., 2017. Axial suspension plasma spraying of Al_2O_3 coatings for superior tribological properties. *Surface and Coatings Technology*, *315*, pp.80–87.
59. Ganvir, A., Joshi, S., Markocsan, N. and Vassen, R., 2018. Tailoring columnar microstructure of axial suspension plasma sprayed TBCs for superior thermal shock performance. *Materials & Design*, *144*, pp.192–208.
60. Ganvir, A., Calinas, R.F., Markocsan, N., Curry, N. and Joshi, S., 2019. Experimental visualization of microstructure evolution during suspension plasma spraying of thermal barrier coatings. *Journal of the European Ceramic Society*, *39*(2–3), pp.470–481.
61. Björklund, S., Goel, S. and Joshi, S., 2018. Function-dependent coating architectures by hybrid powder-suspension plasma spraying: Injector design, processing and concept validation. *Materials & Design*, *142*, pp.56–65.
62. Mistri, M., Joshi, S., Kar, K.K. and Balani, K., 2020. Tribomechanical insight into carbide-laden hybrid suspension-powder plasma-sprayed Tribaloy T400 composite coatings. *Surface and Coatings Technology*, *396*, p.125957.
63. Mahade, S., Björklund, S., Govindarajan, S., Olsson, M. and Joshi, S., 2020. Novel wear resistant carbide-laden coatings deposited by powder-suspension hybrid plasma spray: Characterization and testing. *Surface and Coatings Technology*, *399*, p.126147.
64. Gopal, V., Goel, S., Manivasagam, G. and Joshi, S., 2019. Performance of hybrid powder-suspension axial plasma sprayed Al_2O_3—YSZ coatings in bovine serum solution. *Materials*, *12*(12), p.1922.
65. Murray, J.W., Leva, A., Joshi, S. and Hussain, T., 2018. Microstructure and wear behaviour of powder and suspension hybrid Al2O3–YSZ coatings. *Ceramics International*, *44*(7), pp.8498–8504.
66. Cipri, F., Marra, F., Pulci, G., Tirillò, J., Bartuli, C. and Valente, T., 2009. Plasma sprayed composite coatings obtained by liquid injection of secondary phases. *Surface and Coatings Technology*, *203*(15), pp.2116–2124.
67. Tejero-Martin, D., Rezvani Rad, M., McDonald, A. and Hussain, T., 2019. Beyond traditional coatings: A review on thermal-sprayed functional and smart coatings. *Journal of Thermal Spray Technology*, *28*(4), pp.598–644.

68. Memon, H., Romero, A.R., Derelizade, K., Venturi, F. and Hussain, T., 2022. A new hybrid suspension and solution precursor thermal spray for wear-resistant silicon carbide composite coatings. *Materials & Design*, *224*, p.111382.
69. Fauchais, P.L., Heberlein, J.V. and Boulos, M.I., 2014. Wire arc spraying. In Pierre L. Fauchais, Joachim V. R. Heberlein, Maher I. Boulos (Eds.) *Thermal Spray Fundamentals: From Powder to Part* (pp. 577–629). Springer US. 184 pages.
70. Wilden, J., Bergmann, J.P., Jahn, S., Reich, S., Knapp, S., Van Rodijnen, F. and Fischer, G., 2007. A new method for the production of particle reinforced coatings. Thermal Spray 2007: Proceedings from the International Thermal Spray Conference, Paper No: itsc2007p0359, pp. 359-364; 6 pages https://doi.org/10.31399/asm.cp.itsc2007p0359.
71. Tillmann, W., Hagen, L. and Kokalj, D., 2017. Embedment of eutectic tungsten carbides in arc sprayed steel coatings. *Surface and Coatings Technology*, *331*, pp.153–162.
72. Kumar, N., Mahade, S., Ganvir, A. and Joshi, S., 2021. Understanding the influence of microstructure on hot corrosion and erosion behavior of suspension plasma sprayed thermal barrier coatings. *Surface and Coatings Technology*, *419*, p.127306.
73. Wang, Q., Sun, Q., Zhang, M.X., Niu, W.J., Tang, C.B., Wang, K.S., Rui, X., Zhai, L. and Wang, L., 2018. The influence of cold and detonation thermal spraying processes on the microstructure and properties of Al-based composite coatings on Mg alloy. *Surface and Coatings Technology*, *352*, pp.627–633.
74. Ma, Y.D., Guo, M.Y., Li, W., Yang, Y., Gao, P.Y., Cui, Y.H., Sun, W.W., Wang, Y.W., Wang, L. and Dong, Y.C., 2021. Microstructure and properties of Cr7C3-CrSi2 composite coatings prepared by plasma spraying. *Surface and Coatings Technology*, *412*, p.127011.
75. Sun, Z., Zhu, S., Dong, W., Bai, Y., Ding, H., Luo, Y. and Di, P., 2021. Characteristic comparison of stacked WC-based coatings prepared by high-velocity oxygen-fuel spray and electric contact strengthening. *Surface and Coatings Technology*, *421*, p.127289.
76. Fals, H.C., Lourençato, L.A., Orozco, M.S., Belém, M.J.X. and Lima, C.R.C., 2020. Slurry erosion resistance of thermally sprayed Nb2O5 and Nb2O5+ WC12Co composite coatings deposited on AISI 1020 carbon steel. *Ceramics International*, *46*(17), pp.27670–27678.

2 Nanocomposite Thermal Spray Feedstock Materials

Amrinder Mehta, Hitesh Vasudev, and Rajwinder Singh

2.1 INTRODUCTION

Thermal spray technology encompasses a broad spectrum of deposition techniques, wherein molten or semi-molten particles are accelerated toward a substrate to form a solid coating upon rapid solidification. Recent advances have seen the systematic incorporation of nanomaterials into thermal spray feedstocks, yielding nanocomposite coatings with markedly superior physicochemical attributes. Such enhancements are particularly manifested in elevated hardness, augmented wear resistance, and pronounced resistance to corrosive environments. Domains characterized by stringent performance requirements—including aerospace, automotive, energy, and biomedical applications—have begun to adopt these coatings as a strategic enabler of component longevity and reliability [1]. The creation of coatings for military engines started in the 1950s and marked the first advancement of thermal barrier coatings (TBCs) in industry. The first coatings were used to protect engines and parts from corrosion and extreme heat. Many studies have been conducted to develop advanced ceramic-based coatings with better insulation properties. Plasma-sprayed zirconia coatings became available in the 1970s and are used for turbine blades and other engine blades because they are heat-resistant and durable [2]. As with other coating technologies, hot corrosion presents challenges with the TBCs used in gas turbine engines, which come under attack at high temperatures. TBCs undergo hot corrosion when, at high temperatures, molten salt deposits, which consist of sulfates and vanadates, meet the TBC and chemically interact with the TBC material. Under such circumstances, the top layer of TBC, most likely yttria-stabilized zirconia (YSZ) coating, suffers disproportionate damage due to hot corrosion through dissolution-reprecipitation and infiltration along coarse fissures, pores, and other crevices. Thermally grown oxides coating the TBC can be chemically coupled under such conditions, and molten salt can infiltrate the coating, reacting with the layer and the underlying metallic bond coat of oxidized zirconia [3].

Figure 2.1 illustrates the critical failure modes of boiler components resulting from high-temperature operating conditions and aggressive environments. Figure 2.1 (a) shows severe ash deposition on the tube surfaces, creating insulating layers that impede heat transfer and lead to localized overheating and thermal stress cracking; (b) displays a catastrophic water tube failure characterized by rupture and material

DOI: 10.1201/9781003730637-2

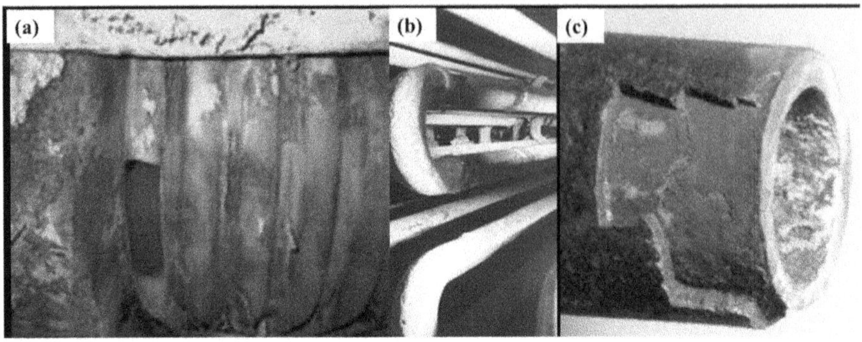

FIGURE 2.1 Failed boiler parts due to: (a) ash deposition, (b) water tube failure, and (c) corroded boiler parts due to hot corrosion [4].

degradation, likely resulting from thermal fatigue, creep deformation, or pressure-induced stress concentration; and (c) demonstrates extensive hot corrosion damage on boiler parts, exhibiting characteristic dark oxidation scales and material wastage caused by sulfur-containing compounds and alkali deposits at elevated temperatures. These failure mechanisms highlight the harsh operating environment of power generation equipment and underscore the critical need for protective thermal barrier coatings and advanced materials to enhance the durability and operational reliability of components [4].

A MCrAlY + AlSiY composite coating was tested for corrosion behaviour at 900°C for 200 h in 75 wt% Na_2SO_4 + 25 wt% K_2SO_4. The coating exhibited excellent resistance to high-temperature oxidation, forming a protective alumina layer on the surface. Microstructural analysis revealed a dense and adherent oxide layer, which effectively prevented further oxidation of the underlying substrate. The addition of AlSiY to MCrAlY coatings significantly enhances their corrosion resistance, potentially extending the service life of components in harsh environments [5].

The ceramic layer undergoes spallation, leaving vital engine parts vulnerable to extreme heat exposure. To address the issue of hot corrosion, scientists have been investigating new TBC materials that offer enhanced chemical stability, such as gadolinium zirconate ($Gd_2Zr_2O_7$). Notably, $Gd_2Zr_2O_7$, with a critical radius ratio of 1.46, has garnered considerable attention because of its low thermal conductivity, excellent insulating capabilities, and corrosion resistance [6]. Nanocomposite materials are currently in the spotlight of both industry and research groups. Researchers are interested in these composites because they can create special materials. These materials combine the beneficial properties of ceramic nanoparticles, such as electrical conductivity and facilitation of chemical reactions with light, with the advantageous properties of polymers, such as flexibility and ease of shaping. Nanostructured ceramics are conventionally defined as inorganic materials composed of structural units with a size scale of less than 100 nm in any dimension. Nanoceramics are classified as zero-dimensional nanocrystals, one-dimensional nanowires and nanotubes, two-dimensional nanofilms and nanowalls, and three-dimensional bulk materials with at least one nanocrystalline phase [7].

Figure 2.2 shows a comprehensive classification system centred around three primary categories for nanocomposite feedstock materials. Metal matrix nanocomposites (MMNCs) [8] utilize metallic matrices reinforced with nanoscale particles, while Ceramic matrix nanocomposites (CMNCs) [9] employ ceramic-based matrices with similar nanoreinforcements. Polymer matrix nanocomposites (PMNCs) [10] complete the triad using polymeric matrices. Each category branches into specific matrices and nanoreinforcements, demonstrating the hierarchical organization of these materials [11].

Granulated nano Al_2O_3 powders were heated at 200°C and sprayed over a standard TBC using atmospheric plasma spraying (APS). The TBC/nano Al_2O_3 coatings were permitted to develop a thermally grown oxide (TGO) layer on top of the bond coat (NiCrAlY layer). The granulated nano Al_2O_3 powders exhibited very high flowability, which, in addition to the increased apparent density, substantially reduced the microcracks and interconnected porosities in the coating [12, 13].

Figure 2.3 illustrates the fundamental principles of TSC technology, wherein a powder feedstock is introduced into a high-temperature combustion environment using a powder feeder system. Fuel gases, typically acetylene or hydrogen, combine with oxygen in the combustion chamber to generate an extremely hot flame. The carrier gas transports the powder particles through this particle heating zone, where they become molten or semi-molten. The heated particles are then accelerated through a converging nozzle to form a concentrated spray stream that impacts the pre-treated substrate surface. Upon collision, the molten particles deform, solidify rapidly, and build successive layers to form a dense, adherent coating with desired properties for protection or functionality [14–16].

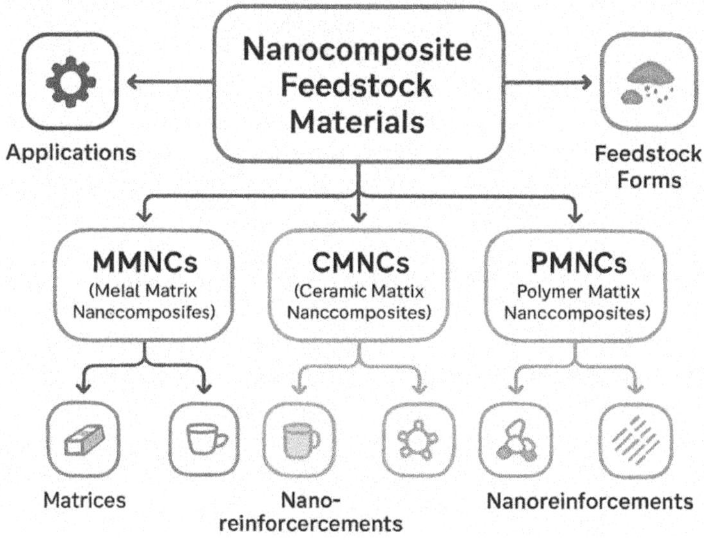

FIGURE 2.2 Nanocomposite feedstock materials are typically classified by their matrix and reinforcement types [8–10].

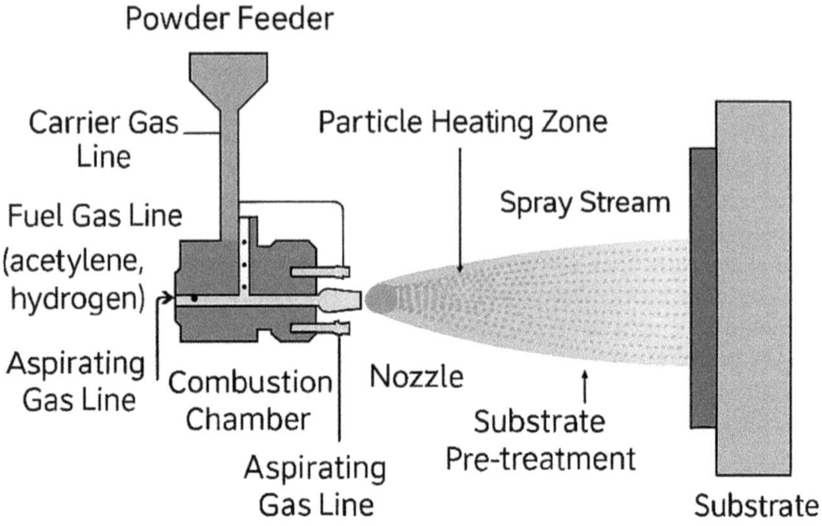

FIGURE 2.3 Principles of the thermal spray technique [14–16].

Thermal spray coating methods, such as APS [17], high-velocity oxy-fuel (HVOF) [18], flame spray [19], detonation gun (D-gun) [20], and high-velocity air fuel (HVAF) [21], offer flexible solutions for applying protective and functional layers to various substrates. These coating techniques have gained significant attention in various industries because of their ability to enhance surface properties and extend the lifespan of components. The selection of an appropriate thermal spray method depends on factors such as the desired coating properties, substrate material, and application requirements [22]. These techniques employ materials like metals, such as nickel and aluminium alloys, ceramics, such as alumina and zirconia, carbides and cermets, like WC-CoCr [23] and Cr_3C_2-NiCr, as well as polymers including polyethylene or polyamide. Each technique uses distinct heat sources and spray speeds to apply the materials, creating coatings with specific microstructures and characteristics. HVOF-sprayed Cr_3C_2-NiCr coatings are extensively utilized to protect engineering components from intense wear and corrosion, especially in challenging industrial settings. The widespread availability of these techniques allows for improved surface durability [24].

Figure 2.4 illustrates the operational parameter ranges for different thermal spray technologies by plotting the flame temperature against the particle velocity. APS operates at the highest temperatures (12,000°C) with moderate velocities (150–400 m/s), whereas vacuum plasma spray/low-pressure plasma spray (VPS/LPPS) achieves similar temperatures with higher velocities (150–600 m/s). HVOF and D-gun systems provide moderate temperatures (3,000°C) but exceptional particle velocities (400–1,000 m/s), optimizing the kinetic energy for dense coatings. Cold gas spray (CGS) operates at the lowest temperatures (300–1,000°C) and achieves the highest velocities (800–1,200 m/s), enabling solid-state deposition. Wire arc and combustion flame spray represent traditional methods with lower velocities and moderate temperatures, suitable for basic coating applications [25].

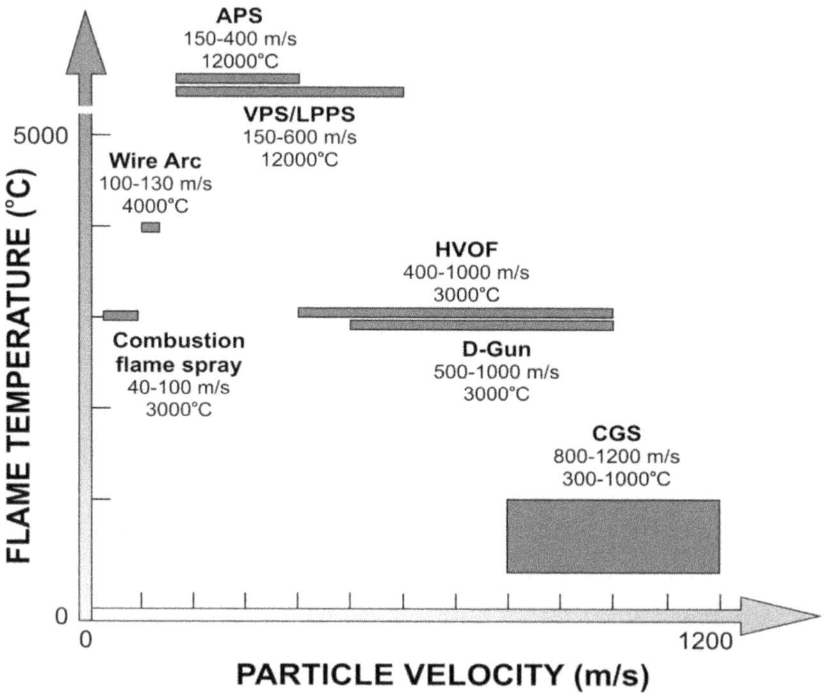

FIGURE 2.4 The typical flame temperature and particle velocity operation ranges for various thermal spray systems [25].

WC powder is frequently utilized to bolster the strength of metal coatings, and the development of WC/Co-based coatings has garnered significant research attention. The incorporation of WC particles into the Stellite-6 matrix resulted in a notable enhancement of the coating's mechanical properties. Microstructural analysis revealed a uniform distribution of WC particles within the cobalt-based alloy, leading to improved wear resistance and overall durability. Furthermore, the presence of WC reinforcement contributed to a significant increase in microhardness values, suggesting potential applications in harsh operating environments where both strength and corrosion resistance are crucial. To extend the durability of high-value-added components, WC particle-reinforced Stellite-6 cladding layers were examined for changes in coating structure, microhardness, and corrosion resistance [26].

Figure 2.5 presents tribological performance data for different coating compositions, comparing systems with and without carbon nanotube (CNT) reinforcement. Figure 2.5 (a) shows coefficient of friction evolution over a 2000-metre sliding distance for six different compositions: high power (HP), medium power (MP), and low power (LP) conditions, each tested with and without CNT addition. The HP coating exhibits the highest friction coefficient, reaching approximately 0.55 and showing considerable fluctuation throughout the test. In contrast, both MP and LP coatings demonstrate significantly lower and more stable friction behaviour, with values ranging between 0.15–0.35. The addition of CNT reinforcement generally reduces

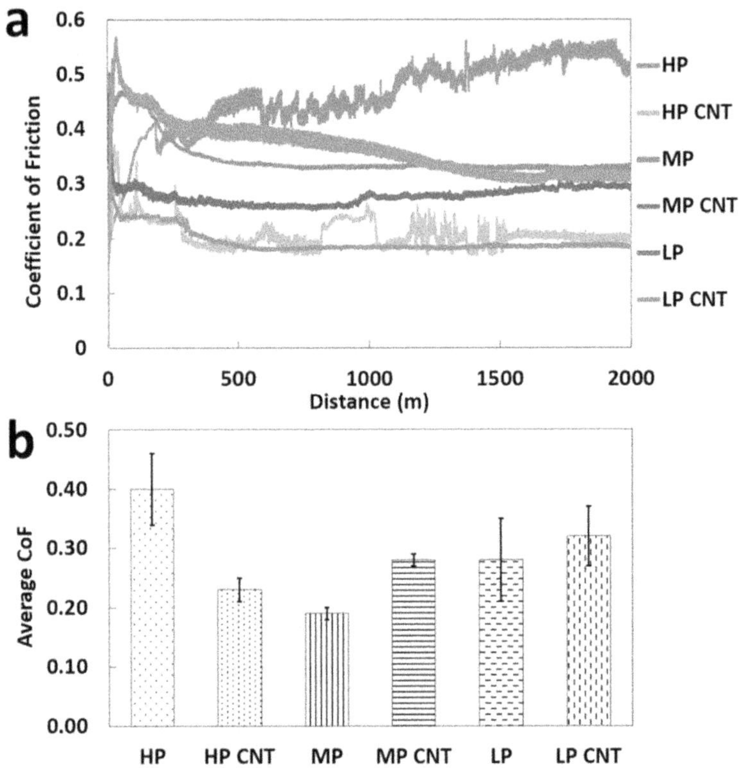

FIGURE 2.5 Wear tests and coefficient of friction: (a) coefficient of friction as a function of distance of the first wear test and (b) coefficient of friction averaged over the whole distance of the two tests with standard deviation as associated uncertainty [27].

friction coefficients across all power conditions, with HP CNT and MP CNT showing particularly pronounced improvements. Figure 2.5 (b) provides statistical validation through averaged friction coefficients with standard deviations, confirming that MP coatings achieve the lowest friction values, around 0.19, while HP coatings maintain the highest, at approximately 0.39, regardless of CNT incorporation [27].

The development of these feedstocks involves advanced preparation methods, including mechanical milling, suspension/solution precursor techniques, and hybrid feedstock strategies, each of which influences the resulting coating microstructure and performance. During the thermal spray process, the behaviour of nanocomposite particles, such as melting, phase evolution, and in-flight dynamics, plays a critical role in determining the final coating properties [28]. Recent research has explored the use of liquid feedstocks and hybrid approaches to overcome the challenges of feeding and dispersing nanoparticles, leading to coatings with tailored architectures for specific industrial applications. This review synthesizes the latest advances in the development, preparation, particle behaviour, and performance of nanocomposite feedstock materials in thermal spray coatings, highlighting key findings, challenges, and future directions.

2.2 PREPARATION METHODS AND FEEDSTOCK DEVELOPMENT

Nanocomposite feedstock preparation encompasses diverse methodologies tailored to achieve optimal particle distribution and microstructural characteristics for TBC [29]. Mechanical milling and high-energy ball milling techniques utilize mechanical forces to reduce particle size and promote intimate mixing of nanoscale reinforcements with matrix materials, creating homogeneous powder blends through repeated deformation and fracture processes. Spray drying converts liquid suspensions containing dispersed nanoparticles into free-flowing granular powders by rapid moisture evaporation, preserving compositional uniformity while enhancing flowability for subsequent processing. Suspension and solution precursor techniques involve dispersing nanomaterials in liquid media, enabling precise control over particle distribution and chemical homogeneity through careful selection of solvents and stabilizing agents [30]. Hybrid powder-suspension approaches combine conventional micron-sized powders with nanosuspensions during processing, facilitating multi-scale microstructures that leverage the benefits of both feedstock types. These preparation methods enable tailoring of nanocomposite feedstocks to specific application requirements, optimizing properties such as thermal conductivity, mechanical strength, and wear resistance in the resulting coatings [31].

Figure 2.6 shows that the novel HVAF spraying approach employs a hybrid feedstock that combines conventional powder (IN-625) and chromium carbide (Cr_3C_2) nanoparticle suspension to produce coatings with superior wear and corrosion performance. The powder and suspension streams merged in a T-joint mixer before entering the HVAF torch, enabling the simultaneous deposition of metallic and ceramic phases. Microstructural analysis revealed a homogeneous distribution of chromium carbide particles embedded within the nickel-based IN-625 matrix, as confirmed by elemental mapping of Cr and C. Wear tests showed that the hybrid coatings exhibited lower specific wear rates and reduced friction coefficients compared to the conventional IN-625 coating, which was attributed to the hard Cr_3C_2 phases arresting splat boundary failure and blunting abrasive particles. Electrochemical polarization measurements indicated enhanced corrosion resistance, with the hybrid layers demonstrating higher breakdown potentials and lower corrosion current densities. Wear mechanism diagrams illustrate that, unlike conventional coatings, where sliding wear leads to micro-cutting and oxide debris formation, hybrid coatings confine wear to reduced-width tracks, with carbide reinforcements mitigating substrate damage. This powder-suspension hybrid strategy enables fine control of the coating composition and microstructure, delivering multifunctional surfaces optimized for demanding tribological and corrosive environments [32].

Suspension and solution precursor methods allow the use of ultrafine or nanoscale particles, overcoming feeding challenges and enabling the formation of unique microstructures. Hybrid feedstocks, which combine powder and suspension, have been shown to enhance the retention of nanoreinforcements and improve coating properties [33].

2.3 COATING PERFORMANCE AND APPLICATIONS

The remarkable multifunctional characteristics of carbon-based nanocomposite films have made them a subject of frontier research. Within the aerospace industry, the

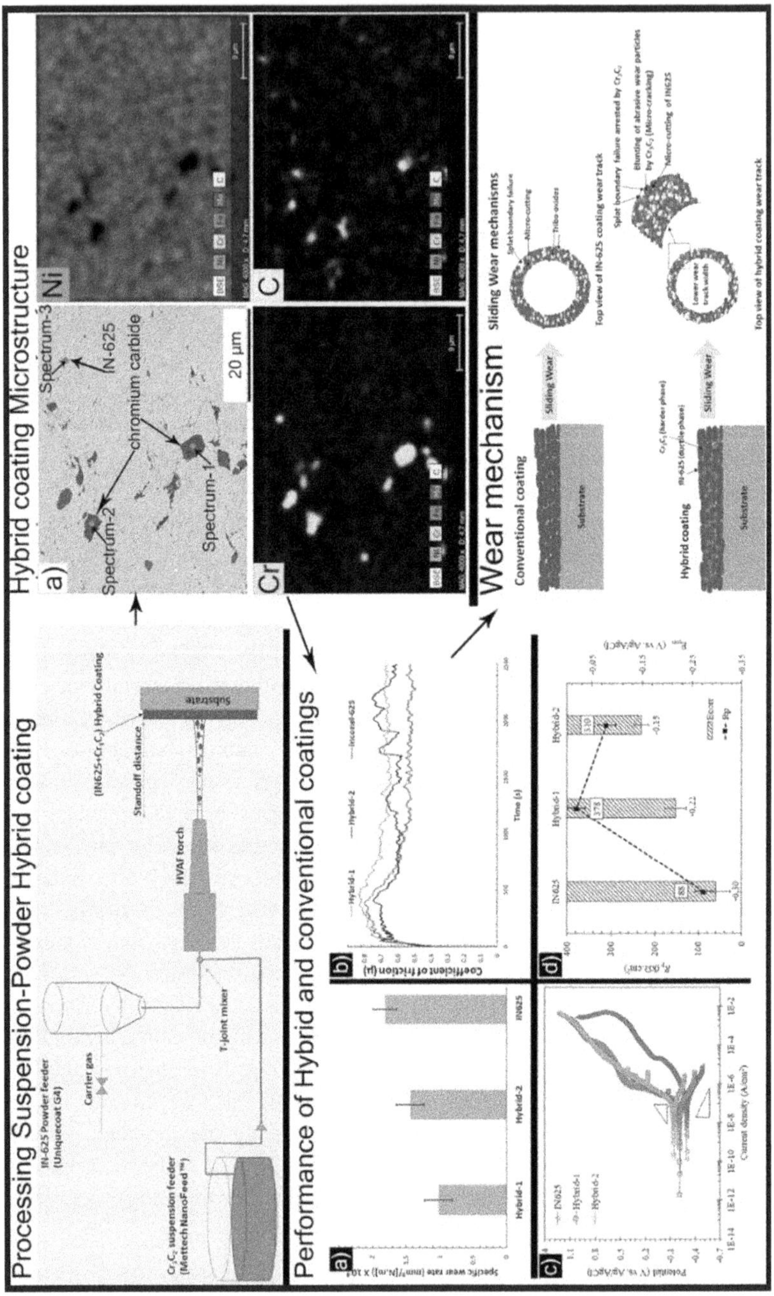

FIGURE 2.6 Novel utilization of powder-suspension hybrid feedstock in HVAF spraying [32].

focus for the fabrication of such nanocomposite films is not only to improve upon the engineered materials of the sector but also to obtain a substantial weight reduction, all while exceeding the electrical, thermal, and mechanical properties of the traditional materials [34]. This results in capturing the unadsorbed zone unreachable for classic substrates like alumina, and all variations of contemporary coatings merge for their alumina base. The result was dense coatings with markedly better wear resistance and much less friction [35].

Figure 2.7 shows that thermal spray coating processes offer versatile solutions for depositing functional layers across diverse applications, from aerospace to renewable energy technology. In hydrogen production via water splitting, critical components such as bipolar plates, electrodes, and corrosion-prone parts in proton exchange membrane (PEM), and alkaline, solid oxide, thermochemical, and photoelectrochemical electrolyzers can benefit from protective or catalytic coatings. Thermal spray methods are categorized based on the energy source. Combustion-based techniques, including flame spraying, wire flame spraying, D-gun, HVOF, and HVAF, deliver robust, dense coatings from powders or suspensions/solutions, with suspension feeds enabling nanostructured or composite layers. Cold spray provides solid-state deposition through high-velocity kinetic impact, preserving feedstock chemistry. Electric processes, such as plasma spraying and arc spraying under atmospheric or vacuum/low-pressure (VPS/LPPS) conditions, allow for fine control over the microstructure and composition. By selecting an appropriate spray technology and feedstock form, engineers can tailor the surface properties, such as catalytic activity, corrosion resistance, electrical conductivity, and mechanical integrity, to meet the specific demands of electrolyzer components, aircraft parts, automotive systems, and industrial infrastructure. This integrated framework underscores the strategic role of thermal spraying in enhancing the performance and longevity of cutting-edge technologies [36].

The incorporation of nanoparticles, such as CNTs, graphene, Al_2O_3, TiO_2, and SiC, leads to increased hardness, improved fracture toughness, reduced coefficient of friction, and enhanced resistance to wear and corrosion. These coatings are used in aerospace, automotive, energy, and biomedical industries, where high-performance surfaces are required. Additionally, multifunctional and smart coatings with self-cleaning, photocatalytic, and bioactive properties are being developed using advanced nanocomposite feedstocks [37].

2.4 CHALLENGES OF NANOMATERIAL-BASED TSC

Performance discrepancies may arise from the non-homogeneous structure of nanostructured coatings. Thus, they may be less suitable for applications requiring consistent corrosion resistance. This intricacy may lengthen production times and raise upfront costs. Despite significant progress, challenges remain in achieving homogeneous nanoparticle dispersion, controlling agglomeration, optimizing feedstock flowability, and scaling up production for industrial applications of these materials [38].

Figure 2.8 presents the SEM fracture sections of the coatings from the conventional (a), B12 (b), and C24 (c) powders with high-magnification views. The conventional

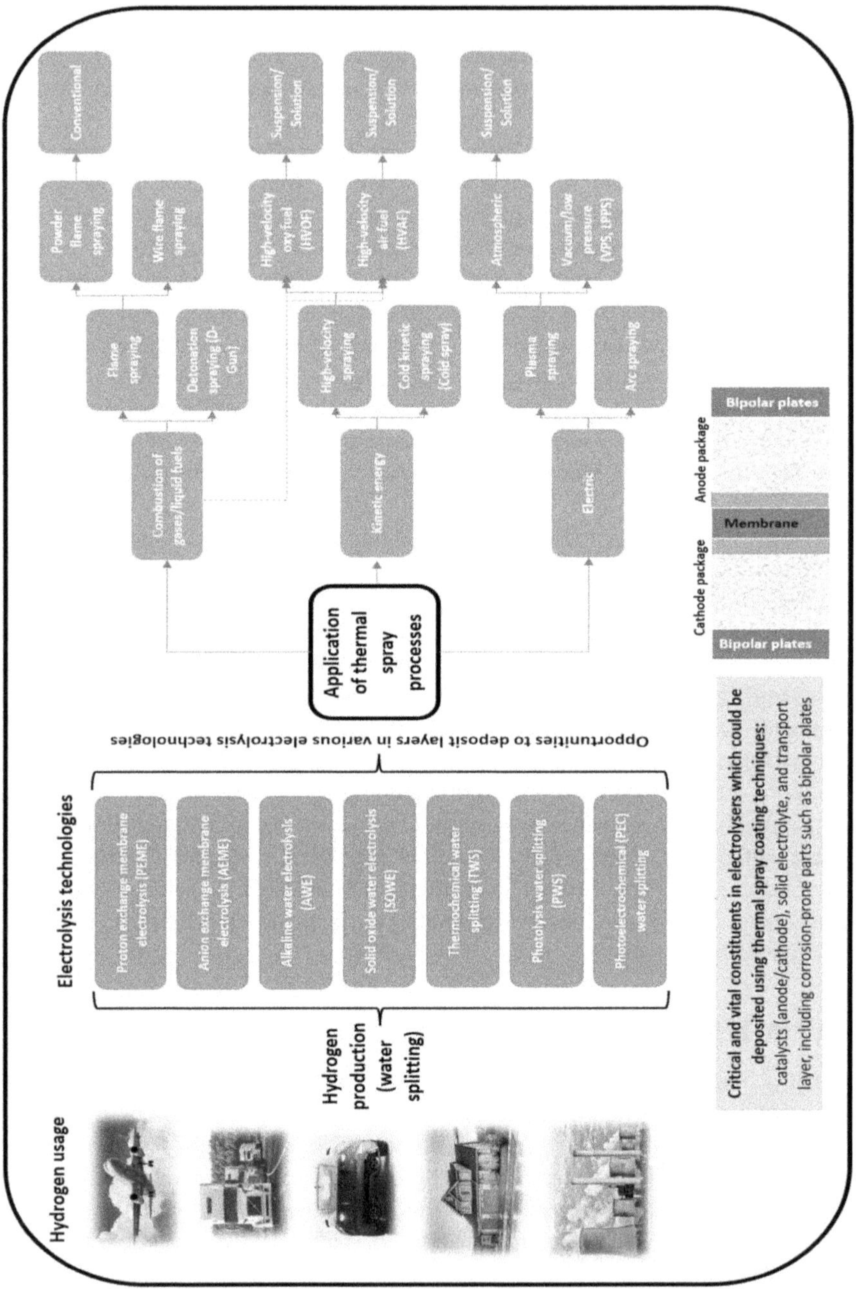

FIGURE 2.7 Applications of thermal spray coating processes for layer deposition in various technologies [36].

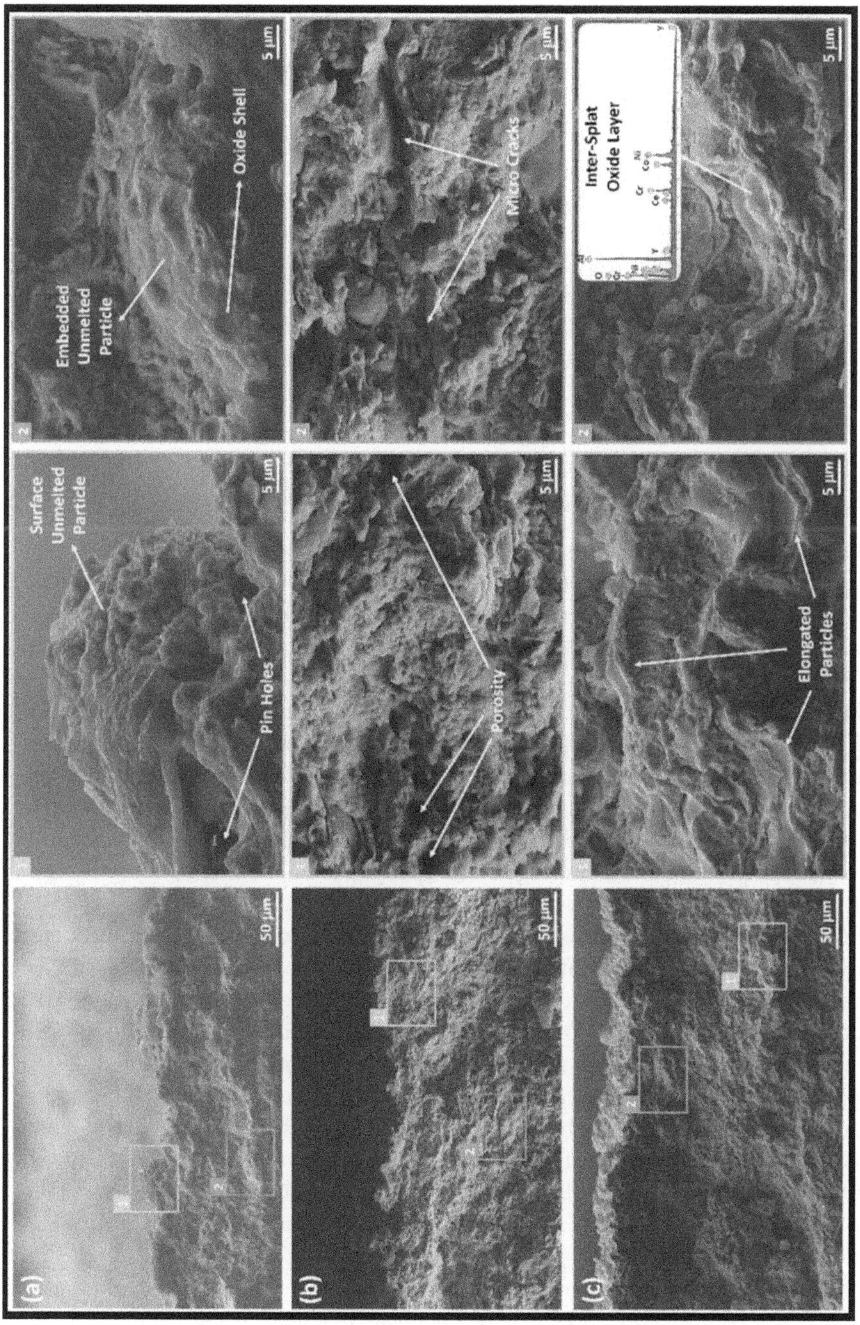

FIGURE 2.8 SEM fracture-sections of the coatings prepared from (a) conventional, (b) B12, and (c) C24 powders alongside their higher magnification images [30].

powder exhibited unmelted particles and oxide shells, B12 revealed pronounced porosity and microcracks, and C24 displayed elongated particles and an inter-splat oxide layer. These microstructural differences underscore the influence of powder type on coating defects and fracture behaviour [30].

Achieving a homogeneous nanoparticle dispersion remains challenging because of the strong tendency of nanoparticles to agglomerate, driven by their high surface energy and van der Waals forces. Effective dispersion requires careful selection of mixing techniques, surfactants, or surface modifications, yet even advanced methods may not fully prevent aggregation, especially at higher concentrations. Scaling up production introduces further complexity, as processes that work at laboratory scale often result in inconsistent particle size, distribution, or stability when applied industrially. Maintaining uniformity, reproducibility, and quality control during large-scale manufacturing is difficult, making the transition from lab to commercial production a significant hurdle. The particle dynamics during spraying and the fragility of the associated parameters necessitate the development of sophisticated diagnostics and modelling tools [39].

2.5 FUTURE DIRECTION

The next step for TBCs that involve nanocomposites is tackling the growing need for withstanding higher temperature, increased thermal insulation, and longevity in extreme conditions. These coatings are important for safeguarding parts in gas turbines, aircraft engines, and other applications that are exposed to elevated temperatures [40].

Researchers are developing novel nanofibre membranes, aerogels, and composite structures that maintain stability and low thermal conductivity at temperatures exceeding 1300°C. New innovations include the integration of microphases into the microstructure of the material, hierarchical porous designs, and self-ceramifiable materials that gain strength upon heating. These developments not only enhance the insulation performance but also provide strength and durability under prolonged operation at elevated temperatures [41].

Figure 2.9 illustrates four distinct TBC configurations on Inconel 718 substrates with NiCrAlY bond coats. Configuration (a) features a traditional single-layer YSZ topcoat approximately 250μm thick, while (b) shows a gadolinium zirconate (GZ) topcoat of similar thickness. The bilayer system (c) combines both materials with 125μm layers of GZ and YSZ stacked sequentially. The most sophisticated design (d) presents a functionally graded coating, where the material composition transitions gradually from pure YSZ at the bottom through mixed ratios of 33%GZ/67%YSZ and 67%GZ/33%YSZ to pure GZ at the top. This graded approach creates smooth property transitions, potentially reducing thermal stress concentrations and improving coating durability compared to the abrupt interfaces found in conventional single-layer or bilayer systems [42].

In situ synthesis during thermal spray processes represents a revolutionary approach to coating fabrication, where chemical reactions occur within a high-temperature plasma environment, creating unique phases and microstructures that are not achievable using conventional feedstock materials. This process involves the

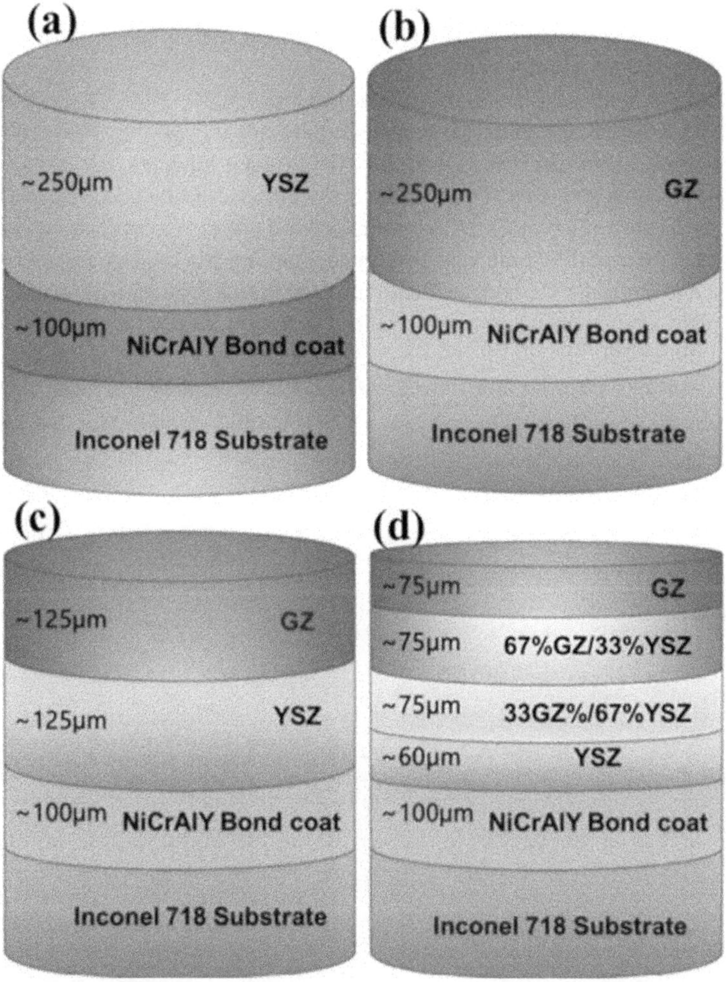

FIGURE 2.9 Schematic representation of TBCs: (a) YSZ, (b) GZ, (c) YSZ/GZ bilayer, and (d) FGC of YSZ/GZ [42].

simultaneous formation and consolidation of reinforcing phases, such as carbides, nitrides, and intermetallics, directly within the coating matrix during deposition. The extreme thermal conditions in plasma jets provide sufficient energy for decomposition reactions, atomic diffusion, and phase transformations that enable the formation of submicron Ti(C,N) particles, TiC reinforcements, and other ceramic phases in situ [43]. The resulting microstructures exhibit distinct characteristics, including bimodal particle distributions, core–shell structures, and refined grain sizes, which significantly enhance the mechanical properties. These coatings exhibit lamellar architectures with embedded reinforcing particles uniformly distributed throughout the metallic matrix, creating interfaces with specific crystallographic orientation relationships that improve load transfer and bonding strength. The in situ formed phases typically range from nanoscale to sub-micron dimensions, providing effective strengthening mechanisms through grain refinement, precipitation hardening, and enhanced interfacial bonding [44].

Figure 2.10 shows the formation of two distinct NbB_2–NbC–Al_2O_3 composite coatings via thermal spraying: (a) feedstock particles comprising NbB_2, NbC, and Al_2O_3 are heated and partially melted; the Al_2O_3 phase liquefies first, encapsulating the boride–carbide particles. Upon impact, the eutectic droplets spread and solidified into a layered microstructure with intact NbB_2/NbC reinforcements within an alumina-rich matrix. (b) Composite powders containing Nb_2O_5, B_4C, and Al undergo reduction and melting; Al liquefies, facilitating the in situ formation of mixed Nb–Al–O–C–B phases, whereas the remaining oxides and carbides react to produce complex nanostructures [45].

Nanostructures improve ion movement and make electrolytes more accessible. When paired with TB, which is known for its metallic conductivity and structural

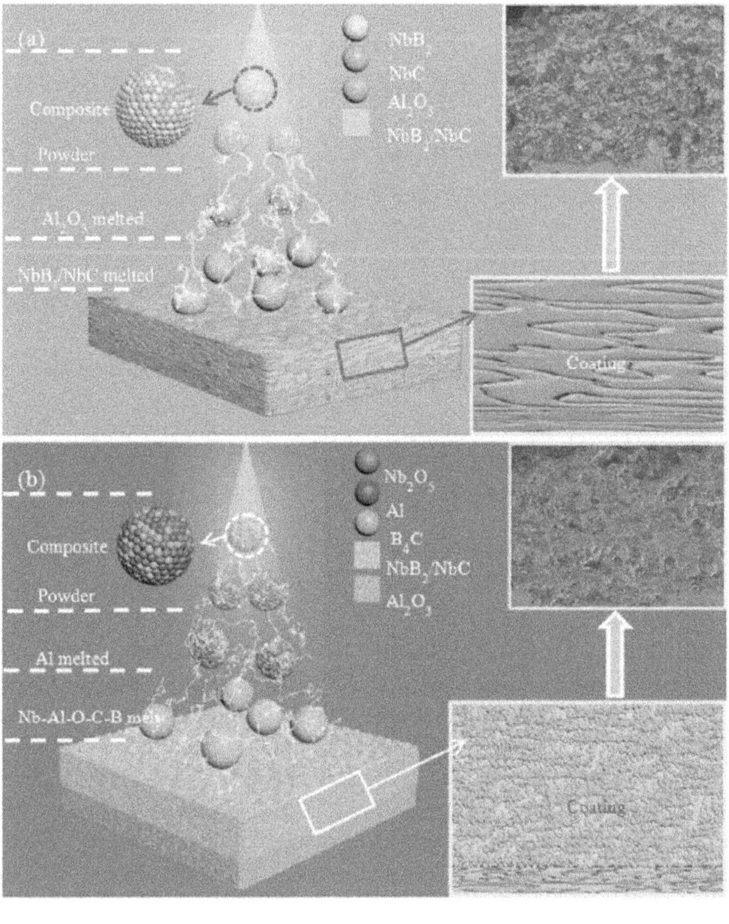

FIGURE 2.10 Schematic graphs of the formation mechanism of two NbB_2–NbC–Al_2O_3 coatings: (a) NbB_2–NbC–Al_2O_3, and (b) Nb_2O_5–B4C–Al [45].

strength, the TB/Co_3O_4 composite benefits from synergistic effects, such as enhanced charge storage and faster electron transport. Co_3O_4, in comparison to other oxides such as MnO_2, NiO, and Fe_2O_3, provides better cycling stability, electrical conductivity, and compatibility with TiB_2 at the interface. This combination results in a superior electrode material with an improved electrochemical performance. The TB/Co_3O_4 composite exhibited a higher specific capacitance, better rate capability, and longer cycle life than its individual components. Furthermore, the unique nanostructure of this composite facilitates rapid ion diffusion and electron transfer, making it a promising candidate for high-performance supercapacitors [46].

In situ synthesized coatings show remarkable property enhancements relative to industry-standard thermal spray deposits. The microhardness increases 3.5 times and wear resistance is 7.66 times better than the substrate to which the material is case bonded. Tunable multilayered structures can be designed, where Fe_2Ti Laves phases and Ti(C,N) particles are interspaced with various oxide phases, providing maintenance-free high-wear surfaces with corrosion-resistant coatings and exceptional thermal stability [47]. Furthermore, the incorporation of specific oxide compounds can be leveraged to enhance the oxidation resistance and thermal barrier properties, which are crucial for components operating in extreme environments. Thermal spray coatings along with other surface engineering technologies have increased within the fields of biomedicine, slurry, and high-temperature corrosion systems most recently in conjunction with microwave processing, additive manufacturing, and magnetron sputtering thin-film coatings [28, 48–60].

Coatings also allow for the prediction and simulation of surface roughness, strength in tension, compression, bending, and performance in solar applications. In addition, coatings allow flow systems and the simulation of more intricate structures like chairs considering ergonomics [12, 61–72].

2.6 CONCLUSION

Nanocomposite thermal spray coatings are sophisticated protective layers formed by integrating nanoparticles into base materials. The preparation techniques include mechanical milling, suspension/solution precursor methods, and hybrid feedstock approaches. These coatings exhibit enhanced hardness, wear resistance, corrosion resistance, and multifunctional properties. The behaviour of nanocomposite particles in-flight, particularly their temperature, speed, and melting conditions during thermal spray processes, is vital in determining the final characteristics of the coating. Achieving the correct particle temperature and velocity is crucial for proper melting and flattening upon impact, resulting in dense, well-bonded microstructures with improved hardness and adhesion and reduced porosity. However, insufficient or excessive particle heating can lead to unmelted particles, increased defects, or unwanted phases, which can undermine the mechanical performance. Thus, precise control and monitoring of in-flight particle dynamics are critical for customizing the microstructure and attaining superior functional properties in nanocomposite thermal spray coatings.

Achieving a uniform dispersion of nanoparticles is a persistent challenge owing to their strong inclination to clump together, driven by high surface energy and van der Waals forces. To effectively disperse them, one must carefully choose the mixing methods, surfactants, or surface modifications; however, even the most advanced techniques may not completely prevent aggregation, particularly at higher concentrations. Scaling up production adds another layer of complexity, as processes that are successful in the laboratory often lead to inconsistent particle size, distribution, or stability when scaled to industrial levels. Ensuring uniformity, reproducibility, and quality control during large-scale manufacturing is difficult, making the transition from laboratory to commercial production a significant challenge. Future developments in high-temperature materials and coatings will focus on achieving greater resistance to extreme temperatures, improved thermal insulation, and longer service lives. Researchers are creating innovative nanofibre membranes, aerogels, and composite structures that remain stable and maintain low thermal conductivity at temperatures above 1300°C. Innovations include the use of multi-phase microstructures, hierarchical porous designs, and self-ceramifiable materials that become stronger when exposed to heat. These strategies not only enhance insulation efficiency but also ensure mechanical strength and durability during extended high-temperature use. The ultimate aim is to develop lightweight, flexible, and reusable materials suitable for demanding industrial and aerospace applications.

REFERENCES

[1] Nelson WA, Orenstein RM. TBC experience in land-based gas turbines. *Journal of Thermal Spray Technology*. 1997;6:176–80.

[2] Brandon JR, Taylor R. Phase stability of zirconia-based thermal barrier coatings part I. Zirconia-yttria alloys. *Surface and Coatings Technology*. 1991;46 :75–90.

[3] Karaoglanli AC, Ogawa K, Turk A, Ozdemir I. Thermal shock and cycling behavior of thermal barrier coatings (TBCs) used in gas turbines. *Progress in Gas Turbine Performance*. 2013;2013:237–60.

[4] Lakkannavar V, Yogesha K, Prasad CD, Phanden RK, Srinivasa G, Prasad C. Thermal spray coatings on high-temperature oxidation and corrosion applications–A comprehensive review. *Results in Surfaces Interfaces*. 2024:100250.

[5] Jiang SM, Peng X, Bao ZB, Liu SC, Wang QM, Gong J, et al. Preparation and hot corrosion behaviour of a MCrAlY+AlSiY composite coating. *Corrosion Science*. 2008;50:3213–20.

[6] Guo Y, He W, Guo H. Thermo-physical and mechanical properties of Yb2O3 and Sc2O3 co-doped Gd2Zr2O7 ceramics. *Ceramics International*. 2020;46:18888–94.

[7] Xu Z, Xie Y, Ebrahimnia M, Dang H. Effect of B4C nanoparticles on microstructure and properties of laser cladded IN625 coating. *Surface and Coatings Technology*. 2021;416:127154.

[8] He F, Han Q, Jackson MJ. Nanoparticulate reinforced metal matrix nanocomposites: A review. *International Journal of Nanoparticles*. 2008;1:301–9.

[9] Omanović-Mikličanin E, Badnjević A, Kazlagić A, Hajlovac M. Nanocomposites: A brief review. *Health and Technology*. 2020;10:51–9.

[10] Dinari M, Soltani R, Mohammadnezhad G. Kinetics and thermodynamic study on novel modified–mesoporous silica MCM-41/polymer matrix nanocomposites: Effective adsorbents for trace CrVI removal. *Journal of Chemical & Engineering Data*. 2017;62:2316–29.

[11] Gujba AK, Mahdipoor MS, Medraj M. Water droplet impingement erosion performance of WC-based coating sprayed by HVAF and HVOF. *Wear*. 2021;484:203904.
[12] Verma PK, Vasudev H, Kumar V. Bioactivity and corrosion performance of low-velocity oxy-fuel-sprayed hydroxyapatite-based coatings on a surgical-grade stainless steel. *Transactions of the Indian Institute of Metals*. 2025;78(8):185.
[13] Duan B, Niu H, Zhang W, Ma Y, Yuan Y, Liu C. Microporous density-mediated response of MSCs on 3D trimodal macro/micro/nano-porous scaffolds via fibronectin/integrin and FAK/MAPK signaling pathways. *Journal of Materials Chemistry B*. 2017;5:3586–99.
[14] Bocklisch F, Bocklisch SF, Grimm M, Lampke T, Joshi S. Hybrid decision-making in atmospheric plasma spraying enables human–machine teaming. *The International Journal of Advanced Manufacturing Technology*. 2024;132:4941–63.
[15] Chade D, Berlouis L, Infield D, Cruden A, Nielsen PT, Mathiesen T. Evaluation of Raney nickel electrodes prepared by atmospheric plasma spraying for alkaline water electrolysers. *International Journal of Hydrogen Energy*. 2013;38:14380–90.
[16] Chade D, Berlouis L, Infield D, Nielsen PT, Mathiesen T. Deactivation mechanisms of atmospheric plasma spraying Raney nickel electrodes. *Journal of The Electrochemical Society*. 2016;163:F308.
[17] Bellippady M, Björklund S, Li X-H, Frykholm R, Kjellman B, Joshi S, et al. Performance of atmospheric plasma-sprayed thermal barrier coatings on additively manufactured super alloy substrates. *Coatings*. 2024;14:626.
[18] Abhijith N, Kumar D, Kalyansundaram D. Development of single-stage TiNbMoMnFe high-entropy alloy coating on 304L stainless steel using HVOF thermal spray. *Journal of Thermal Spray Technology*. 2022;31:1032–44.
[19] Bergant Z, Trdan U, Grum J. Effect of high-temperature furnace treatment on the microstructure and corrosion behavior of NiCrBSi flame-sprayed coatings. *Corrosion Science*. 2014;88:372–86.
[20] Karaoglanli AC, Ozgurluk Y, Doleker KM. Comparison of microstructure and oxidation behavior of CoNiCrAlY coatings produced by APS, SSAPS, D-gun, HVOF and CGDS techniques. *Vacuum*. 2020;180:109609.
[21] Alroy RJ, Pandey R, Kamaraj M, Sivakumar G. Role of process parameters on microstructure, mechanical properties and erosion performance of HVAF sprayed Cr3C2-NiCr coatings. *Surface Coatings Technology*. 2022;449:128941.
[22] Gao X, Li C, Han X, Chen X, Zhao X. Numerical simulation and parameter sensitivity analysis of multi-particle deposition behavior in HVAF spraying. *Surface Coatings Technology*. 2022;441:128569.
[23] Lee SS. Wear behaviors of WC-CoCr and WC-CrC-Ni coatings sprayed by HVOF. *Journal of the Korea Academia-Industrial cooperation Society*. 2020;21:204–11.
[24] Du J, Li Y, Li F, Lu H, Ran X, Zhang X, et al. Numerical and experimental analysis of temperature field and mechanical analysis for the multi-layer Cr3C2-NiCr coating deposited by HVOF. *Journal of Alloys and Compounds*. 2021;889:161656.
[25] Oksa M, Turunen E, Suhonen T, Varis T, Hannula S-P. Optimization and characterization of high velocity oxy-fuel sprayed coatings: Techniques, materials, and applications. *Coatings*. 2011;1:17–52.
[26] Bartkowski D, Młynarczak A, Piasecki A, Dudziak B, Gościański M, Bartkowska A. Microstructure, microhardness and corrosion resistance of Stellite-6 coatings reinforced with WC particles using laser cladding. *Optics & Laser Technology*. 2015;68:191–201.
[27] Venturi F, Kamnis S, Hussain T. Internal diameter HVOAF thermal spray of carbon nanotubes reinforced WC-Co composite coatings. *Materials & Design*. 2021;202:109566.

[28] Verma PK, Vasudev H, Singh G. Improving the corrosion performance of SS 304 stainless steel using HVOF-sprayed WC-Co-Cr coatings in chloride environments. *Journal of Bio-and Tribo-Corrosion*. 2025;11(4):1–3.

[29] Keyvani A, Mahmoudinezhad P, Jahangiri A, Bahamirian M. Synthesis and characterization of ((La1-xGdx) 2Zr2O7; x= 0, 0.1, 0.2, 0.3, 0.4, 0.5, 1) nanoparticles for advanced TBCs. *Journal of the Australian Ceramic Society*. 2020;56:1543–50.

[30] Zakeri A, Ghadami F, Rouhaghdam AS, Saeedi B. Study on production of modified MCrAlY powder with nano oxide dispersoids as HVOF thermal spray feedstock using mechanical milling. *Materials Research Express*. 2020;7:015030.

[31] Björklund S, Goel S, Joshi S. Function-dependent coating architectures by hybrid powder-suspension plasma spraying: Injector design, processing and concept validation. *Materials Design*. 2018;142:56–65.

[32] Mahade S, Baiamonte L, Sadeghimeresht E, Björklund S, Marra F, Joshi S. Novel utilization of powder-suspension hybrid feedstock in HVAF spraying to deposit improved wear and corrosion resistant coatings. *Surface and Coatings Technology*. 2021;412:127015.

[33] Carnicer V, Orts MJ, Moreno R, Sánchez E. Microstructure assessment of suspension plasma spraying coatings from multicomponent submicronic Y-TZP/Al2O3/SiC particles. *Ceramics International*. 2018;44:12014–20.

[34] Laurenzi S, Clausi M, Zaccardi F, Curt U, Santonicola MG. Spray coating process of MWCNT/epoxy nanocomposite films for aerospace applications: Effects of process parameters on surface electrical properties. *Acta Astronautica*. 2019;159:429–39.

[35] Goyal R, Sidhu BS, Chawla V. Multi-Walled CNT reinforcement to thermal spray coatings. *Asian Journal of Engineering and Applied Technology*. 2018;7:93–8.

[36] Faisal NH, Prathuru A, Ahmed R, Rajendran V, Hossain M, Venkatachalapathy V, et al. Application of thermal spray coatings in electrolysers for hydrogen production: Advances, challenges, and opportunities. *ChemNanoMat*. 2022;8:e202200384.

[37] Julkapli NM, Bagheri S, Sapuan SM. Multifunctionalized carbon nanotubes polymer composites: Properties and applications. In: Thakur VK, Thakur MK, editors. *Eco-friendly Polymer Nanocomposites: Processing and Properties*. New Delhi: Springer India; 2015. pp. 155–214.

[38] Abdesselam Y, Belloufi A, Rezgui I, Abdelkrim M, Catalin T, Chiriță B. Composite coatings based on nickel and Y2O3 nanoparticles: A comprehensive analysis of developments in electrodeposition and functional property optimization. *The International Journal of Advanced Manufacturing Technology*. 2025;137:3273–332.

[39] Alkunte S, Fidan I, Naikwadi V, Gudavasov S, Ali MA, Mahmudov M, et al. Advancements and challenges in additively manufactured functionally graded materials: A comprehensive review. *Journal of Manufacturing Materials Processing*. 2024;8:23.

[40] Fathi R, Wei H, Saleh B, Radhika N, Jiang J, Ma A, et al. Past and present of functionally graded coatings: Advancements and future challenges. *Applied Materials Today*. 2022;26:101373.

[41] Lima RS. Porous APS YSZ TBC manufactured at high powder feed rate (100 g/min) and deposition efficiency (70%): Microstructure, bond strength and thermal gradients. *Journal of Thermal Spray Technology*. 2022;31:396–414.

[42] Ariharan S, Parchovianský M, Singh P, Rani P, Maurya R, Sekar A, et al. Hot corrosion behavior of La2Ce2O7-based plasma-sprayed coating. *High Temperature Corrosion of Materials*. 2024;101:779–88.

[43] Antou G, Montavon G, Hlawka F, Cornet A, Coddet C, Machi F. Modification of thermal barrier coating architecture by in situ laser remelting. *Journal of the European Ceramic Society*. 2006;26:3583–97.

[44] Cai Y, Xiong J, Chen H, Zhang G. A review of in-situ monitoring and process control system in metal-based laser additive manufacturing. *Journal of Manufacturing Systems*. 2023;70:309–26.
[45] Wang X, Yang Y, Jia S, Wang Y, Ma Y, Cui Y, et al. In-situ synthesis, microstructure, and properties of NbB2-NbC-Al2O3 composite coatings by plasma spraying. *Journal of Advanced Ceramics*. 2022;11:1263–78.
[46] Xu W, Li T-T, Zheng Y-Q. Porous Co 3 O 4 nanoparticles derived from a Co (ii)-cyclohexanehexacarboxylate metal–organic framework and used in a supercapacitor with good cycling stability. *RSC Advances*. 2016;6:86447–54.
[47] Wang M, Cui H, Wei N, Ding L, Zhang X, Zhao Y, et al. A new design of in situ Ti (C, N) reinforced composite coatings and their microstructures, interfaces, and wear resistances. *ACS Applied Materials & Interfaces*. 2018;10:4250–65.
[48] Dutta V, Thakur L, Singh B, Vasudev H. A study of erosion–corrosion behaviour of friction stir-processed chromium-reinforced NiAl bronze composite. *Materials*. 2022;15(15):5401.
[49] Singh M, Vasudev H, Kumar R. Corrosion and tribological behaviour of BN thin films deposited using magnetron sputtering. *International Journal of Surface Engineering and Interdisciplinary Materials Science (IJSEIMS)*. 2021;9(2):24–39.
[50] Sharanabasava H, Raviprakash M, Prasad CD, Ramesh MR, Phanibhushana MV, Vasudev H, Kumar S. Microstructure, mechanical and wear properties of SiC and Mo reinforced NiCr microwave cladding. *Advanced Materials and Processing Technologies*. 2023;10(4):3620–33.
[51] Singh M, Vasudev H, Singh M. Surface protection of SS-316L with boron nitride based thin films using radio frequency magnetron sputtering technique. *Journal of Electrochemical Science and Engineering*. 2022;12(5):851–63.
[52] Prashar G, Vasudev H, Thakur L, Bansal A. Performance of thermally sprayed hydroxyapatite coatings for biomedical implants: A comprehensive review. *Surface Review and Letters*. 2023;30(1):2241001.
[53] Vasudev H. Wear characteristics of Ni-WC powder deposited by using a microwave route on mild steel: Microwave cladding of Ni-WC. *International Journal of Surface Engineering and Interdisciplinary Materials Science (IJSEIMS)*. 2020;8(1):44–54.
[54] Parkash J, Saggu HS, Vasudev H. A short review on the performance of high velocity oxy-fuel coatings in boiler steel applications. *Materials Today: Proceedings*. 2022;50:1442–6.
[55] Singh P, Vasudev H, Bansal A. Effect of post-heat treatment on the microstructural, mechanical, and bioactivity behavior of the microwave-assisted alumina-reinforced hydroxyapatite cladding. *Proceedings of the Institution of Mechanical Engineers, Part E: Journal of Process Mechanical Engineering*. 2024;238(3):1060–71.
[56] Verma PK, Vasudev H, Singh G, Singh S. A review on the microstructural and biomedical properties of implants manufactured using additive manufacturing. *Journal of the Institution of Engineers (India): Series D*. 2024;106:1587–601.
[57] Singh SP, Verma PK, Singh S, Kapoor M, Balu R. Microstructural characterization and electrochemical behaviour of HVOF sprayed WC-Co-Cr coatings on 316L boiler steel stainless steel. *Results in Surfaces and Interfaces*. 2024;16:100248.
[58] Verma PK, Minhas AS, Singh P, Kumar S. Slurry erosion behaviour of thermal sprayed Al2O3 and Cr2O3 coatings for turbine steels. *AIP Conference Proceedings*. 2024;2986(1):020012.
[59] Verma PK, Kumar V, Vasudev H. Bioactivity and corrosion analysis of thermally sprayed hydroxyapatite based coatings. *Journal of Electrochemical Science and Engineering*. 2024;14(4):419–39.

[60] Verma PK, Vasudev H, Kumar V. Biocompatibility and corrosion performance of HA-based LVOF sprayed and post-annealed coatings on SS 254 surgical-grade bio-implant material. *Ceramics International*. 2025;51(10):13207–25.

[61] Kulkarni SD, Manjunatha, Chandrasekhar U, Manjunath KV, Prasad CD, Vasudev H. Design and optimization of polyvinyl-nitride rubber for tensile strength analysis. *International Journal on Interactive Design and Manufacturing (IJIDeM)*. 2024;18(8):5893–908.

[62] Kattimani MA, Venkatesh PR, Masum H, Math MM, Bahadurdesai VN, Mustafkhadri S, Prasad CD, Vasudev H. Design and numerical analysis of tensile deformation and fracture properties of induction hardened inconel 718 superalloy for gas turbine applications. *International Journal on Interactive Design and Manufacturing (IJIDeM)*. 2024;18(8):5939–49.

[63] Poojari M, Hanumanthappa H, Prasad CD, Jathanna HM, Ksheerasagar AR, Shetty P, Shanmugam BK, Vasudev H. Computational modelling for the manufacturing of solar-powered multifunctional agricultural robot. *International Journal on Interactive Design and Manufacturing (IJIDeM)*. 2024;18(8):5725–36.

[64] Prakash C, Gupta LR, Mehta A, Vasudev H, Tominov R, Korman E, Fedotov A, Smirnov V, Kesari KK. Computing of neuromorphic materials: an emerging approach for bioengineering solutions. *Materials Advances*. 2023;4:5882–919.

[65] Singh J, Gill HS, Vasudev H. Computational fluid dynamics analysis on role of particulate shape and size in erosion of pipe bends. *International Journal on Interactive Design and Manufacturing (IJIDeM)*. 2023;17(5):2631–46.

[66] Rao GS, Mukkamala U, Hanumanthappa H, Prasad CD, Vasudev H, Shanmugam B, KishoreKumar KC. Evaluating and optimizing surface roughness using genetic algorithm and artificial neural networks during turning of AISI 52100 steel. *International Journal on Interactive Design and Manufacturing (IJIDeM)*. 2024;18(8):6151–60.

[67] Singh J, Singh R, Singh S, Vasudev H, Kumar S. Reducing scrap due to missed operations and machining defects in 90PS pistons. *International Journal on Interactive Design and Manufacturing (IJIDeM)*. 2023;17(5):2527–39.

[68] Nagabhushana P, Ramprasad S, Prasad CD, Vasudev H, Prakash C. Numerical investigation on heat transfer of a nano-fluid saturated vertical composite porous channel packed between two fluid layers. *International Journal on Interactive Design and Manufacturing (IJIDeM)*. 2024;18(5):2927–44.

[69] Verma PK, Vasudev H, Kumar V. Effect of electro-corrosion behavior of different grades of stainless steel used in bio-implants: A review. In Hitesh Vasudev, Chander Prakash, Mohammad Khalid, Kavindra Kesari (Eds.) *Advanced Manufacturing Processes*. 2025. pp. 269–83, CRC Press, Boca Raton.

[70] Verma PK, Vasudev H. Importance of hydroxyapatite coatings in improving new bone osteogenesis: A review. In Hitesh Vasudev, Chander Prakash, Mohammad Khalid, Kavindra Kesari (Eds.) *Advanced Manufacturing Processes*, 2025. pp. 284–99, CRC Press, Boca Raton.

[71] Mahantesh MM, Rao KR, Chandra AP, Vijayakumar MN, Nandini B, Prasad CD, Vasudev H. Design and modeling using finite element analysis for the sitting posture of computer users based on ergonomic perspective. *International Journal on Interactive Design and Manufacturing (IJIDeM)*. 2024;18(8):5875–91.

[72] Verma PK, Vasudev H, Kumar V. Dental implant materials and designs and failure during implant placement and loading: A review. In Hitesh Vasudev, Chander Prakash, Mohammad Khalid, Kavindra Kesari (Eds.) *Advanced Manufacturing Processes*, 2025. pp. 138–50, CRC Press, Boca Raton.

3 Microstructure and Tribology of Plasma-Sprayed Al-Si-Cu Composite Coatings

Praveen Kumar Verma, Hitesh Vasudev, and Shivam Kumar

3.1 INTRODUCTION

Aluminium alloys, with their characteristic properties of lightweight structure, high specific strength, and good machinability, have found widespread usage in several industries, e.g. food packaging, automobile manufacturing, ships, and aviation. However, the weakness of conventional aluminium-based alloys, with low strength and poor wear resistance, completely limits their application in the field of friction and wear [1]. Thus, the need to increase the wear resistance of conventional aluminium-based alloys has attracted significant interest from scientists globally. High-silicon aluminium (HSA) alloys containing a silicon content above 17 wt% are famous for their light weight, low thermal expansion coefficient, and excellent wear resistance. They have widespread use in the field of wear resistance, especially in the manufacture of elements such as pistons and cylinder liners of automobile engines [2]. The mechanical properties of binary HSA alloys are controlled by the silicon content, size, morphology, and distribution of the silicon phases. But with increased silicon content, coarse and irregular primary silicon and needle-like eutectic silicon phases could occur in the microstructure. These phases tend to cut through the α-Al matrix and increase stress concentration, thus drastically weakening the mechanical properties [3].

The continuous demand for materials with superior wear resistance, mechanical stability, and cost-effectiveness has accelerated the development of composite coatings for automotive, aerospace, and industrial applications. Aluminium-based alloys, particularly Al-Si-Cu systems, are widely employed in structural and tribological components due to their favourable combination of light weight, strength, and corrosion resistance. However, under severe frictional loading, conventional Al-Si-Cu alloys often suffer from adhesive wear and surface degradation. Plasma spraying offers an effective means to deposit dense Al-Si-Cu composite coatings with modified microstructure and enhanced surface performance [4].

With the aim to realize the potential application of HSA alloys, the utilization of low-cost and high-efficiency surface modification technologies for the fabrication of

DOI: 10.1201/9781003730637-3

HSA alloy coatings is crucial. This method not only preserves the lightweight and high-strength feature of aluminium alloy components but also truly enhances the surface wear-resistant performance of conventional aluminium-based alloys. It is one of the successful ways to speed up the rapid progress of HSA alloys. Laser cladding, plasma spraying [5-6], and magnetron sputtering have been utilized to prepare high-performance coatings for HSA coatings. Of these, plasma spraying technology offers advantages, including a low process, low cost, high efficiency, and high cooling speed, making it an efficient way to prepare HSA coatings [7]. At the same time, the incorporation of the alloying element Cu to create hard second-phase particles (Al_2Cu) improves coating wear resistance; however, the improvement is quite limited [8]. One of the effective and dependable methods to enhance the wear resistance of the coating is by adding hard ceramic phases [9]. TiC, possessing a high melting point, high hardness, and superior wear and corrosion resistance, is one of the widely applied ceramic reinforcing phases for thermal spraying coatings. This paper reviews the microstructural evolution, tribological behaviour, and potential applications of plasma-sprayed Al-Si-Cu coatings [10].

3.2 PLASMA SPRAYING FUNDAMENTALS OF AL-SI-CU COATINGS

Plasma spraying, a branch of thermal spray technologies, has become one of the most advanced and versatile surface engineering processes for the production of high-performance metallic and composite coatings. The fundamental principle is based on the creation of a very high-temperature plasma jet, formed by maintaining a direct current (DC) electric arc between an anode and a cathode with the assistance of a plasma-forming gas mixture (most commonly argon, nitrogen, hydrogen, or helium). Temperatures within this plasma plume can be up to 10,000–15,000 K, hot enough to cause total or partial melting of injected feedstock particles. These molten or semi-molten particulates are then accelerated towards a substrate at supersonic speeds, where they flatten rapidly, quench, and solidify to create lamellar splats that overlap to construct a coating microstructure [11].

Applicable to Al-Si-Cu feedstock systems, the plasma spray process allows the production of microstructures impossible by traditional casting or powder metallurgy paths. The high plasma enthalpy, coupled with the abrupt thermal gradients encountered during deposition, induces novel solidification routes and phase selection mechanisms. The resultant microstructural characteristics—grain refinement, metastable phase retention, intermetallic formation, and refined eutectic morphologies—are a determining factor in deciding the ultimate tribological and mechanical performance of the coating [12]. Figure 3.1 shows a schematic of the setup for plasma spraying.

The effectiveness of plasma spraying as a means of Al-Si-Cu coating design is directly reliant on a set of processing parameters. Plasma arc power, controlled by voltage and current, directly affects the particle thermal history, melting extent, and in-flight oxidation effects. The gas composition controls the plasma enthalpy and velocity profile, with hydrogen additions frequently used for augmenting reducing capacity and particle heating efficiency. Stand-off distance, or the distance between substrate and plasma nozzle, controls thermal and kinetic energy transferred to

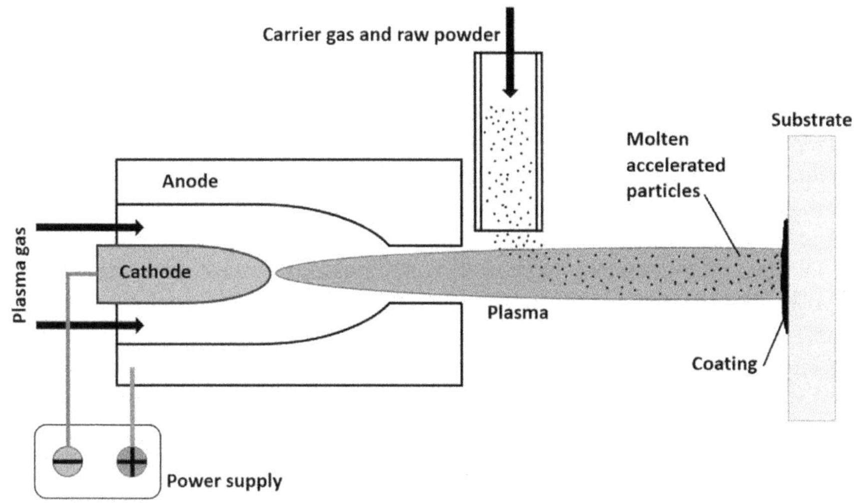

FIGURE 3.1 Schematic representation of plasma spraying setup [13].

striking particles; very large distances favour premature solidification and porosity, but short distances enhance splashing and overheating of substrate. In the same way, powder feed rate controls particle–plasma interaction time and directly influences coating thickness uniformity, deposition efficiency, and porosity content. The optimization of these mutually dependent parameters is of prime importance for the deposition of dense, adherent, and microstructurally tailored Al-Si-Cu composite coatings. One of the characteristic features of plasma-sprayed Al-Si-Cu coatings is their intrinsic lamellar microstructure. When deposited, molten droplets strike the substrate, spread under hydrodynamic pressure, and solidify in microseconds to form splats of thickness normally in the range of 1–5 μm. The overlapping of adjacent splats creates a heterogeneous microstructure of inter-splat boundaries, inter-lamellar porosity, and the presence of unmelted or partially melted particles. These morphological factors, although inherent to the process, have intense effects on the density of coating, thermal conductivity, hardness, and wear behaviour [14, 15].

From the phase-evolution point of view, the Al-Si-Cu system is of special interest under plasma spraying conditions. Al matrix solidifies quickly, encasing Si particles that are finely dispersed, and favours the development of Al-Si eutectic colonies. The addition of Cu also alters solidification routes further, with the precipitation of intermetallic phases like Al_2Cu and intricate Al-Cu-Si ternary compounds. These intermetallics, though being brittle in bulk form, play a valuable role in improving the microhardness and load-carrying capacity of the coating when dispersed as nanoscale or submicron reinforcements in the ductile Al matrix. In addition, the fast solidification kinetics associated with plasma spraying can stabilize metastable Al-rich solid solutions and inhibit coarsening of Si and Cu-rich phases, thereby conferring greater hardness and tribological performance than conventionally processed Al-Si-Cu alloys. An important determinant of the quality of Al-Si-Cu plasma coatings is interfacial adhesion. In contrast to diffusion-bonded or metallurgically continuous coatings, adhesion within plasma-sprayed systems is mainly created through a hybrid process of mechanical interlocking and localized metallurgical interactions. The substrate surface is usually roughened by grit blasting and develops asperities that provide anchoring of splats during solidification. Besides, localized metallurgical bonding also results from partial remelting of the substrate surface or interdiffusion across the interface, where higher plasma energies are used. The strength of adhesion is therefore closely related to substrate pre-treatment, particle kinetic energy, and local thermal gradients. Optimized conditions can provide adhesion strengths above 35–50 MPa, making the coatings feasible for demanding tribological environments [16].

Although the merits of plasma spraying are clear, there are some inherent difficulties that need to be recognized. High quenching and the non-equilibrium conditions of deposition tend to cause residual stresses within the coating, which can appear as microcracks or delamination during service. In addition, the lamellar structure, while useful in sustaining strain, can serve as a pathway for crack extension and wear particle generation. Porosity, oxides, and unmelted particles—characteristic in plasma-sprayed coatings—further compound these constraints if not effectively managed by parameter optimization and post-treatment. Solutions like laser remelting, hot isostatic pressing, and controlled atmosphere spraying are thus increasingly investigated to counteract these microstructural deficiencies and improve coating performance [17].

3.3 MICROSTRUCTURAL CHARACTERISTICS

The microstructural makeup of plasma-sprayed Al-Si-Cu composite coatings is the strongest single factor governing their mechanical and tribological properties. Contrary to conventionally processed alloys, wherein equilibrium phase transformations prevail, the plasma spray process brings about highly non-equilibrium solidification, with very steep thermal gradients (10^5–10^6 K/s) and very short solidification times (microseconds). This imparts to the coating a distinctive hierarchy of structural characteristics, from nanoscale precipitates to lamellar topographies, each of which plays a pivotal role in mediating hardness, adhesion, and wear behaviour [18]. Figure 3.2 shows the Al-25Si-4Cu-0.9Mg powder and inner pore coating microscopic morphologies: (a) powder OM image; (b) powder SEM picture following corrosion; (c) powder OM image; and (d) coating SEM image following corrosion. SEM pictures of coating surfaces with different Al contents (a) 0%, (b) 20%, (c) 40%, and (d) 60% at ×500 magnification are shown in Figure 3.3. Increasing Al content significantly influences surface morphology, reducing porosity, refining splat boundaries, and enhancing coating density, thereby improving wear resistance and mechanical integrity.

Grain refinement is a primary consequence of the rapid quenching inherent in plasma spraying. After impact, molten Al-Si-Cu drops experience hydrodynamic spreading and instantaneous solidification, inhibiting long-range diffusion and preventing coarsening of solidification fronts. This effect leads to the stabilization of ultrafine equiaxed grains within the Al matrix, often pushing into the nanocrystalline regime. Such refinement increases the Hall–Petch strengthening effect, thus increasing

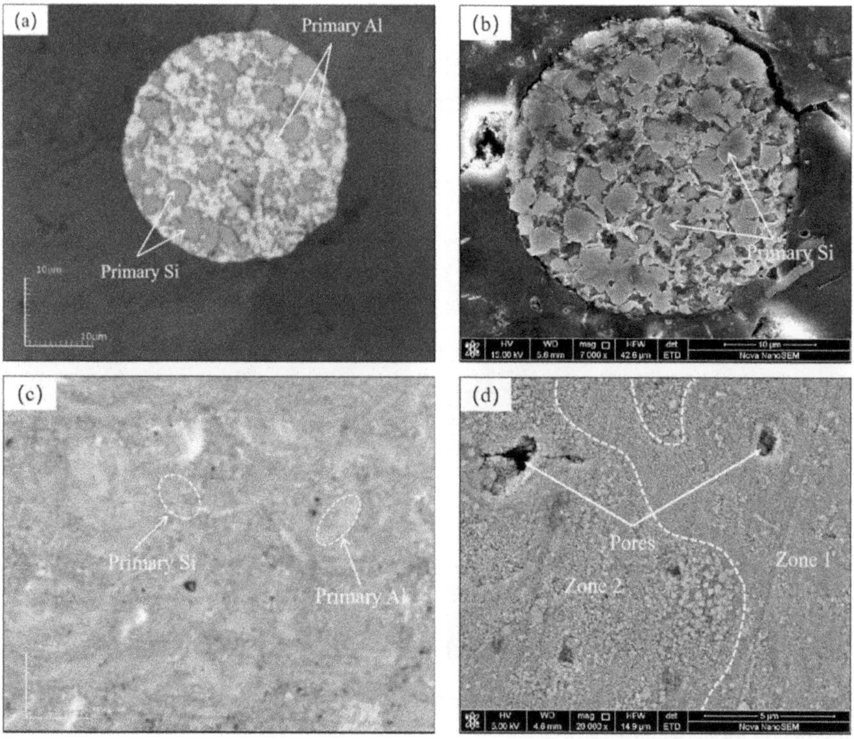

FIGURE 3.2 Microscopic morphologies of the Al-25Si-4Cu-0.9Mg powder and inner pore coating: (**a**) OM image of the powder; (**b**) SEM image of the powder after the corrosion; (**c**) OM image of the coating; (**d**) SEM image of the coating after corrosion [19].

hardness and resistance to localized plastic deformation. Refinement, though, can become too much and create high-angle grain boundaries that serve as nucleation sites for cracks under cyclic or high-stress wear conditions, illustrating the bimodal function of refinement in strengthening and embrittlement [21].

The eutectic structure in Al-Si-Cu coatings is a characteristic feature. Silicon, being virtually insoluble in Al under ambient conditions, precipitates as discrete or connected particles within the Al matrix, frequently occurring in a well-refined eutectic morphology because of rapid solidification. The Si-rich phases are intrinsic hard reinforcements and contribute abrasive wear resistance by supporting load and mitigating direct metallic contact during sliding. Their jagged morphology, though, can lead to stress concentrations at the Al/Si interface, facilitating crack nucleation under harsh tribological loading. The potential to control the eutectic network by tuning process parameters is therefore critical: coarse Si leads to brittle failure, but ultra-fine Si particles increase hardness but can embrittle the coating if the volume content is beyond the regime for effective load partitioning [22]. Optical micrographs of the as-cast A354 alloy are shown in Figure 3.4. Eutectic silicon particles around the dendritic α-Al matrix are shown in Figure 3.4(a), while a higher magnification view

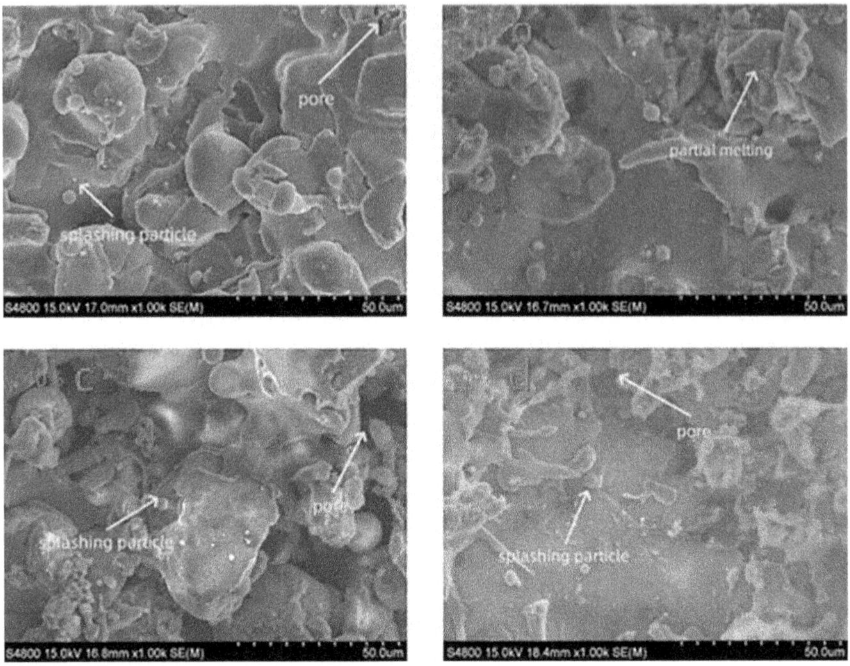

FIGURE 3.3 SEM morphology of coating surfaces with different Al content ×500. (**a**) 0%, (**b**) 20%, (**c**) 40%, (**d**) 60% [20].

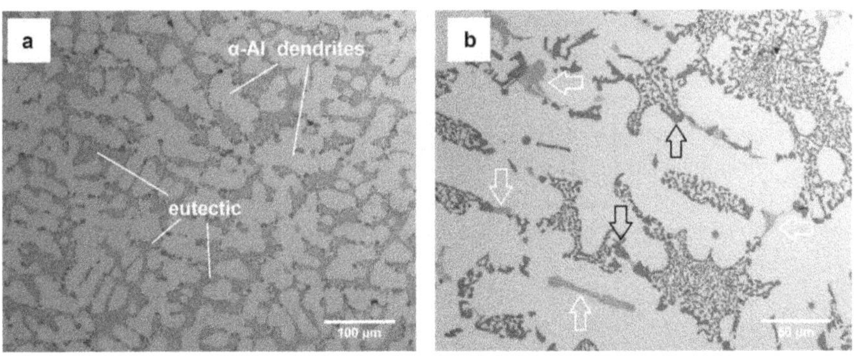

FIGURE 3.4 Optical micrographs of the as cast A354 alloy: (**a**) dendritic structure surrounded by eutectic silicon particles and (**b**) higher magnification image showing coarse intermetallics formed during solidification (white arrows) and eutectic Si particles, partially unmodified (black arrows) [23].

(Figure 3.4(b)) shows largely unaltered eutectic Si particles (black arrows), which indicate incomplete modification during casting, and coarse intermetallic phases (white arrows) generated during solidification.

Also vital is the microdistribution of Cu-containing intermetallics, including Al_2Cu and intricate Al-Cu-Si ternaries. These phase elements impart improved load-bearing properties and heat stability, both of which are vital in coatings used in friction-demanding environments. In finely dispersed forms, they act synergistically with the Si reinforcements to enhance hardness and reduce adhesive wear. On the other hand, uncontrolled intermetallic network growth promotes brittleness, decreases toughness, and makes the coating susceptible to catastrophic delamination under dynamic loading conditions. Non-equilibrium states imposed by processing often tend to stabilize metastable intermetallic variants, which, although imparting additional hardness, have the potential to transform under stress conditions during service, leading to structural instability [24].

The occurrence of defects, i.e. pores, oxides, and microcracks, is an inevitable but central factor that determines Al-Si-Cu coating integrity. Porosity results from incomplete melting, lack of particle acceleration, or turbulence upon droplet impingement, and excessive porosity (>5%) will significantly reduce hardness, adhesion, and wear resistance by acting as sites of crack initiation and propagation. The controlled plasma spray process parameters can keep porosity below 5%, which is usually acceptable for tribological purposes. Al_2O_3 oxide inclusions are produced by the high reactivity of Al in the plasma jet and tend to be found along lamellar interfaces. Although individually they might add to local hardness, their brittleness lessens toughness and promotes delamination under cyclic stresses. Microcracks, frequently produced through thermal mismatch between the substrate and coating or due to quick quenching stresses, are precursors for spallation when sliding wear occurs [25].

Hence, Al-Si-Cu coatings' performance is inextricably linked with the engineering of its microstructural characteristics. Maintaining an optimum balance means minimizing defects while harnessing grain refinement, eutectic strengthening, and constrained intermetallic dispersion simultaneously. Customization of the spray parameters—plasma power, gas composition, stand-off distance, and powder feed morphology—is essential to arrive at this balance. For example, greater plasma enthalpy enhances particle melting and densification but poses the danger of oxidation and residual stress buildup; lower enthalpy decreases oxidation but causes greater porosity and unmelted inclusions [26]. The microstructure of plasma-sprayed Al-Si-Cu coatings is a fine balance between desirable strengthening mechanisms and undesirable defect formation. Grain refinement, proper distribution of eutectic Si and Cu-rich intermetallics, and porosity and crack suppression all determine the wear resistance, hardness, and adhesion strength of the coatings. In the future, progress depends on accurate control of microstructural development through optimized parameters and post-deposition treatments with a focus on avoiding harmful heterogeneities while maintaining the reinforcing effectiveness of metastable and eutectic microphases [27].

3.4 TRIBOLOGICAL BEHAVIOUR

3.4.1 Wear Resistance

The plasma-sprayed Al-Si-Cu coatings exhibit significantly higher wear resistance compared to their bulk alloy counterparts through the synergistic strengthening mechanisms obtained from microstructural refinement and heterogeneity in phases. Ultrafine Al grains, Si particulates distributed uniformly, and Cu-rich intermetallic phases obtained due to rapid solidification in plasma spraying provide the architecture that facilitates efficient load partitioning, inhibits localized plastic deformation, and contributes to the improved abrasive and adhesive wear resistance of the coatings. The prevailing wear mechanisms depend on the severity of contact and testing conditions. Under moderate loads, abrasive wear is the dominant process controlled by the hard Si reinforcements and Al_2Cu phases' capacity to resist micro-cutting by counterface asperities. Hard phases function as load-bearing phases, while the ductile Al matrix accommodates the redistribution of stress. With rising loading or sliding speed, adhesive wear is enhanced since the softened Al matrix deforms plastically and smears, transferring the material on to the counterbody. Oxidative wear at high temperatures plays a major role, as the quick development of an oxide film (mainly Al_2O_3) acts as a solid lubricant that minimizes metallic contact. Repeated fracture and re-formation of these oxide layers, however, can induce transient wear acceleration [28]. Surface material under impacts (a) normal impact; (b) oblique impact) is shown in Figure 3.5.

The reduction in wear volume is credited to the dispersion of Si and intermetallics, which slow down crack propagation and delamination. Nevertheless, residual porosity and microcracks inherent in plasma spraying may serve as stress concentrators, aggravating wear in sub-optimally treated coatings. Therefore, the optimization of spray parameters to reduce defect density continues to be critical to obtaining reliable wear resistance in high-friction service. With the progress of aerospace, equipment manufacturing, and other high-tech industries, it is necessary to develop solid lubricating coatings with good performance at a low friction coefficient and wear rate over a broad temperature range to increase the service life of machine parts under severe conditions. NiAl intermetallic alloys were extensively explored for their application in the protection of structural materials against corrosion and high-temperature oxidation owing to their high melting point, high thermal conductivity, and remarkable oxidation and corrosion resistance at high temperature [30].

Besides, the Ni and Al components of the NiAl alloy will also oxidize at elevated temperatures to produce NiO and A_2O_3 and be utilized as solid lubricants at some temperatures to lower the friction coefficients of materials. Thus, NiAl intermetallic compounds also have much potential for application as wear-resistant coatings. Nevertheless, the clean NiAl coating normally experiences intense wear, which narrows its application in anti-wear areas. In recent years, numerous studies have aimed at employing NiAl intermetallic compounds as the matrix phase of high-temperature self-lubricating coatings to enhance the wear-resistance of the coatings at a broad temperature range by incorporating various lubricants. Among all kinds of lubricants, nano-diamond particles as a carbon source are the best material for tribological applications because their good chemical stability, high thermal conductivity, and high hardness have an important effect on the tribological properties of the materials. It is demonstrated that nano-diamond particles exhibit a self-lubricating effect and are stable for lubrication at even a high temperature of 727°C owing to the sp^3 hybridization and graphite-like carbon shell structure [31].

In comparison, plasma spraying is a low-cost and easy coating preparation method by which a coating can be deposited with a uniform and controlled thickness on any substrate material. It is thus extensively utilized for the preparation of lubricant coatings. The graphitization transformation of nano-diamond particles can occur

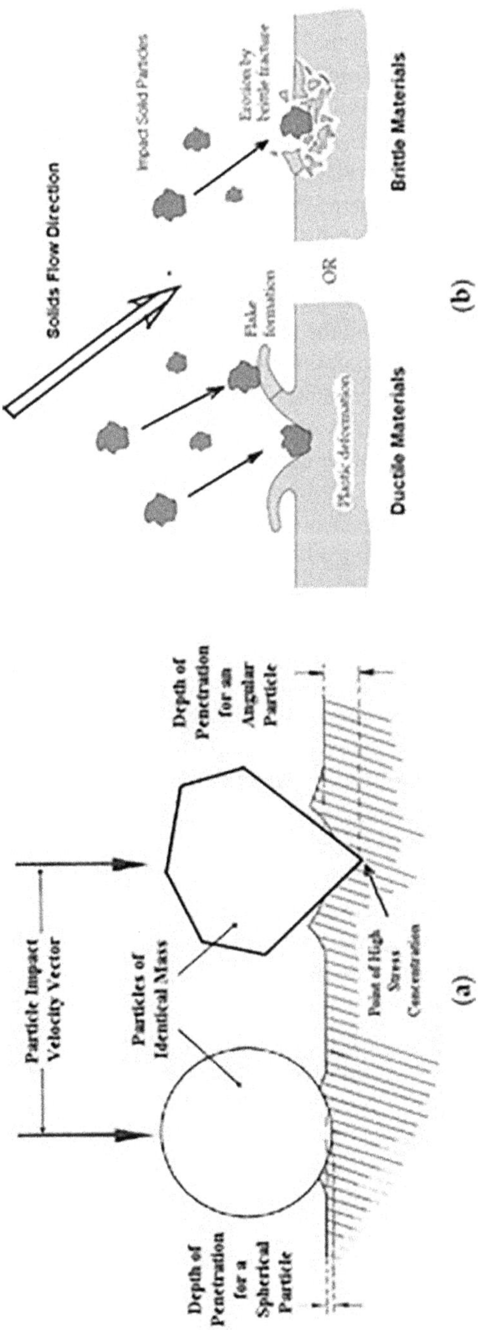

FIGURE 3.5 Surface material under impacts: (**a**) normal impact; (**b**) oblique impact [29].

during plasma spraying due to the thermal cycling up to a high temperature above their melting points undergone by individual spray particles. Graphite is a good lubricant material and the application of it as a lubricant phase is anticipated to enhance the friction and wear properties of the coatings from room temperature to 500°C. It is known that plasma-sprayed coatings commonly exhibit a standard lamellar structure, and both the coating/substrate and inter-splat interface bonding in the coating are the key factors influencing the performance of the coating. For metal powders, particularly those with Al, severe oxidation of the coating during plasma spraying in an ambient atmosphere always happens due to its high oxygen affinity, leading to high oxide inclusion content in the resulting coatings. The oxide inclusions cause generally poor coating/substrate interface bonding and inter-splat bonding between the coatings, leading to premature coating spalling along the inter-splat interface during wear, which significantly hampers the application of APS technology used in metal-based wear-resistant coating preparation [32].

The adhesion strength of the coating to the substrate is a very important parameter, which depends on substrate roughness and spray energy. Published adhesion strengths are in excess of 35 MPa for optimized coatings. Vickers hardness (200–350 HV) values are higher than those of normal Al alloys due to rapid solidification and intermetallic strengthening. Adhesive strength is a significant factor that influences the wear resistance of plasma-sprayed coatings [33].

3.5 FACTORS INFLUENCING TRIBOLOGY

The tribological behaviour of Al–Si–Cu coatings produced by plasma spraying is controlled not only by inherent alloy chemistry but also by external parameters like spraying conditions, powder shape, post-deposition processing, and addition of reinforcement. The parameters control the microstructural hierarchy, defect structure, and phase content in the coating, which together regulate hardness, adhesion, and wear resistance to varied loading regimes [34].

Plasma conditions are the key regulators of coating density and strength of interfacial bonding. Arc power, plasma gas composition, and standoff distance regulate melting efficiency, droplet velocity, and cooling rate. Increased plasma power is responsible for thorough melting of feedstock particles to result in well-flattened splats with excellent interlamellar bonding, while subthreshold energy input results in porous deposits with incomplete particle fusion that act as sites for the initiation of cracks on wear. Standoff distance has a twofold influence: very short distances result in oxidation and overheating of Al-rich microphases, while longer spray distances lead to early solidification and inferior splat adhesion. Regulation of these parameters can reduce porosity to less than 5%, effectively reducing third-body abrasive mechanisms and enhancing long-term wear resistance. Feedstock morphology is another determining factor in the quality of the coating. Spherical powders, conventionally made through gas atomization, exhibit better flowability and melting homogeneity than angular mechanically crushed powders. This guarantees homogenous splat deposition, decreased inter-splat porosity, and homogeneous dispersion of Si and Cu phases within the Al matrix. In contrast to angular feedstock particles, angular particles are apt to shadow during spraying, creating heterogeneous microstructures susceptible to adhesive wear and fatigue crack initiation. Therefore, spherical feedstocks provide isotropic tribological behaviour in cases of multidirectional sliding [35]. An overview of the many simulation methods frequently employed in tribologically related applications is shown in Figure 3.6.

Post-treatment methods like heat treatment, laser remelting, and hot isostatic pressing are commonly used to improve tribological stability. Thermal annealing may precipitate Cu-rich intermetallics and stabilize networks of eutectic Si, thus increasing hardness, although over-coarsening could lower toughness. Laser remelting, for instance, re-melts the surface layer, removing pores and oxide inclusions and enhancing metallurgical continuity across splat boundaries. The densification leads to smoother surfaces, decreased abrasive ploughing, and enhanced fatigue endurance in cyclic wear. In addition, post-treatments reduce residual stresses created through rapid quenching, enhancing adhesion and preventing spallation in adverse wear environments. Reinforcement additions are vital in the adaptation of tribological behaviour. Addition of ceramic particulates like SiC, Al_2O_3, and TiC converts Al–Si–Cu coatings into hybrid composites with enhanced hardness and wear resistance. SiC additions inhibit plastic deformation of the Al matrix and also retard abrasive wear, whereas Al_2O_3 enhances oxidative stability at high temperatures. TiC reinforcement also increases hardness and resists adhesive wear because of its high chemical stability and load-carrying capacity. Nonetheless, inadequate interfacial bonding between ceramics and the metal matrix tends to enhance interfacial debonding and crack growth. Properly optimized reinforcement dispersion guarantees effective load transfer, inhibits wear debris formation, and reduces the coefficient of friction, hence increasing coating life in high-friction conditions [37].

3.6 APPLICATIONS IN HIGH-FRICTION ENVIRONMENTS

The use of plasma-sprayed Al–Si–Cu composite coatings in harsh frictional environments has attracted growing industrial interest based on their novel capability to integrate light weight with excellent tribological performance. The intrinsic lamellar microstructure, combined with the presence of hard silicon phases and reinforcing Cu-rich intermetallics, confers a specific combination of hardness, wear resistance, and intermediate frictional response. These types of properties render such coatings extremely desirable in industries where components experience prolonged sliding, cyclic loading, and thermal stresses, yet are also needed to be reduced in weight for improved efficiency and performance [38].

In the automobile sector, Al–Si–Cu plasma-sprayed coatings are most applicable to high-performance engine and braking parts. Cylinder liners and piston rings, which function under boundary lubrication conditions, experience intense adhesive and abrasive wear. The existence of nanoscale Si dispersions in the coating matrix performs as a hard load-supporting skeleton that does not microplough, and the Al matrix offers toughness to prevent crack development. Moreover, Cu-rich intermetallics help provide enhanced load-bearing capability, promoting piston–liner interface durability under conditions of high-pressure combustion cycles. Brake rotors too are improved by such coatings, as the composite design resists combined effects of high-temperature oxidative degradation and abrasive wear from brake pads, thus

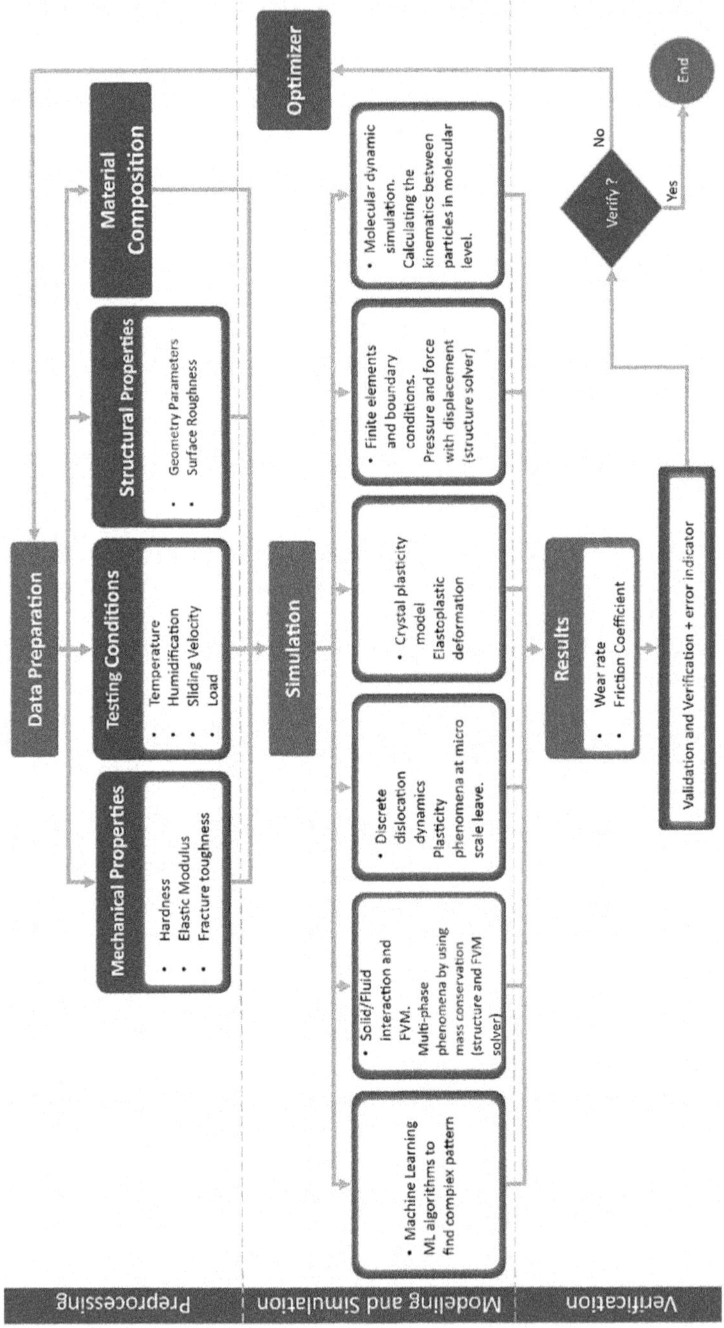

FIGURE 3.6 Overview of different simulation techniques commonly used in tribological-related applications [36].

improving service intervals and lowering particulate emissions. These improvements lead directly to improved fuel efficiency and lower maintenance costs, which are both vital to the contemporary automotive industry where there is strict regulation for sustainability [39].

In aerospace technology, the combination of light weight and wear resistance is particularly useful in enhancing fuel economy and reliability. Plasma-sprayed Al–Si–Cu coatings are being studied more and more for turbine parts subject to severe sliding contact, including abradable seals, compressor vanes, and structural joints. The coatings' fine microstructure reduces adhesive wear and restrains friction-induced material loss, thus ensuring dimensional integrity and aerodynamic efficiency throughout extended service lifetimes. Sliding seals of jet engines, for example, are under constant exposure to temperature and cyclic loads; the Al–Si–Cu system, with eutectic Si reinforcement and Cu-rich microstructure, provides stability and load-carrying ability in these adverse environments. Structural joints in aerospace fittings also appreciate the increased adhesion strength and friction control contributed by these coatings, enabling increased reliability during dynamic fluctuations in loads. This combination of wear resistance and lower mass results in Al–Si–Cu coatings being significantly oriented towards the aerospace industry's continued move towards lightweight materials while not sacrificing functional performance [40].

The microstructure and electrochemical behaviour of WC-Co-Cr coatings deposited by High-Velocity Oxy-Fuel (HVOF) spraying on 316L boiler steel were studied. The coatings had a high-density microstructure with negligible porosity, which was responsible for their wear and corrosion resistance. Electrochemical experiments showed enhanced corrosion resistance in high-temperature conditions due to the stable carbide phases and the presence of cobalt and chromium. Subsequent heat treatment after coating also further optimized the microstructure, which resulted in superior performance in aggressive conditions. These results indicate that HVOF-sprayed WC-Co-Cr coatings are suitable for use in corrosive high-temperature environments [41]. A study assessed the erosion resistance of Al_2O_3 and Cr_2O_3 coatings deposited on turbine steels by thermal spraying. The work determined that Cr_2O_3 coatings showed better erosion resistance than Al_2O_3 coatings, which was due to the greater hardness and stronger microstructure of the Cr_2O_3 coatings. The performance of the coatings under different impingement angles and velocities was tested, and the findings relate to their potential application in turbines where erosion is a major issue. The findings emphasize the significance of coating materials selection to maximize the reliability and lifespan of turbine components [42].

A study investigated the improvement of corrosion resistance of SS 304 stainless steel by coating it with HVOF-sprayed WC-Co-Cr coatings. The research proved that the coatings drastically enhance the resistance of the material against corrosion due to chlorides, thus qualifying them for marine and chemical processing uses. The electrochemical analyses showed that the coatings serve to prevent corrosion by lowering the corrosion rate and lengthening the working life of the steel under aggressive conditions. The research proves the efficiency of the WC-Co-Cr coatings in preventing corrosion problems in environments rich in chlorides [43]. An investigation focused on the bioactivity and corrosion performance of hydroxyapatite

(HA)-based coatings deposited on surgical-grade stainless steel by the Low-Velocity Oxy-Fuel (LVOF) spraying method. The study contrasted pure HA coatings and HA composites with micrometric and nanometric Al_2O_3. The findings showed that the HA-Al_2O_3 composites possess higher mechanical properties and anticorrosion behaviour, and thus are better suited for use in biomedical applications. The bioactivity tests further showed that the coatings enhance osteointegration, which is vital for the success of orthopedic implants. From the research, HA-Al_2O_3 composite coatings are potential candidates for enhancing the performance and lifespan of surgical implants [44].

The industrial machine sector is another significant area in which Al–Si–Cu coatings prove to be highly beneficial. In pump shafts, gear parts, and sliding bearings, tribological effectiveness dictates system reliability and operational life. These parts often undergo mixed lubrication conditions, with metal-to-metal contact causing adhesive wear as well as fatigue. Al–Si–Cu coatings deposited by plasma spraying overcome this problem through the formation of a dense, fine microstructure with enhanced hardness and wear resistance. The hard Si phases promote abrasion resistance against particulates present in fluids, while the Al–Cu matrix is retained for toughness to oppose fatigue cracking. Additionally, the coatings' capability to maintain moderate friction coefficients in dry or lubricated conditions decreases energy losses in machinery systems, ultimately enhancing overall operational efficiency. In industrial gear boxes, the coatings minimize scuffing and pitting, effects most evident under high-load operation, while elongating service life in harsh environments such as mining, oil production, and heavy industry [41].

3.7 FUTURE PERSPECTIVES

Although plasma-sprayed Al–Si–Cu composite coatings have shown encouraging tribological performance, there are a number of challenges that still curtail their applicability in industry on a large scale. Next-generation research will need to transcend traditional coating optimization and adopt novel design paradigms, nanostructured precursors, and computational-based methods in order to fully realize the potential of these systems in extreme friction conditions [42].

Microcrack and porosity control is one of the main challenges. Plasma spraying intrinsically creates fast solidification, thermal gradients, and splat boundaries that create localized stresses, usually leading to crack generation and inter-splat porosity. The defects reduce coating toughness, lower adhesion strength, and serve as stress concentrators during cyclic wear. Although optimized plasma conditions and post-spray heat treatments have alleviated these problems to some extent, future research should focus on defect engineering through tighter control of droplet solidification dynamics. Enhanced in-flight diagnostics and real-time particle temperature and velocity monitoring may offer predictive control of splat formation and hence low defect density. Further, integration of such monitoring with closed-loop feedback mechanisms might allow for adaptive optimization of spray parameters during deposition. FGCs can bypass this drawback by composition and microstructure transition from ductile Al-rich regions close to the substrate to hard, Si- and Cu-rich surface phases gradually. These architectures would enhance thermal stability and wear resistance while, at the same time, lowering interfacial stresses. Investigations into such developments should consider examining both powder mixing methods and novel spray processing routes that enable compositional modulation during deposition. The application of nanostructured feedstocks is another revolutionary direction. Nanostructured Al–Si–Cu powders, processed by cryomilling or high-energy ball milling, can substantially reduce grain size, increase solid solution strengthening, and enhance uniform dispersion of Si and Cu phases in the Al matrix. Deposited by plasma spraying, such powders can maintain their nanocrystalline nature, and the resulting coatings will have significantly enhanced hardness, wear resistance, and crack propagation resistance [43].

One promising area is combining computational modelling with machine learning for predictive design of coating systems. Plasma spraying represents an extremely nonlinear coupling of plasma dynamics, particle trajectory, melting, solidification, and coating build-up. Conventional experimental optimization is time-consuming and usually inadequate to describe the intricate parameter space. Multiscale computational models ranging from particle-scale thermal-fluid models to macroscale coating stress analysis can offer an understanding of defect inception and microstructure development. When integrated with machine learning algorithms developed from large experimental and simulation datasets, predictive models could be built to optimally expedite spray parameters, powder morphology, and feedstock composition at high rates. This paradigm revolution towards data-driven coatings engineering will dramatically expedite the development of high-performance Al–Si–Cu coatings. Post-spray treatments are also at the fore of future developments. Although laser remelting has been shown to be effective in minimizing porosity and encouraging metallurgical continuity, future research must investigate hybrid methods combining remelting with surface alloying to design microstructural gradients and intermetallic phase distributions. Just as hot isostatic pressing (HIP) is a route towards internal porosity elimination and interlamellar bonding improvement, cost and scalability are yet to be overcome in its industrial implementation. Localized HIP and pulsed laser treatments are emerging technologies that can make cost-effective alternatives a reality, allowing defect mitigation without sacrificing throughput [44].

Most recently, in conjunction with microwave processing, additive manufacturing, and magnetron sputtering thin-film coatings, thermal spray coatings and other surface engineering technologies have expanded within the fields of biomedicine, slurry, and high-temperature corrosion systems [45–52].

Predicting and simulating surface roughness, tensile, compressive, and bending strength, and performance in solar applications is possible with coatings. Coatings facilitate the design of flow systems and the simulation of more complex structures, such as ergonomically designed chairs [53–63].

3.8 CONCLUSION

Plasma-sprayed Al–Si–Cu composite coatings are notable for a distinctive synergy of light weight, sophisticated microstructure, and improved tribological performance, rendering them highly desirable for high-friction and high-load situations.

Their excellent hardness, wear resistance, and moderate friction coefficients are attributable to rapid solidification-induced grain refinement, silicon phase dispersion, and intermetallic Cu-rich formation. However, inherent limitations like porosity, microcracking, and residual stresses remain to hinder long-term reliability. Improving plasma spray parameters, embracing ceramic reinforcements, and utilizing post-spray treatment are the main strategies for addressing these limitations. With continuing developments in nanostructured powders, hybrid composites, and predictive process models, Al–Si–Cu coatings are set to dominate next-generation automotive, aerospace, and industrial parts with longer service life, lower maintenance expenses, and better energy efficiency.

REFERENCES

1. Zhang J, Fu B, Jiang Y, Dong T, Li G. Microstructure and tribological properties of Al-30Si-5Cu composite coating reinforced with ceramic phases synthesized in situ by plasma spraying. *Surface and Coatings Technology*. 2024 Jun 30;486:130946.
2. Peng Q, Liu M, Huang Y, Zhou X, Ma G, Wang H, Xing Z. Microstructure, phase composition, mechanical properties and tribological properties of plasma sprayed Al-25Si wear-resistant coatings. *Surfaces*. 2022 Jul 13;5(3):350–64.
3. Xu HB, Ma GZ, He PF, Li GL, Song Y, Zhu XY, Liu M, Zhao HC, Wang HD. Structural characteristics and tribological properties of an ultrafine-grained Al–40Si–5Fe coating prepared by supersonic plasma spraying. *Journal of Materials Research*. 2024 Jan 28;39(2):220–30.
4. Zhang J, Fu B, Jiang Y, Dong T, Li G. Preparation of in-situ synthesized TiC/AlSi30Cu5 coating by high velocity oxy-fuel (HVOF) spraying: Microstructure evolution, mechanical properties, and tribological behavior. *Tribology International*. 2025 Jan 1;201:110254.
5. Altunpak Y, Akbulut H, Üstel F. The effect of plasma spraying on the microstructure and aging kinetics of the Al-Si Matrix Alloy and Al-Si/SiC composites. *Journal of materials engineering and performance*. 2010 Feb;19(1):116–22.
6. Kucita P. *The development of a wear resistance aluminium bronzes (Cu-Al-Fe) coating* (Doctoral dissertation, University of Southampton).
7. Zavareh MA, Doustmohammadi E, Sarhan AA, Karimzadeh R, Nia PM, Al RS, Singh K. Comparative study on the corrosion and wear behavior of plasma-sprayed vs. high velocity oxygen fuel-sprayed Al8Si20BN ceramic coatings. *Ceramics International*. 2018 Aug 1;44(11):12180–93.
8. Zaguliaev D, Konovalov S, Ivanov Y, Gromov V. Microstructure and micro-hardness behavior of Ti–Y2O3–Al–Si composite coatings prepared in electron-plasma alloying. *Materials Characterization*. 2019 Dec 1;158:109934.
9. Subbiah R, Arun A, Lakshmi AA, Harika AN, Ram N, Sateesh N. Experimental study of wear behaviour on Al-2014 alloy coated with thermal spray HVOF (High Velocity Oxy-Fuel) and plasma spray process: A review. *Materials Today: Proceedings*. 2019 Jan 1;18:5151–7.
10. Arendarchuck BE, Bertuol K, Rivadeneira F, de Castilho BC, Barnett B, Moreau C, Stoyanov P. Unveiling the influence of nickel on the erosion and the tribological performance of AlSi-based abradable coatings. *Journal of Materials Science*. 2024 Dec;59(48):22266–82.

11. Dahotre NB, Nayak S. Nanocoatings for engine application. *Surface and Coatings Technology.* 2005 Apr 20;194(1):58–67.
12. Deuis RL, Yellup JM, Subramanian C. Metal-matrix composite coatings by PTA surfacing. *Composites Science and Technology.* 1998 Jan 1;58(2):299–309.
13. Benčina M, Resnik M, Starič P, Junkar I. Use of plasma technologies for antibacterial surface properties of metals. *Molecules.* 2021 Mar 5;26(5):1418.
14. Fu Y, Chen X, Zhang B, Gong Y, Zhang H, Li H. Fabrication of nanodiamond reinforced aluminum composite coatings by flame spraying for marine applications. *Materials Today Communications.* 2018 Dec 1;17:46–52.
15. Song Y, Jin G, Ma G, Guo W, He P, Liu M, Wang H, Mou H. Formation mechanism of nano planar defects in SAPS Al-Si-Ce coatings. *Materials Today Communications.* 2025 Mar 1;44:112145.
16. Li C, Li S, Liu C, Zhang Y, Deng P, Guo Y, Wang J, Wang Y. Effect of WC addition on microstructure and tribological properties of bimodal aluminum composite coatings fabricated by laser surface alloying. *Materials Chemistry and Physics.* 2019 Aug 1;234:9–15.
17. Morks MF, Gao Y, Fahim NF, Yingqing FU, Shoeib MA. Influence of binder materials on the properties of low power plasma sprayed cermet coatings. *Surface and Coatings Technology.* 2005 Sep 1;199(1):66–71.
18. Yang G, Fu B, Dong T, Li G. Effect of laser remelting on the microstructure and properties of the aluminum high silicon alloy coating. *Journal of Materials Processing Technology.* 2024 Mar 1;324:118278.
19. Mehta, A., Vasudev, H., Sabareshwaran, S. *et al.* Future research directions and applications of artificial intelligence in tribology. *Discov Mechanical Engineering* **5**, 68 (2026). https://doi.org/10.1007/s44245-026-00249-0
20. Dai P, Zhang Y, Wang X, Yan J, Zhao L, Yi X. Study on the microstructure and wear properties of Al2O3-3% TiO2-xAl composite coatings prepared by plasma spraying. *Crystals.* 2025 Sep 15;15(9):809.
21. Han BY, Gao XH, Chen SY, Cong MQ, Cui FF, Hang WX, Wu HD, Chu JJ. Investigation tribological behavior of plasma sprayed Ni-based coating prepared on aluminum alloy under different temperatures. https://dx.doi.org/10.2139/ssrn.4399113
22. Kumar D, Pandey KN. Experimental investigations of sol-gel process parameters for wear reduction on thermal barrier coated AA2024 aluminum alloys with the use of Taguchi-based optimization. *Sādhanā.* 2020 Dec;45(1):187.
23. Toschi S. Optimization of A354 Al-Si-Cu-Mg alloy heat treatment: Effect on microstructure, hardness, and tensile properties of peak aged and overaged alloy. *Metals.* 2018 Nov 17;8(11):961.
24. Javidani M, Larouche D. Application of cast Al–Si alloys in internal combustion engine components. *International Materials Reviews.* 2014 Apr 1;59(3):132–58.
25. Rajaram G, Kumaran S, Rao TS, Kamaraj M. High temperature dry sliding wear behaviour of Al-Si/graphite composites processed by stir casting. In TMS 2011 140th annual meeting and exhibition, materials fabrication, properties, characterization, and modeling, 2011 Feb 28 (Vol. 2, p. 191). Wiley.
26. MacGregor OS. *Evaluating the wear and corrosion resistance of plasma-sprayed tungsten carbide coatings on aluminium-6082 alloy* (Master's thesis, University of the Witwatersrand, Johannesburg (South Africa)).
27. Feng LP, Fleury E, Zhang GS. Preparation of Al-Cu-Fe-(Sn, Si) quasicrystalline bulks by laser multilayer cladding. *International Journal of Minerals, Metallurgy, and Materials.* 2012 May;19(5):434–40.

28. Ram SC, Chattopadhyay K, Bhushan A. A literature review on Al-Si alloy matrix based in situ Al-Mg2Si FG-composites: Synthesis, microstructure features, and mechanical characteristics. *Proceedings of the Institution of Mechanical Engineers, Part C: Journal of Mechanical Engineering Science*. 2023 Feb;237(4):919–40.
29. Deng T. Erosive wear mechanisms of materials—A review of understanding and progresses. *Materials*. 2025 Apr 2;18(7):1615.
30. Quazi MM, Fazal MA, Haseeb AS, Yusof F, Masjuki HH, Arslan A. Laser-based surface modifications of aluminum and its alloys. *Critical Reviews in Solid State and Materials Sciences*. 2016 Mar 3;41(2):106–31.
31. Zhang L, Liao XJ, Chen R, Luo XT, Li CJ. Tribological behavior of NiAl coating deposited by plasma spraying of diamond-contained Ni/Al composite powder at a wide range temperature from 25 °C to 700 °C. *Transactions of Materials Research*. 2025 Sep 9;1:100100.
32. Zaguliaev D, Gromov V, Rubannikova Y, Konovalov S, Ivanov Y, Romanov D, Semin A. Structure and phase states modification of AL-11SI-2CU alloy processed by ion-plasma jet and pulsed electron beam. *Surface and Coatings Technology*. 2020 Feb 15;383:125246.
33. Min ZE, Hong YA, Bao-biao YU, Zhi HU. Microstructure, microhardness and corrosion resistance of laser cladding Ni–WC coating on AlSi5Cu1Mg alloy. *Transactions of Nonferrous Metals Society of China*. 2021 Sep 1;31(9):2716–28.
34. Ordóñez Jiménez AF. Propiedades mecánicas y tribológicas de recubrimientos nanoestructurados de TiAlCrN-Si por co-sputtering con fuentes Hipims. https://redcol.minciencias.gov.co/Record/REPOUPTC2_05c1a9730424ad2fff6f3f9042ec4c7d/Details
35. Li K, Li W, Zhang G, Zhu W, Zheng F, Zhang D, Wang M. Effects of Si phase refinement on the plasma electrolytic oxidation of eutectic Al-Si alloy. *Journal of Alloys and Compounds*. 2019 Jun 25;790:650–6.
36. Kurdi A, Alhazmi N, Alhazmi H, Tabbakh T. Practice of simulation and life cycle assessment in tribology—A review. *Materials*. 2020 Aug 7;13(16):3489.
37. Sundaramoorthy R, Sasidhar G, Akhilesh AP, Vardhan AH. Tribological study of modified Al-6013 aluminum alloys. *Journal of Bio-and Tribo-Corrosion*. 2024 Jun;10(2):26.
38. Chengwen S, Jun KD. Microstructure and electrochemical performance of plasma sprayed Cu11. 85Al3. 2Mn0. 1Ti alloy coating in different corrosive solutions. *Canadian Metallurgical Quarterly*. 2023 Oct 2;62(4):734–44.
39. AVAL HJ. Effects of aging heat treatment on microstructure and corrosion behavior of friction surfacing treated Al–Zn–Mg–Cu matrix composite. *Transactions of Nonferrous Metals Society of China*. 2023 Aug 1;33(8):2303–13.
40. Feng HK, Yu SR, Li YL, Gong LY. Microstructure and properties of Sip/Al–Si surface composites prepared by ultrasonic method. *Materials & Design*. 2009 Aug 1;30(7):2420–4.
41. Singh SP, Verma PK, Singh S, Kapoor M, Balu R. Microstructural characterization and electrochemical behaviour of HVOF sprayed WC-Co-Cr coatings on 316L boiler steel stainless steel. *Results in Surfaces and Interfaces*. 2024 Aug 1;16:100248.
42. Verma PK, Minhas AS, Singh P, Kumar S. Slurry erosion behaviour of thermal sprayed Al2O3 and Cr2O3 coatings for turbine steels. *AIP Conference Proceedings*. 2024 Feb 20;2986(1):020012.
43. Verma PK, Vasudev H, Singh G. Improving the corrosion performance of SS 304 stainless steel using HVOF-sprayed WC-Co-Cr coatings in chloride environments. *Journal of Bio-and Tribo-Corrosion*. 2025 Dec;11(4):1–3.

44. Verma PK, Vasudev H, Kumar V. Bioactivity and corrosion performance of low-velocity oxy-fuel-sprayed hydroxyapatite-based coatings on a surgical-grade stainless steel. *Transactions of the Indian Institute of Metals*. 2025 Aug;78(8):185.
45. Dutta V, Thakur L, Singh B, Vasudev H. A study of erosion–corrosion behaviour of friction stir-processed chromium-reinforced NiAl bronze composite. *Materials*. 2022;15(15):5401.
46. Singh M, Vasudev H, Kumar R. Corrosion and tribological behaviour of BN thin films deposited using magnetron sputtering. *International Journal of Surface Engineering and Interdisciplinary Materials Science (IJSEIMS)*. 2021;9(2):24–39.
47. Sharanabasava H, Raviprakash M, Prasad CD, Ramesh MR, Phanibhushana MV, Vasudev H, Kumar S. Microstructure, mechanical and wear properties of SiC and Mo reinforced NiCr microwave cladding. *Advanced Materials and Processing Technologies*. 2023;10(4): 3620–33.
48. Singh M, Vasudev H, Singh M. Surface protection of SS-316L with boron nitride based thin films using radio frequency magnetron sputtering technique. *Journal of Electrochemical Science and Engineering*. 2022;12(5):851–63.
49. Prashar G, Vasudev H, Thakur L, Bansal A. Performance of thermally sprayed hydroxyapatite coatings for biomedical implants: A comprehensive review. *Surface Review and Letters*. 2023;30(1):2241001.
50. Vasudev H. Wear characteristics of Ni-WC powder deposited by using a microwave route on mild steel: microwave cladding of Ni-WC. *International Journal of Surface Engineering and Interdisciplinary Materials Science (IJSEIMS)*. 2020;8(1):44–54.
51. Parkash J, Saggu HS, Vasudev H. A short review on the performance of high velocity oxy-fuel coatings in boiler steel applications. *Materials Today: Proceedings*. 2022;50:1442–6.
52. Singh P, Vasudev H, Bansal A. Effect of post-heat treatment on the microstructural, mechanical, and bioactivity behavior of the microwave-assisted alumina-reinforced hydroxyapatite cladding. *Proceedings of the Institution of Mechanical Engineers, Part E: Journal of Process Mechanical Engineering*. 2024;238(3):1060–71.
53. Prakash C, Gupta LR, Mehta A, Vasudev H, Tominov R, Korman E, Fedotov A, Smirnov V, Kesari KK. Computing of neuromorphic materials: An emerging approach for bioengineering solutions. *Materials Advances*. 2023;4:5882–19.
54. Singh J, Gill HS, Vasudev H. Computational fluid dynamics analysis on role of particulate shape and size in erosion of pipe bends. *International Journal on Interactive Design and Manufacturing (IJIDeM)*. 2023;17(5):2631–46.
55. Rao GS, Mukkamala U, Hanumanthappa H, Prasad CD, Vasudev H, Shanmugam B, KishoreKumar KC. Evaluating and optimizing surface roughness using genetic algorithm and artificial neural networks during turning of AISI 52100 steel. *International Journal on Interactive Design and Manufacturing (IJIDeM)*. 2024;18(8):6151–60.
56. Singh J, Singh R, Singh S, Vasudev H, Kumar S. Reducing scrap due to missed operations and machining defects in 90PS pistons. *International Journal on Interactive Design and Manufacturing (IJIDeM)*. 2023;17(5):2527–39.
57. Nagabhushana P, Ramprasad S, Prasad CD, Vasudev H, Prakash C. Numerical investigation on heat transfer of a nano-fluid saturated vertical composite porous channel packed between two fluid layers. *International Journal on Interactive Design and Manufacturing (IJIDeM)*. 2024;18(5):2927–44.
58. Mahantesh MM, Rao KR, Chandra AP, Vijayakumar MN, Nandini B, Prasad CD, Vasudev H. Design and modeling using finite element analysis for the sitting posture of computer users based on ergonomic perspective. *International Journal on Interactive Design and Manufacturing (IJIDeM)*. 2024;18(8):5875–91.

59. Verma PK, Vasudev H, Kumar V. Dental implant materials and designs and failure during implant placement and loading: A review. In Hitesh Vasudev, Chander Prakash, Mohammad Khalid, Kavindra Kesari (Eds.) *Advanced manufacturing processes*, 2025. pp. 138–50, Boca Raton: CRC Press.
60. Kulkarni SD, Manjunatha, Chandrasekhar U, Manjunath KV, Prasad CD, Vasudev H. Design and optimization of polyvinyl-nitride rubber for tensile strength analysis. *International Journal on Interactive Design and Manufacturing (IJIDeM)*. 2024;18(8):5893–908.
61. Kattimani MA, Venkatesh PR, Masum H, Math MM, Bahadurdesai VN, Mustafkhadri S, Prasad CD, and Vasudev H. Design and numerical analysis of tensile deformation and fracture properties of induction hardened inconel 718 superalloy for gas turbine applications. *International Journal on Interactive Design and Manufacturing (IJIDeM)*. 2024;18(8):5939–49.
62. Poojari M, Hanumanthappa H, Prasad CD, Jathanna HM, Ksheerasagar AR, Shetty P, Shanmugam BK, Vasudev H. Computational modelling for the manufacturing of solar-powered multifunctional agricultural robot. *International Journal on Interactive Design and Manufacturing (IJIDeM)*. 2024;18(8):5725–36.
63. Verma PK, Vasudev H. Importance of hydroxyapatite coatings in improving new bone osteogenesis: A review. In Hitesh Vasudev, Chander Prakash, Mohammad Khalid, Kavindra Kesari (Eds.) *Advanced manufacturing processes*, 2025. pp. 284–299, Boca Raton, CRC Press.

4 Microstructure and Properties of Molybdenum Composite Coatings

Jeyaprakash Natarajan, Praveen Kumar Verma, Hitesh Vasudev, and Harminder Singh Saggu

4.1 INTRODUCTION

Molybdenum (Mo) is a refractory metal with a high melting point (2623°C), a very high modulus of elasticity (~320 GPa), and high resistance to creep and wear at harsh high temperatures and aggressive chemical environments. This renders it essential for surface engineering, where components face extreme thermal, mechanical, and corrosive stresses. However, bulk molybdenum has adverse characteristics that limit its use. These include brittle behaviour at room temperatures, high susceptibility to oxidation above 400°C, and difficulty of machining due to its high melting temperature. For the last 30 years, the majority of research aimed at meeting these challenges has been focused on developing molybdenum composite coatings—engineered composites where molybdenum is the principal matrix or constituent phase, with added ceramic or metallic particles for performance multifunctionality. Molybdenum-based coatings exhibit low-friction, wear resistant, and high-temperature strength which makes them versatile and indispensable in tribological, aerospace, energy, and biomedical applications. The addition of secondary phases, such as carbides (SiC, WC, TiC), oxides (Al_2O_3, ZrO_2, Cr_2O_3), and metals (Ni, Cr, Cu, Ti), has significantly improved the mechanical and functional properties. The molybdenum (Mo) composite principle coating synergistic effect refers to enhancing the performance of pure Mo coatings utilized for wear resistant and high-temperature applications by reinforcing a second phase (like carbides, or oxides) to overcome disadvantages such as lower microhardness and low oxidation resistance. Molybdenum composite coatings are made using thermal spraying, physical vapour deposition (PVD), chemical vapour deposition (CVD), electrodeposition, and laser cladding. Different coating processes yield different microstructures due to the process conditions, energy input, substrate interaction, and feedstock morphology. Plasma spraying and laser cladding are considered high-energy processes. They offer excellent coating adhesion and density which enable the production of fine-scale composite microstructures with low porosity [1, 2].

The structural performance of a molybdenum composite coating is strongly associated with its microstructure—the structure, morphology, and interface of

DOI: 10.1201/9781003730637-4

constituent phases. Microstructural refinement usually increases hardness and wear resistance by grain boundary strengthening, load transfer effects, and constrained plasticity. In nanocomposite variants, where reinforcements exist at submicron or nanoscale dimensions, further strengthening arises from Orowan looping and dislocation pinning mechanisms. As a result, the systematic study of microstructural evolution and its correlation to coating properties is essential to optimize material performance. In industrial practice, Mo-based coatings have been used in piston rings, turbine blades, cutting tools, crucibles, and thermal protection systems. For instance, Mo–Ni and Mo–Cr coatings serve as protective barriers in gas turbines, while Mo–SiC and Mo–Al_2O_3 coatings are applied in aerospace and biomedical devices for wear and oxidation resistance. The development of molybdenum disilicide ($MoSi_2$) and molybdenum carbide (Mo_2C) systems has further expanded the high-temperature capabilities of Mo-based coatings [3].

This chapter provides a comprehensive overview of the microstructure–property relationships in molybdenum composite coatings. It begins with an overview of various processing routes, followed by an examination of microstructural features, phase formation, and interface characteristics. Subsequent sections discuss mechanical, tribological, and corrosion behaviours, highlighting the influence of reinforcement type, volume fraction, and processing parameters. The chapter concludes with emerging trends, challenges, and future research directions in this field.

4.2 PROCESSING TECHNIQUES FOR MOLYBDENUM COMPOSITE COATINGS

4.2.1 Thermal Spraying Techniques

Thermal spraying involves a collection of processes—plasma spraying, high-velocity oxy-fuel (HVOF) spraying, and detonation gun (D-gun) spraying—that utilize high-temperature jets to drive molten or semi-molten particles onto a substrate. Atmospheric plasma spraying (APS) is especially appropriate for Mo and its composites because of its capability to process refractory materials above temperatures of 10,000 K. In this process, Mo and reinforcement powders like Al_2O_3, TiC, or SiC are mixed and injected into a plasma plume, which is melted and later solidifies upon impact to create a lamellar microstructure. Morphology and porosity of splats are highly influenced by parameters like arc current, gas composition, stand-off distance, and powder feed rate [4]. Figure 4.1 shows a schematic of the thermal spray coating process, illustrating the stages of feedstock powder injection, particle melting, acceleration through a high-velocity gas stream, and eventual deposition onto the substrate to produce a dense coating layer.

HVOF spraying produces denser coatings with higher bond strength compared to APS, as the lower flame temperature and higher particle velocities minimize oxidation and enable better consolidation. Mo–Ni, Mo–Cr, and Mo–Al_2O_3 composites fabricated by HVOF exhibit superior wear and corrosion resistance due to reduced oxide inclusions and finer splat interfaces. For instance, a HVOF-deposited Mo–NiCr coating demonstrated a porosity below 2% and microhardness exceeding 900 $HV_{0.3}$, significantly higher than conventional plasma-sprayed coatings [1]. Detonation gun spraying offers even higher particle velocities (700–900 m/s), producing dense and well-adhered coatings. Mo–Cr_3C_2 and Mo–Ni–Al_2O_3 composites fabricated by D-gun have shown excellent abrasion resistance and thermal stability up to 800°C [6].

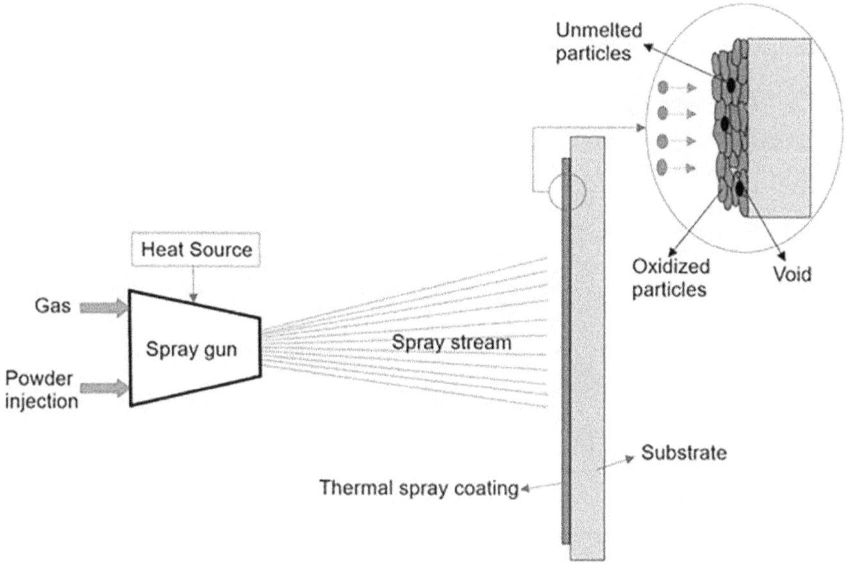

FIGURE 4.1 Schematic illustration of the thermal spray coating [5].

4.2.2 Physical and Chemical Vapour Deposition (PVD and CVD)

PVD and CVD techniques enable the deposition of highly pure, fine-grained molybdenum coatings under controlled vacuum environments. In PVD processes, such as magnetron sputtering, co-deposition of Mo with metallic or ceramic targets (e.g. Ni, TiN, or Al_2O_3) can create multilayer or nanocomposite architectures with nanometric control over layer thickness. The resulting coatings exhibit high hardness, low surface roughness, and superior adhesion, suitable for precision components and cutting tools. In contrast, CVD coatings are formed through gas-phase reactions, often employing molybdenum hexacarbonyl ($Mo(CO)_6$) as a precursor. When combined with silane or methane, Mo–Si or Mo–C composite coatings can be produced. Although CVD offers excellent coating uniformity and conformal coverage, the high deposition temperatures (typically 800–1100°C) restrict its use to thermally stable substrates [7]. Figure 4.2 shows a schematic of the different stages involved in the coating application using PVD. The figure presents the stages of target material evaporation, transport of vapour species in a vacuum chamber, and the subsequent condensation of vapour onto a substrate to create a thin, uniform coating layer.

4.2.3 Electrodeposition and Electroplating

Electrodeposition of molybdenum-based coatings is challenging due to the refractory nature of Mo, which does not deposit easily from aqueous electrolytes. However, composite electrodeposition, where Mo is co-deposited with Ni or Cu from mixed electrolytes containing molybdate ions and suspended reinforcements, has shown promising results. For instance, Mo–Ni–SiC and Mo–Cu–Al_2O_3 coatings exhibit enhanced hardness and corrosion resistance due to the uniform dispersion of

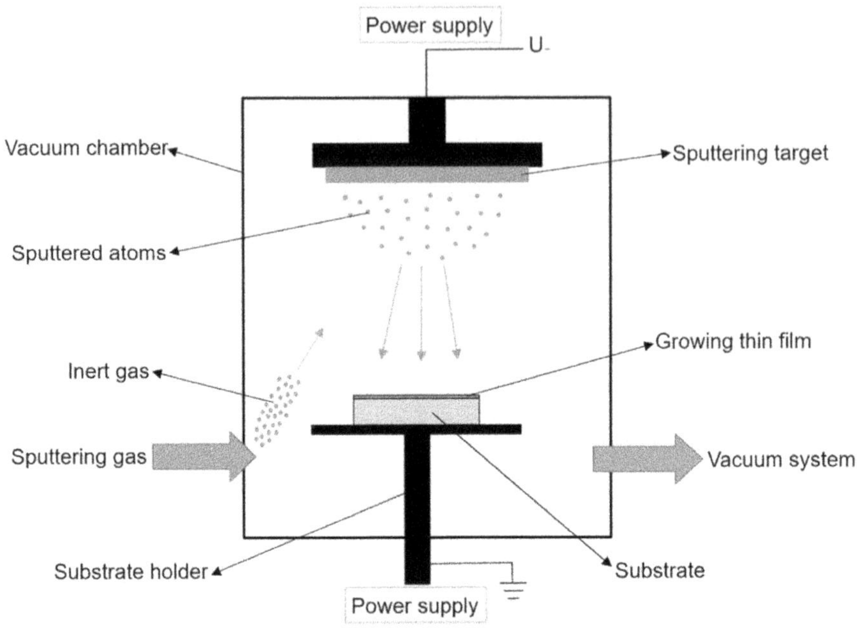

FIGURE 4.2 Physical vapour deposition coating technique [5].

reinforcing particles within a ductile metallic matrix. The process is cost-effective and suitable for coating components with complex geometries [8].

4.2.4 Laser Cladding and Additive Manufacturing

Laser cladding is increasingly recognized as a superior route for fabricating dense and metallurgically bonded Mo composite coatings. A high-energy laser beam melts a pre-placed or simultaneously fed powder mixture of Mo and reinforcements, forming a coating with minimal dilution and excellent interfacial bonding. The rapid solidification associated with laser cladding results in fine microstructures, improved hardness, and reduced cracking. Studies on Mo–NiCrBSi and Mo–Al_2O_3 composites have demonstrated microhardness values exceeding 1000 HV and wear rates an order of magnitude lower than bulk Mo coatings. With the rise of additive manufacturing (AM), techniques such as laser engineered net shaping (LENS) and directed energy deposition (DED) have enabled the fabrication of functionally graded Mo composites with controlled composition gradients [9]. As illustrated in Figure 4.3 the laser powder bed fusion (LPBF) method employs a stepwise construction technique in which a high-power laser focuses on and fuses discrete areas of a metallic powder bed in accordance with a CAD blueprint and subsequently re-coats the powder to create the additively manufactured part.

Each coating method yields a distinct microstructure and property set. In general, thermal spraying offers high productivity and thickness control but can introduce porosity; PVD/CVD produce fine-grained, pure coatings with limited thickness;

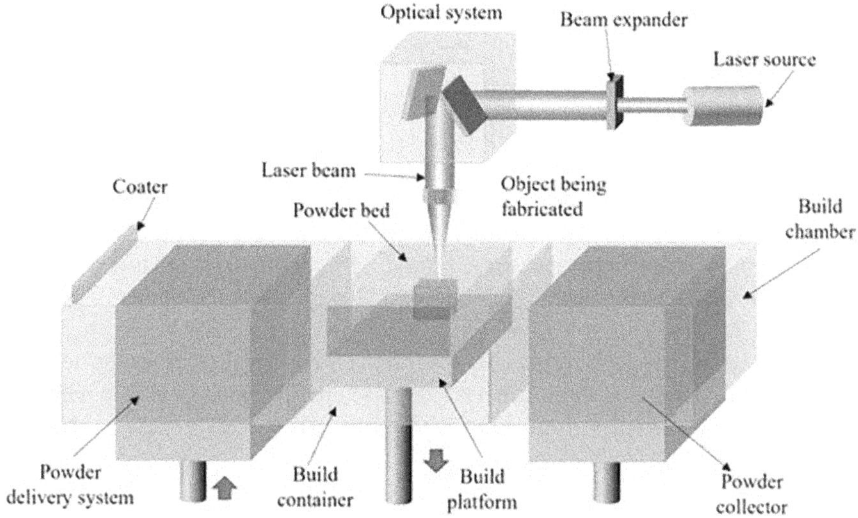

FIGURE 4.3 Schematic diagram of laser powder bed fusion (LPBF) [10].

electrodeposition provides cost-effective coatings with moderate performance; and laser cladding/AM deliver dense, metallurgically bonded coatings ideal for high-performance applications. The choice of method depends on the desired combination of wear resistance, oxidation stability, and cost considerations [11].

4.3 MICROSTRUCTURAL EVOLUTION AND CHARACTERIZATION

The microstructure of molybdenum composite coatings plays a central role in dictating their mechanical, tribological, and thermal properties. Microstructural evolution during coating formation depends strongly on the deposition technique, cooling rate, feedstock morphology, and the thermodynamic compatibility of the constituent phases. In Mo-based composites, the microstructure is typically composed of a metallic Mo matrix containing dispersed reinforcements such as oxides, carbides, or other metals. The characteristics of splat formation, grain refinement, and interfacial bonding largely determine the coating's performance [12].

4.3.1 Microstructural Development during Deposition

During thermal spray or laser cladding deposition, the molten or semi-molten droplets of Mo and reinforcement particles impact the substrate at high velocities. Upon impact, these droplets flatten, spread, and solidify rapidly, forming lamellar structures commonly known as "splats." The successive accumulation of splats leads to the formation of a layered coating microstructure. The cooling rate during splat solidification can exceed 10^6 K/s, resulting in fine-grained or even amorphous regions at the splat interfaces. In plasma-sprayed coatings, the high temperature of the plasma plume leads to partial melting of the Mo particles, while the reinforcements may remain solid or partially molten, depending on their melting point. For instance, Al_2O_3 particles tend to remain solid and become embedded within the molten Mo matrix, creating a heterogeneous composite structure. In contrast, carbides such as SiC or WC may partially decompose or react with the molten Mo to form intermediate phases like Mo_2C or $MoSi_2$. These interfacial reactions can significantly modify the mechanical behaviour of the coating. Controlled plasma parameters are therefore crucial to prevent excessive decomposition of reinforcements. In laser-cladded coatings, the localized melting and rapid solidification of the melt pool generate metallurgical bonding between the coating and the substrate. The cooling rates are typically lower than those in plasma spraying, allowing for the development of columnar or dendritic Mo grains. Reinforcements are uniformly distributed in the interdendritic regions or along grain boundaries. Laser scanning speed and energy density govern the degree of dilution and microstructural refinement. Excessive energy input can lead to reinforcement dissolution, whereas insufficient melting results in weak interfacial bonding [13]. Figure 4.4 is a depiction of a thermally grown oxide (TGO) system with (a) a pre-oxidation state showing the ceramic top coat, the bond coat, and the substrate, and (b) a post-oxidation state showing the TGO layer formed between the bond coat and top coat illustrating thermally grown oxide. Scanning Electron Microscopy (SEM) images in Figure 4.5 depict the typical surface morphology of an atmospheric plasma-sprayed chromium oxide (Cr_2O_3) coating on Q235 steel. The coating microstructure displays lamellar splat morphology, microcracking, and porosity consistent with the plasma spraying technique. The observed microstructural characteristics also reveal the coating's adhesion and densification porosity behaviour.

4.3.2 Phase Composition and Reaction Products

Phase formation in Mo composite coatings depends on both thermodynamic compatibility and processing temperature. For example, Mo readily reacts with SiC to form molybdenum silicides (Mo_5Si_3, $MoSi_2$) and carbides (Mo_2C). These phases contribute to increased hardness and oxidation resistance but may embrittle the coating if present in excess. Similarly, Mo–Al_2O_3 coatings can exhibit spinel-type phases (e.g. $MoAl_2O_4$) at high deposition temperatures, especially in oxygen-rich

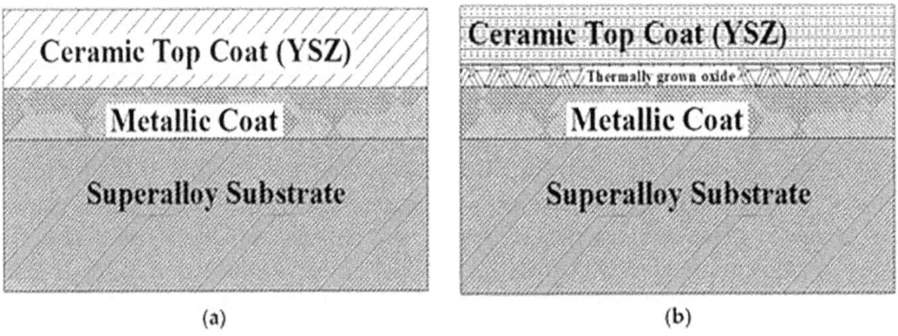

FIGURE 4.4 The components of thermal barrier coating (TBC): (**a**) before oxidation, and (**b**) after oxidation with the inclusion of thermally grown oxide (TGO) [14].

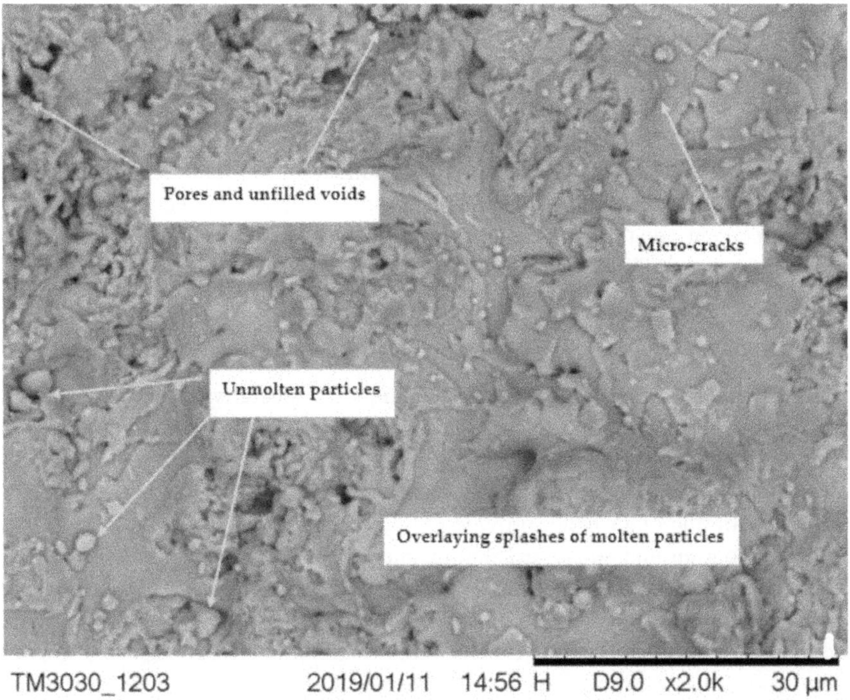

FIGURE 4.5 Scanning electron microscopy (SEM) images showing typical atmospheric plasma-sprayed coating morphology for chromium oxide coating on Q235 steel [14].

environments. In contrast, metallic reinforcements such as Ni, Cr, or Cu are generally soluble in Mo and promote the formation of solid solutions or intermetallic compounds. Mo–Ni and Mo–Cr systems show improved ductility and adhesion due to the formation of coherent phase boundaries. The presence of Ni also facilitates self-healing of microcracks by acting as a ductile buffer layer during service. X-ray diffraction (XRD) studies typically reveal strong reflections corresponding to body-centred cubic (bcc) Mo, accompanied by peaks of secondary phases such as Mo_2C, $MoSi_2$, or Al_2O_3. Depending on processing temperature, amorphous oxide films may also be observed, particularly on the splat boundaries of plasma-sprayed coatings. Post-deposition heat treatment often promotes the crystallization of these amorphous regions, improving coating stability and cohesion [15].

4.3.3 Grain Structure and Size Refinement

One of the most striking features of molybdenum composite coatings is the grain size refinement achieved during high-energy processing. In thermally sprayed coatings, the rapid solidification and repeated splat stacking result in grain sizes in the range of 50–300 nm. In PVD or sputtered coatings, the grain size can be even smaller, often below 50 nm, forming nanocrystalline or amorphous structures. Grain refinement enhances hardness and yield strength according to the Hall–Petch relationship.

Reinforcements further contribute to grain refinement through Zener pinning, where the presence of dispersed particles inhibits grain boundary migration. For example, nano-Al_2O_3 additions to Mo have been shown to reduce the average grain size from 400 nm to below 150 nm, resulting in a hardness increase of nearly 30%. The effectiveness of grain refinement depends on particle size, distribution, and interfacial energy. In laser-cladded coatings, epitaxial solidification from the substrate can lead to columnar grains extending along the heat flow direction. However, the presence of reinforcements disrupts this columnar growth, promoting equiaxed grains that enhance isotropy and reduce crack susceptibility [16].

4.3.4 Porosity, Defects, and Interfacial Features

Porosity is inevitable in many coating techniques, especially in thermal spraying, because of incomplete splat coalescence and gas entrapment. Mo coatings produced with plasma spraying techniques usually separate 5–10% porosity, whereas HVOF porosity-optimized coating streams record under 2%. Laser cladding and PVD techniques, on the other hand, provide closed, dense coatings with under 1% porosity. The coating and substrate coating interface is significant for adhesion and mechanical integrity. For thermal spray coatings, the interstructural bonds are formed mechanically and are merely metallurgical diffusion, whereas with laser cladding, metallurgical bonding is achieved. A bonded interface consists of shifting compositions, thus eliminating thermal stresses. Thermal "inter" adding of metals is done using Ni or Cu, which can strengthen adhesion further. Metallo-organic adhesion builders are interlayers. Microcracks, unmelted particles, and oxide inclusions are other defects that result in the concentration of stress and fatigue weakening. The coating surface may be treated with hot isostatic pressing (HIP) techniques and then followed by annealing in order to relax residual stress and reinforce the occluded surface. Residual compressive stress in PVD coatings is beneficial to a certain degree and serves to enhance the resistance to wear and fractures; if unmodified or uncontrolled, compressive stress leads to delamination of the coatings [17].

4.3.5 Effect of Post-processing and Heat Treatment

Post-deposition heat treatment profoundly influences the microstructure of Mo composite coatings. Controlled annealing promotes recrystallization, stress relaxation, and phase homogenization. In Mo–Al_2O_3 systems, annealing at 800–1000°C enhances interfacial bonding by reducing amorphous oxide films. In Mo–Ni or Mo–Cr composites, diffusion of alloying elements results in solid-solution strengthening and improved ductility. However, excessive heat treatment can lead to grain coarsening, which may degrade hardness and wear resistance. Therefore, an optimized balance between stress relaxation and grain growth must be maintained. In certain cases, laser remelting has been used as an in-situ post-treatment step to refine surface microstructure and close surface pores without significant grain growth [18].

4.4 MECHANICAL PROPERTIES OF MOLYBDENUM COMPOSITE COATINGS

The mechanical properties of molybdenum composite coatings determine their use in advanced applications including cutting tools, turbine parts, and surfaces resistant to wear. These properties consist of hardness, elastic modulus, adhesion strength, fracture toughness, and residual stress. They stem from the interaction among the microstructural features—such as grain size, porosity, and phase distribution—and the processing conditions. The addition of secondary reinforcements, whether metallic or ceramic, enables significant adjustments of these properties and can enhance load transfer, grain refinement, and dislocation interactions [19].

4.4.1 Hardness and Elastic Modulus

Hardness is one of the most critical parameters for assessing wear resistance and surface durability. The hardness of pure molybdenum coatings varies widely with deposition technique, ranging from ~300 HV for plasma-sprayed coatings to over 900 HV for dense HVOF or laser-cladded layers. In contrast, composite coatings incorporating hard ceramic reinforcements such as Al_2O_3, SiC, or WC can exceed 1000–1200 HV, depending on reinforcement volume fraction and distribution [4]. According to the Hall–Petch relationship, hardness increases with decreasing grain size. Rapid solidification and the presence of dispersoids restrict grain boundary motion, resulting in refined microstructures. Reinforcement particles bear a portion of the applied load, effectively transferring stress away from the softer Mo matrix. The efficiency of load transfer depends on the interfacial bonding quality and reinforcement volume fraction. Nano-sized reinforcements obstruct dislocation motion, forcing dislocations to bow between particles. This mechanism dominates in nanocomposite coatings where reinforcement spacing is in the range of tens of nanometres [20].

Metallic additions such as Ni, Cr, or Ti can dissolve into the Mo lattice, generating lattice distortions that impede dislocation motion. Complementary to other methods, Mo-based coatings' elastic modulus values in the range of 180 to 320 GPa can be examined using nanoindentation techniques. Incorporation of rigid ceramic reinforcements of Al_2O_3 or SiC increases the modulus by 25%. Ductile metallic components, such as Ni, compromise stiffness but enhance toughness. Consequently, the construction of Mo composite coatings is primarily concerned with the compromise between the hardness and elasticity [21].

4.4.2 Adhesion Strength

For applications involving load-bearing components, the adhesion of the coatings to the substrate is of utmost importance. Weak adhesion could cause coatings to spall or delaminate when the components experience mechanical or thermal stress. With molybdenum composite coatings, adhesion is usually a function of the deposition technique, which is often the most important parameter, and interface characteristics. Molybdenum thermal spray coatings have adhesion values of

20–60 MPa. Although adhesion is primarily attributed to mechanical interlocking caused by the surface roughness, the role of the metallurgical interface is minimal. Bond coats, of which NiCr or Cu are examples, appreciably improve adhesion due to the presence of a ductile intermediate layer that balances the difference in thermal expansion along the composite and coating. Laser-cladded coatings display superior adhesion (often >80 MPa) due to metallurgical fusion with the substrate. PVD coatings, particularly those deposited by magnetron sputtering, achieve excellent adhesion on metallic or ceramic substrates when appropriate bias voltage and pre-cleaning are used. The use of metallic reinforcements like Ni or Cr helps improve adhesion through diffusion bonding, but too many ceramics can lead to brittleness and localized detachment. Adhesion strength is often assessed using scratch tests and pull-off adhesion tests. The failure modes reported include cohesive cracking of the coating and adhesive failure at the coating–substrate interface [22].

4.4.3 Fracture Toughness and Crack Resistance

Though molybdenum shows good intrinsic toughness relative to the other refractory metals, it is still the case that coatings—particularly those that are reinforced with ceramics—may display brittle behaviour due to the presence of hard, non-deforming constituents. Therefore, fracture toughness (K_IC) and crack resistance are key design considerations. Interfaces between the Mo matrix and reinforcement particles act as barriers that deflect or branch propagating cracks, increasing the energy absorption during fracture. Ductile metallic phases or well-bonded particles can bridge cracks and retard their propagation. Controlled microcracking around hard particles helps dissipate fracture energy, delaying catastrophic failure. Typical fracture toughness values for Mo composite coatings range between 3–6 $MPa\cdot m^{1/2}$, depending on the reinforcement type and deposition process. Laser-cladded Mo–Ni–Al_2O_3 coatings, for example, have demonstrated improved fracture toughness compared with pure Mo due to the dual contribution of ductile Ni and fine Al_2O_3 dispersion [23].

4.4.4 Residual Stress and Elastic Strain

Residual stresses arise from thermal expansion mismatch between the coating and substrate as well as from rapid solidification during deposition. For Mo coatings on steel substrates, compressive residual stresses typically develop because of the higher thermal expansion coefficient of steel compared to Mo. Moderate compressive stresses are beneficial, as they enhance resistance to crack initiation and delamination. However, excessive stress can lead to coating spallation or substrate deformation. The magnitude of residual stress can be tailored by controlling process parameters such as substrate temperature, spray distance, and cooling rate. Post-deposition annealing or laser remelting is often applied to relieve residual stresses and improve coating integrity. X-ray diffraction-based $\sin^2\psi$ methods or curvature measurements are commonly used to quantify these stresses [24].

4.4.5 Wear Resistance and Load-bearing Capacity

Wear performance is the most widely studied property of molybdenum composite coatings. Pure Mo coatings already exhibit low friction and moderate wear resistance due to the formation of lubricious MoO_3 films during sliding. Hard ceramic particles act as load-bearing reinforcements, reducing the real contact area between sliding surfaces and suppressing plastic deformation. Metallic binders promote ductility, which accommodates stress concentrations and prevents brittle fracture. Lubricating oxides formed in-service (MoO_3, $NiMoO_4$) reduce friction and minimize adhesive wear. Wear tests such as pin-on-disk or ball-on-flat configurations typically reveal that Mo–Al_2O_3 and Mo–SiC composites exhibit wear rates one to two orders of magnitude lower than pure Mo coatings. The improvement is attributed to a combination of high hardness and stable oxide tribofilms formed during sliding. The coefficient of friction of Mo composites generally falls in the range of 0.25–0.45 under dry conditions, and even lower in lubricated environments [25].

Molybdenum's superior high-temperature strength makes its composites attractive for applications above 800°C. At elevated temperatures, pure Mo softens due to grain boundary sliding and diffusion creep. The inclusion of stable ceramic reinforcements such as SiC or Al_2O_3 effectively retards these processes by pinning grain boundaries and reducing grain boundary mobility. Laser-cladded Mo–SiC coatings have exhibited stable hardness values up to 1000°C, while retaining good oxidation resistance. Moreover, fine dispersions of molybdenum carbides (Mo_2C) or silicides (Mo_5Si_3, $MoSi_2$) formed during processing contribute to secondary-phase strengthening. These phases act as thermally stable barriers to dislocation motion, thereby enhancing creep resistance [26].

4.5 TRIBOLOGICAL BEHAVIOUR OF MOLYBDENUM COMPOSITE COATINGS

The tribological behaviour of molybdenum composite coatings has been widely investigated because these materials are specifically developed to function in severe sliding, rolling, and abrasive conditions where both high hardness and low friction are essential. Tribological performance is a complex function of hardness, surface roughness, interfacial chemistry, oxide film formation, and the stability of reinforcement–matrix bonding under load. Among refractory metals, molybdenum exhibits an intrinsically low friction coefficient due to its ability to form a lubricious molybdenum trioxide (MoO_3) layer during sliding. However, pure Mo coatings often suffer from rapid oxidation, poor adhesion, and limited wear life at high temperature or under oscillating stresses. Reinforcement by ceramics or ductile metals is therefore employed to stabilize friction, reduce wear rate, and extend service life across a range of operating conditions. During sliding contact, the frictional heat generated at asperity junctions can locally raise temperatures beyond several hundred degrees Celsius, triggering oxidation, diffusion, and tribochemical reactions [27].

The formation of a stable tribo-oxide film is a key mechanism governing the frictional response of Mo composites. Under dry sliding, oxygen from the environment reacts with the surface to produce MoO_2 and MoO_3, both of which possess layered crystal structures that shear easily along basal planes. This self-lubricating film can drastically lower the coefficient of friction to values between 0.25 and 0.45, depending on sliding speed and environment. When metallic reinforcements such as nickel or chromium are present, mixed oxides such as $NiMoO_4$ or $Cr_2(MoO_4)_3$ may form, further stabilizing the tribofilm and preventing the volatilization of MoO_3 at high temperature. In vacuum or inert atmospheres where oxide formation is suppressed, friction tends to increase because metallic contact dominates. In such cases, the presence of solid reinforcements still mitigates wear through load sharing and micro-abrasion suppression. The wear mechanisms in molybdenum composite coatings vary with the sliding conditions and the hardness ratio between the coating and its counterface. In general, micro-abrasion, oxidative wear, and delamination are the dominant modes. Pure Mo coatings exhibit pronounced ploughing and material transfer, resulting in higher wear rates. In contrast, Mo–Al_2O_3 and Mo–SiC composites show smooth worn surfaces with fine, compacted oxide debris layers. These oxide layers act as dynamic lubricants, continuously forming and regenerating during sliding, thereby maintaining low friction over extended durations. The incorporation of ductile metal phases such as nickel or cobalt improves adhesion of the oxide debris to the surface, preventing spallation [28].

Studies employing ball-on-disk and pin-on-flat tribometers have shown that wear rate reductions of one to two orders of magnitude are achievable when 5–15 wt% ceramic reinforcement is added to molybdenum coatings. For instance, Mo–10 wt% Al_2O_3 coatings prepared by high-velocity oxy-fuel spraying exhibited a wear rate of 2×10^{-6} $mm^3\,N^{-1}\,m^{-1}$, compared to 3×10^{-4} $mm^3\,N^{-1}\,m^{-1}$ for unreinforced Mo under similar conditions. The reduction in wear rate arises not only from increased hardness but also from the stabilization of the oxide tribofilm and the suppression of severe adhesive wear. In laser-cladded Mo–SiC coatings, wear rates remain nearly constant even after long-duration tests up to 10 km of sliding distance, confirming the superior endurance of composite architectures [29].

The tribological response of molybdenum composites is also sensitive to temperature. At ambient conditions, the friction coefficient is governed by asperity deformation and adhesive junction formation, while at elevated temperatures (>500°C), oxidation becomes dominant. The transition from metallic to oxide-controlled friction often leads to a decrease in coefficient of friction with increasing temperature until volatilization of MoO_3 occurs above 700–800°C. Reinforcement phases play a critical role in maintaining the stability of the oxide film at these temperatures. For example, alumina is chemically inert and thermally stable, acting as a diffusion barrier that slows oxygen transport and MoO_3 evaporation. Similarly, carbides and silicides formed in Mo–SiC systems exhibit high thermal stability, preventing catastrophic oxidation while providing microstructural rigidity. Consequently, these composites retain favourable friction behaviour even in oxidizing atmospheres at temperatures up to 1000°C [30].

Surface roughness and coating density also influence tribological outcomes. High porosity and rough surfaces lead to abrasive wear and oxide film rupture. Densification attained through HVOF or laser cladding mitigates these effects by stiffening weak splat interfaces and augmenting the load-bearing contact area. Additionally, post-deposition polishing or laser remelting enhances the surface finish, which promotes more stable friction. Therefore, the surface topography and microstructure interplay define tribofilm initiation and evolution during sliding. The counterpart material is another significant factor for tribological performance. When Mo composite coatings are slid against steel or ceramic counterfaces, the wear behaviour is dictated by the material pairing. When coated surfaces are slid against hard ceramics like alumina, the coating undergoes primarily oxidative wear with minimal wear loss. In contrast, steel bonding results in adhesive wear and the creation of transfer layers. In these situations, the Mo coating's ceramic reinforcements inhibit wear and adhesion. The tribochemical interaction of the coating and the counterface defines whether the oxide film is continuously adherent or undergoes periodic removal and regeneration [31].

Recent advances have explored the development of nanostructured molybdenum composites with ultrafine grain sizes and uniformly dispersed nanoscale reinforcements. Such coatings demonstrate exceptional tribological stability due to homogeneous distribution of stress and the formation of continuous, thin tribofilms. The nanoscale structure enhances the availability of oxide nucleation sites and facilitates self-healing of the film during wear. Additionally, the mechanical compatibility between matrix and reinforcement reduces interfacial cracking, ensuring longer coating life. The cumulative evidence from experimental and modelling studies indicates that the superior tribological performance of molybdenum composite coatings originates from a synergistic combination of mechanical strengthening and tribochemical stabilization. The hard phases reduce plastic deformation and micro-cutting, while the lubricious oxides formed dynamically during sliding lower friction and isolate the substrate from direct wear. The optimal performance is achieved when the microstructure provides both hardness and toughness, enabling the coating to sustain cyclic stresses without fracture or delamination [32].

4.6 CORROSION AND OXIDATION RESISTANCE OF MOLYBDENUM COMPOSITE COATINGS

Molybdenum composite coatings possess exceptional anti-corrosion and anti-oxidation properties as a result of molybdenum's own characteristics and the properties of composite constituents. For metallic parts exposed to aggressive or high-temperature environments, the service life is primarily limited by the rate of corrosion and oxidation. Molybdenum in its pure form does provide reasonable resistance to oxidation through the passive oxide film; however, at high temperatures or in wet environments, the oxide film becomes volatile and unstable, resulting in rapid degradation. Adding ceramics or metallics to molybdenum matrix improves its durability in harsh environments by altering the kinetics of oxidation, blocking diffusion pathways, and stabilizing the protective oxide layers. The environment which includes the pH, temperature, and concentration of chlorides or sulfates, shapes the corrosion behaviour of molybdenum and its composites. The passive layers formed by molybdenum in water as corrosion progresses consist of layers of MoO_2 and MoO_3. These layers provide some protection but under strongly oxidizing conditions, the oxides of molybdenum dissolve and result in pitting. The corrosion behaviour of molybdenum composites can certainly improve due to the addition of passive reinforcement such as alumina, titania, or chromium because they form insoluble barriers which increase the stability of the passive films and corrosion behaviour. For example, Mo–Al_2O_3 composite coatings show considerable improvement by corrosion current density reduction because the dispersed alumina particles impede the diffusion of oxidizing agents, water, and oxygen through the passive film. For Mo–Cr and Mo–Ni composites, the improvement comes from more adherent and less soluble mixed oxides of molybdenum such as Cr_2O_3 or $NiMoO_4$, which increase passivation and result in lower corrosion rates [33].

Molybdenum-based coatings don't always defend perfectly in environments with aggressive chlorides. Because of their small size, chloride ions can breach oxide layers and start pitting at grain boundaries or splat interfaces of thermally sprayed coatings. Composite formulations have porosity and intersplat region sealing, but these are mostly for reduced effects. Coating reinforcements more than help with densifying a coating; they alter its electrochemical behaviour, creating microgalvanic cells for rapid site repassivation for damaged areas. In studies, Mo–Al_2O_3 coatings in 3.5% NaCl solution did shift corrosion potential to more noble values, and passive current density also dropped by an order of magnitude compared to unreinforced Mo coatings. This shows the ceramic phase stabilization in passive film regeneration, thus extending time before pitting and underfilm corrosion start [34].

At elevated temperatures, oxidation resistance becomes the governing factor in determining the long-term performance of molybdenum coatings. Pure molybdenum begins to oxidize rapidly above 400°C, forming MoO_2 and MoO_3 layers. While MoO_2 is relatively stable, MoO_3 is volatile and sublimes at temperatures above 700°C, leaving behind a porous, nonprotective surface. This process leads to accelerated material loss, known as "catastrophic oxidation." The incorporation of thermally stable reinforcements can suppress this behaviour by promoting the formation of adherent, non-volatile oxide layers and by providing mechanical stability to the oxide scale. In Mo–Al_2O_3 systems, the alumina particles remain chemically inert and act as physical diffusion barriers to oxygen ingress. As a result, oxidation rates are significantly lower, and the oxide scale remains compact and adherent even after prolonged exposure. Furthermore, the interfacial bonding between alumina and molybdenum improves scale adhesion, preventing spallation during thermal cycling [35].

Another effective strategy for improving oxidation resistance is the addition of elements such as silicon, chromium, or boron, which form highly stable oxides and silicides. In Mo–SiC coatings, oxidation produces a composite oxide film containing SiO_2 and MoO_3. The silica component forms a viscous glassy layer that seals surface pores and retards further oxidation. Even though MoO_3 can volatilize, the underlying SiO_2 network traps it, leading to the formation of a self-healing film that can reform after thermal cracking. The net result is a substantial decrease in mass loss rate and surface roughening. Likewise, in Mo–Cr composites, a thin Cr_2O_3 layer forms preferentially and acts as a diffusion barrier, preventing oxygen from reaching the underlying molybdenum. Chromium also increases the thermodynamic stability of the protective oxide film, extending its effectiveness up to 900–1000°C [36].

The performance of molybdenum composites in oxidation environments is closely tied to microstructural features such as porosity, grain boundary distribution, and reinforcement morphology. Dense coatings fabricated by high-velocity oxy-fuel or laser cladding exhibit superior oxidation resistance compared to plasma-sprayed coatings, which often possess microcracks and interconnected porosity. The elimination of open porosity minimizes diffusion pathways for oxygen and volatile oxides, resulting in more uniform oxide layers. Fine-grained microstructures also enhance oxidation resistance by providing numerous short-circuit diffusion paths that facilitate rapid oxide scale formation during initial exposure, producing a thin but continuous film that protects the bulk material from further attack. Cyclic oxidation also has a marked influence on thermal stability. Repeated cycles of heating and cooling can cause spalling and cracking due to the substrate and oxide layer differing thermal expansion. Composite coatings can alleviate some of these problems due to multiple phases that redistribute thermal stresses and tether the oxide layer. In the case of Mo–Al_2O_3 systems, the thermal expansion coefficient of alumina is very close to that of molybdenum, thus interfacial thermal stresses are reduced. Likewise, spallation is limited due to metallic reinforcements, such as nickel and cobalt, that absorb strain through plastic deformation. Retaining a protective coating after several cycles of oxidation thermally is indicative of the coating's suitability for aerospace applications, as well as for turbines and furnaces [37].

The studies involving electrochemical impedance spectroscopy as well as potentiodynamic polarization have indicated that the primary reason the Mo composites possess outstanding corrosion and oxidation resistance is the synergy between the passive film's electronic conductivity and ionic transport. Mixed oxides that form at the interface behave as semiconductors which regulate charge transfer and corrosion reaction rates. Modifications of the reinforcement composition and distribution control the coating-tailoring for desired service conditions. Mo–Ni–Al_2O_3 coatings are an example where conduction pathways of both the metal and ceramic are present and where the balance of passivation and mechanical strength is achieved. Molybdenum is also particularly susceptible to the formation of volatile molybdates and therefore catastrophic attacks in high-temperature corrosive environments that contain sulfur, halides, or molten salts. Composite coatings still provide superior protection by stifling the aggressive ion transport and passivating the oxide-sulfide interface. Adding oxides, such as alumina or yttria, to the surface of rare-earth elements significantly improves the surface film's chemical stability. These oxides hinder further attack from volatile sulfur and halogen compounds by forming non-volatile compounds which lead to further attack. Resistance is especial [38].

4.7 FUNCTIONAL AND HIGH-TEMPERATURE PERFORMANCE OF MOLYBDENUM COMPOSITE COATINGS

The adequacy of molybdenum composite coatings for advanced engineering applications such as turbine engines, furnace hardware, plasma-facing components, and space propulsion systems is determined by their functional and high-temperature performance. These coatings must endure extreme thermal and mechanical stress, coupled with oxidizing or corrosive atmospheres, and at times, temperatures of 1000°C or higher. Besides, coatings must withstand thermal and mechanical stress without losing structural integrity, mechanical strength, and chemical stability. Given the conditions in these applications, the properties of molybdenum are advantageous as a base material, especially its high melting point (2623°C), good thermal conductivity, and low coefficient of thermal expansion. Nevertheless, pure molybdenum has the challenges of low-temperature brittleness and high-temperature rapid oxidation. Composite coatings overcome these challenges by integrating secondary phases that impart additional functionalities such as enhanced oxidation resistance, improved thermal shock tolerance, and tailored electrical or catalytic properties [39].

At elevated temperatures, the primary challenge for molybdenum coatings is to retain mechanical strength while resisting creep and thermal degradation. The incorporation of ceramic reinforcements such as Al_2O_3, SiC, TiC, or ZrO_2 significantly improves the creep strength and hardness retention of the coatings. These hard phases impede dislocation motion and grain boundary sliding, which are the main deformation mechanisms at high temperature. Moreover, fine dispersions of ceramics stabilize the microstructure by inhibiting grain growth, thereby maintaining strength even after long exposure. Studies on Mo–SiC and Mo–Al_2O_3 coatings have shown that after annealing at 1000°C for several hours, the hardness decreases by less than 10%, while unreinforced Mo coatings experience a reduction exceeding 30%. This superior stability is attributed to the pinning effect of the reinforcements and the formation of thermally stable interface structures that resist coarsening. Performance at high temperatures requires understanding thermal shock resistance since it concerns the rapid heating and cooling of materials, as in plasma spraying, thermal barrier systems, and rocket nozzles. If the coating and substrate materials have different thermal expansion coefficients, cracking and delaminating will occur. Unlike other coatings, composite coatings are better able to withstand thermal cycling because the reinforcements are stress-distributing inclusions that inhibit the expansion of fractures. In the case of the coating to which the alumina or yttria particles are added, the particles are within the molybdenum matrix, and they will inhibit cracking, absorbing energy, and increasing the toughness of the coating. This ability is reinforced by nickel and cobalt which are added as ductile bridges to help the coating withstand microcracks and added strain. Consequently, Mo-based composites exhibit stable adhesion and minimal spallation even after hundreds of thermal cycles between room temperature and 1000°C [40].

The oxidation resistance of molybdenum composite coatings under high-temperature exposure has been discussed earlier, but it also directly influences their functional performance. The ability to maintain an intact and adherent oxide film under cyclic conditions is essential for sustained performance in aerospace and energy applications. Coatings containing reactive elements such as silicon or chromium exhibit the best high-temperature oxidation behaviour. Mo–SiC coatings form continuous SiO_2-rich layers that prevent further oxidation, while Mo–Cr systems develop Cr_2O_3 films that provide long-term protection. In addition to acting as diffusion barriers, these oxides can also improve the emissivity of the coating surface, which is beneficial in radiative heat management applications where thermal control is critical. Beyond oxidation resistance, molybdenum composite coatings display exceptional functional properties such as high electrical and thermal conductivity, catalytic activity, and radiation shielding capability. The electrical resistivity of molybdenum is relatively low, and this property can be tuned through composite design. Metallic reinforcements such as nickel or copper enhance electrical connectivity, while ceramic additions introduce resistive interfaces that can be exploited in specialized applications. For instance, Mo–Ni composites are used in contact materials and electrodes where both conductivity and wear resistance are required, whereas Mo–Al_2O_3 systems serve as protective coatings in plasma-facing components where electrical insulation and thermal stability are desired. The balance between conductivity and insulation is controlled by the volume fraction and distribution of the reinforcement phase, allowing the composite to be tailored for specific electronic or thermal applications [41].

Molybdenum oxides and carbides are known for their catalytic performance in hydrogenation and desulfurization reactions. Composite coatings containing Mo and oxides such as Al_2O_3 or TiO_2 can maintain catalytic activity at high temperature without structural degradation. The oxide phase provides a high surface area and anchoring sites for active Mo species, while the metallic Mo ensures electronic conductivity, facilitating redox reactions. This combination allows for durable catalytic layers suitable for energy conversion devices and fuel reformers operating at elevated temperatures. Thermophysical characteristics continue defining how molybdenum composites function. Molybdenum's impressive thermal conductivity (about 138 W m^{-1} K^{-1} at 25°C) facilitates rapid heat dissipation, which is critical in applications that involve high power or high heat flux. Though heat conductivity is slightly diminished when ceramic reinforcements are incorporated due to phonon scattering at the interfaces, this decrease is mitigated by the reduction of oxidation and reinforcement that enhance the oxidation resistance and mechanical strength. In the Mo–SiC or Mo–AlN systems, the composites maintain adequate thermal conductivity for heat sink and heat shield applications, while exhibiting improved oxidation resistance when compared with pure Mo [42].

Molybdenum-based composites perform exceptionally well in highly radioactive places, like nuclear reactors and during space missions. They have a low neutron absorption cross-section and excellent dimensional stability when exposed to irradiation. SiC and ZrO_2 reinforcements not only strengthen the material but defend and capture radiation-induced vacancies and interstitials, enhancing performance. This microstructural resilience and coating stability under neutron bombardment makes these coatings perfect for plasma-facing and protective fusion and fission interfaces. These composites do not lose mechanical integrity under thermal and irradiation stress, showcasing even more potential for advanced energy systems. Mo composites' extended functional versatility stems from the high temperature, friction related, and tribological behaviour. Under high temperatures, tribo-oxidation, wear reducing, lubricious, and friction dissipating films of MoO_3 and other mixed oxides form. In the Mo–Al_2O_3 and Mo–SiC systems, these films have a stabilizing oxide-ceramic phase which preserves structural integrity. Consequently, the coatings can withstand dry or weakly lubricated environments and temperatures up to 800–900°C while operation and stability are maintained [43].

Molybdenum-based composites enhance energy generation systems. In concentrated solar power receivers and in nuclear heat exchanger systems, the combination of high thermal conductivity, oxidation resistance, and mechanical durability of these coatings functions optimally. These coatings are also efficient for long-term service, as they sustain repeated thermal cycling. In fuel cells and high-temperature electrolyzers, Mo composites also function as interconnects or protective layers; they maintain electrical conductivity and protect fuel cells and electrolyzers from oxidation and corrosion due to reactive, corrosive, and primary steam gas hydrates such as oxygen, hydrogen, and steam. For unprecedented levels of performance, the functional design of molybdenum composite coatings relies heavily on nanostructuring. In nanocomposite architectures, where reinforcements are controlled at the nanoscale, microstructures and interfaces are greatly enhanced. As a fused matrix interface, the interfacial zone between matrix and reinforcement becomes diffusion barriers, impeding flow and enhancing load transfer efficiency. Nanoscale reinforcements are also key to enhanced oxide formation, and thus control the patterns of oxidation, forming protective layers. Through advanced methods such as laser cladding, pulsed plasma deposition, and additive manufacturing, efficient control for various functions of molybdenum composite coatings is achievable [44].

4.8 APPLICATIONS OF MOLYBDENUM COMPOSITE COATINGS

Molybdenum composite coatings have found widespread application across diverse industries, owing to their unique combination of high-temperature stability, wear resistance, low friction, and corrosion and oxidation resistance. In aerospace engineering, these coatings are extensively employed on turbine blades, vanes, and combustion chamber components, where they are exposed to extreme temperatures exceeding 1000°C, aggressive oxidizing environments, and cyclic mechanical loads. The high melting point and thermal conductivity of molybdenum, combined with the protective effect of ceramic or metallic reinforcements, allow these coatings to protect underlying superalloy substrates from oxidation, creep, and thermal fatigue, thereby extending the operational lifetime of critical components. Laser-cladded Mo–Al_2O_3 or Mo–SiC coatings on nickel-based superalloys have demonstrated exceptional durability in high-temperature jet engine simulations, where cyclic thermal stresses and erosive particle impacts are prevalent [45]. Figure 4.6 shows scanning electron microscopy (SEM) images depicting the surface and cross-sectional morphologies of (a, c) AlTiN and (b, d) AlTiMoN coatings. Examination of the micrographs shows distinct differences in surface texture, grain structure, and coating density, affirming the impact of Mo addition on microstructural refinement and the density of the coating.

Energy generation and power plant applications benefit significantly from molybdenum composite coatings due to their high-temperature corrosion resistance and thermal stability. In gas turbines, Mo–Al_2O_3 and Mo–SiC coatings protect nozzle guide vanes and turbine blades from oxidative degradation and erosion caused by hot gases and particulates. In concentrated solar power systems, these coatings serve as protective layers on receiver tubes, reducing oxidation and scaling while maintaining efficient heat transfer. Nuclear energy applications exploit the low neutron absorption cross-section of molybdenum and the stability of composite coatings under

FIGURE 4.6 Scanning electron microscopy (SEM) images of the surface and cross-sectional morphologies of the (**a**,**c**) AlTiN and (**b**,**d**) AlTiMoN coatings [46].

radiation exposure. Mo–SiC and Mo–ZrO_2 coatings provide protection to plasma-facing components, heat exchangers, and control rods, where extreme temperatures, high neutron flux, and corrosive coolants create a uniquely challenging environment. The combination of high thermal conductivity, oxidation resistance, and radiation tolerance makes molybdenum composites ideal candidates for next-generation nuclear reactors and space-based power systems [47].

In the automotive industry, molybdenum composite coatings are applied to engine components, piston rings, and valve seats, where high wear resistance and low friction are required. The self-lubricating properties of molybdenum oxides formed during sliding reduce friction and improve fuel efficiency, while the ceramic reinforcements ensure long-term wear resistance under high loads and repeated thermal cycling. Similarly, in heavy machinery, pumps, and compressors, Mo-based composites protect shafts, bearings, and gears from abrasive and adhesive wear, contributing to increased service intervals and reduced maintenance costs. The electronics and semiconductor industries utilize molybdenum composite coatings for functional applications such as thin-film electrodes, contact layers, and diffusion barriers. The low electrical resistivity of molybdenum, combined with the chemical and thermal stability of composite phases, allows coatings to maintain conductivity while resisting oxidation and chemical attack. Mo–Al_2O_3 and Mo–TiC coatings are employed in microelectronic devices and MEMS components, where high-temperature stability and wear resistance are essential for reliable operation. Additionally, these coatings provide protective surfaces for sensors and actuators exposed to harsh chemical or thermal conditions. In emerging energy technologies, molybdenum composites play an important role in hydrogen production and storage, catalysis, and fuel reforming. $Mo–TiO_2$, $Mo–Al_2O_3$, and Mo–SiC composites have been investigated as catalytic layers for hydrogen evolution, hydrodesulfurization, and other redox reactions, where high-temperature stability, electronic conductivity, and chemical durability are required. The composite structure ensures that active molybdenum species remain anchored to the support phase, maintaining catalytic efficiency under thermal and chemical cycling [48].

Molybdenum composite coatings are utilized in protective armour, thermal barriers, and missile parts in the defence and aerospace industries. The unique combination of hardness, high melting points, and oxidation resistance enables these coatings to endure extreme and high-velocity impacts, and endure the intense heat of jet propulsion and re-entry oxidation environments. Specifically, Mo–SiC and $Mo–Al_2O_3$ coatings offer the thermal shock resistance and erosion protection critical for nozzles, combustion chambers, and heat shield covers. The ability to maintain structural integrity during extreme temperature and fast mechanical loading helps the coatings to defend and assist the aerospace industry safely. The last important application area covers chemical processing equipment exposed to corrosive gases, molten salts, and high-temperature liquids. Chemical corrosion and structural integrity are preserved by Mo-based composite coatings on reactor vessels, heat exchangers, and pipes. In these cases, erosion and corrosion-resistant coatings with embedded alumina, chromium, and silicon carbide reduce maintenance intervals and increase operational lifespans. For the processed materials, the coatings' chemical inertness minimizes the risk of cross-contamination, which is particularly important in pharmaceutical, petrochemical, and food-processing industries [49].

4.9 FUTURE TRENDS

Building on the trends of integrating nanostructuring, functional grading, hybrid reinforcement approaches, and computational materials design, the future of molybdenum composite coatings will focus on targeted and advanced efforts to bring next-generation coatings to market. Advanced processing will overcome limitations to the next generation of coatings where unprecedented performance can be achieved in extreme conditions of temperature, pressure, and corrosion. Enhanced efficiency and working lifetimes will be achieved in future critical component cross-sector applications. Emerging applications will also include advanced electronics for extreme environments, advanced propulsion, and components for the new generation of sustainable energy systems. The molybdenum composite coatings evolve on the strategic intersection of materials science, surface engineering, and functional design integrating diverse and advanced multipurpose protective surface technologies for extreme environments [50].

The fabrication of novel material combinations, specifically involving reinforced ceramics and complex layered materials, is driving research, focusing on need for refinement of material properties and the integration of added functional components. Research also focuses on improving materials for new functional capabilities to meet advanced materials. New additions in core layered materials or core and shell advanced layered materials are opening new avenues for self-cleaning and multifunctional advanced materials in the field for extreme environments. The development of layered materials in the field of extreme environments and self-cleaning advanced materials is also an opportunity. Recent research on advanced electronics, aerospace, and nuclear materials is opening up layered advanced materials opportunities for oxide ceramics in aerospace and self-cleaning materials. New materials require added functional capabilities for extreme environments and self-cleaning. Future research aims to mitigate these challenges by exploring innovative deposition methods, integrating real-time process control, and applying sophisticated post-processing techniques like laser remelting, annealing, and hot isostatic pressing. The design and application of molybdenum composite coatings increasingly consider environmental and sustainability impacts. The reduction of hazardous materials, lower energy expenditures during deposition, and longer service lives of coated parts are some ways to minimize material waste. In energy production, chemical processing, and transportation, operational and ecological efficiencies in coatings are essential, thus coatings with high durability and low environmental impacts are becoming highly sought after. The design of self-lubricating and self-healing molybdenum composites provides additional opportunities for resource-efficient and sustainable maintenance of crucial components through reduced maintenance to sustain their longer service lives [51].

Surface engineering technologies such as thermal spray coatings, additive manufacturing, microwave processing, and magnetron sputtering thin-film coatings have lately grown in the biomedical, slurry, and high-temperature corrosion systems industries [52–65].

Coatings allow one to simulate or predict surface roughness, tensile, compressive, and bending strengths, as well as solar application performance. A more complex structure, like an ergonomically built chair, can be more easily modelled and created with the help of coatings [66–78].

4.10 CONCLUSIONS

Composite coatings utilizing molybdenum have attained outstanding results in an array of mechanical, tribological, thermal, and chemical activities, and, of course, the research and development in the area have not stopped. Such advancements have become necessary to respond to the more complex demands in the field of engineering. Anticipated future developments in the area relate to the need for multi-functional coatings. These coatings should retain their effectiveness in extreme and higher temperature settings, when subjected to aggressive chemical attacks, complex mechanical loads, and cyclic thermal and extraordinary radiation stress. One of the foremost endeavours entails the crafting of nanostructured and ultrafine composite materials, comprising nanoscale dispersed reinforcements, which results in superior mechanical strength as well as higher wear resistance, as well as oxidation and corrosion resistance. Their nanocomposite structures also result in the establishment of continuous, adherent tribo-oxide layers and the display self-healing properties in elevated temperatures, which prolong the life of the coatings. Another notable development is in the creation of functionally graded composite molybdenum coatings, wherein the chemical composition and microstructure are in the gradual transition across the coating thickness. This is in response to the problems concerning the mismatch of thermal expansion, residual stress, and interfacial brittleness which predominates at elevated temperatures.

The use of AI together with computational design in materials science will also hasten the advancement of molybdenum composite coatings. By modelling high-temperature oxidation, thermal stress, and wear, researchers ascertain the ideal reinforcements, volume fractions, and particle distributions for particular service conditions. Machine learning automates the correlation of microstructural features and design parameters, which significantly streamlines the process of evaluating composite designs and minimizes the need for experimental iteration and testing. These methodologies will likely yield coatings with ideal attributes for extreme conditions and improved efficiency and reliability for high failure cost applications in aerospace, energy, and other industries. Molybdenum composite coatings are high-performance, adaptive, and able to provide the materials system necessary for contemporary engineering's most critical applications. The unique intrinsic properties of molybdenum—such as high melting point, high thermal conductivity, and low friction—combined with the reinforcing effect of ceramics, carbides, silicides, and metallic phases, provide a customized distribution of hardness, toughness, abrasion resistance, toughness, resistance to oxidation and corrosion, high temperature stability, and high operational temperature stability. The effective utilization of these coatings in the aerospace, energy, automotive, electronics, defence, and chemical processing industries attests to their multifunctionality and reliability.

REFERENCES

1. Wang K, Chang B, Chen J, Fu H, Lin Y, Lei Y. Effect of molybdenum on the microstructures and properties of stainless steel coatings by laser cladding. *Applied Sciences*. 2017 Oct 15;7(10):1065.
2. Laribi M, Vannes AB, Treheux D. Study of mechanical behavior of molybdenum coating using sliding wear and impact tests. *Wear*. 2007 May 10;262(11–12):1330–6.
3. Barrera E, Gonzalez F, Rodriguez E, Alvarez-Ramirez J. Correlation of optical properties with the fractal microstructure of black molybdenum coatings. *Applied Surface Science*. 2010 Jan 1;256(6):1756–63.
4. Hou QY. Influence of molybdenum on the microstructure and properties of a FeCrBSi alloy coating deposited by plasma transferred arc hardfacing. *Surface and Coatings Technology*. 2013 Jun 25;225:11–20.
5. Ramezani M, Mohd Ripin Z, Pasang T, Jiang CP. Surface engineering of metals: Techniques, characterizations and applications. *Metals*. 2023 Jul 20;13(7):1299.
6. Deambrosis SM, Miorin E, Montagner F, Zin V, Fabrizio M, Sebastiani M, Massimi F, Bemporad E. Structural, morphological and mechanical characterization of Mo sputtered coatings. *Surface and Coatings Technology*. 2015 Mar 25;266:14–21.

7. McPherson R. The relationship between the mechanism of formation, microstructure and properties of plasma-sprayed coatings. *Thin Solid Films*. 1981 Sep 18;83(3):297–310.
8. Lyo IW, Ahn HS, Lim DS. Microstructure and tribological properties of plasma-sprayed chromium oxide–molybdenum oxide composite coatings. *Surface and Coatings Technology*. 2003 Jan 30;163:413–21.
9. Xi HH, He PF, Wang HD, Liu M, Chen SY, Xing ZG, Ma GZ, Lv ZL. Microstructure and mechanical properties of Mo coating deposited by supersonic plasma spraying. *International Journal of Refractory Metals and Hard Materials*. 2020 Jan 1;86:105095.
10. Yue W, Zhang Y, Zheng Z, Lai Y. Hybrid laser additive manufacturing of metals: A review. *Coatings*. 2024 Mar 6;14(3):315.
11. Wang XH, Han F, Liu XM, Qu SY, Zou ZD. Effect of molybdenum on the microstructure and wear resistance of Fe-based hardfacing coatings. *Materials Science and Engineering: A*. 2008 Aug 20;489(1–2):193–200.
12. Kazimierczak H, Ozga P, Berent K, Kot M. Microstructure and micromechanical properties of electrodeposited Zn–Mo coatings on steel. *Journal of Alloys and Compounds*. 2015 Jul 5;636:156–64.
13. Liu C, Huang X, Xu R, Mai Y, Zhang L, Jie X. Microstructure and properties of nanocrystalline Ni-Mo coatings prepared by ultrasound-assisted pulse electrodeposition. *Journal of Materials Engineering and Performance*. 2021 Apr;30(4):2514–25.
14. Vaidya A, Streibl T, Li L, Sampath S, Kovarik O, Greenlaw R. An integrated study of thermal spray process–structure–property correlations: A case study for plasma sprayed molybdenum coatings. *Materials Science and Engineering: A*. 2005 Aug 25;403(1–2):191–204.
15. Gilewicz A, Warcholinski B, Murzynski D. The properties of molybdenum nitride coatings obtained by cathodic arc evaporation. *Surface and Coatings Technology*. 2013 Dec 15;236:149–58.
16. Battu AK, Zade VB, Deemer E, Ramana CV. Microstructure-mechanical property correlation in size controlled nanocrystalline molybdenum films. *Advanced Engineering Materials*. 2018 Oct;20(10):1800496.
17. Jiang X, Matejicek J, Sampath S. Substrate temperature effects on the splat formation, microstructure development and properties of plasma sprayed coatings: Part II: Case study for molybdenum. *Materials Science and Engineering: A*. 1999 Nov 15;272(1):189–98.
18. Wang J, Munroe P, Zhou Z, Xie Z. Nanostructured molybdenum nitride-based coatings: Effect of nitrogen concentration on microstructure and mechanical properties. *Thin Solid Films*. 2019 Jul 31;682:82–92.
19. Liao YH, Wu FB. Microstructure evolution and mechanical properties of refractory molybdenum-tungsten nitride coatings. *Surface and Coatings Technology*. 2024 Jan 30;476:130154.
20. Bartkowska A, Bartkowski D, Popławski M, Piasecki A, Przestacki D, Miklaszewski A. Microstructure, microhardness, corrosion resistance and chemical composition of Mo, B and Mo-B coatings produced using laser processing. *Materials*. 2020 Jul 22;13(15):3249.
21. Lee CH, Min KO. Effects of heat treatment on the microstructure and properties of HVOF-sprayed Ni–Cr–W–Mo–B alloy coatings. *Surface and Coatings Technology*. 2000 Oct 1;132(1):49–57.
22. Liu JH, Li WH, Pei ZL, Gong J, Sun C. Investigations on the structure and properties of nanocrystalline Ni-Mo alloy coatings. *Materials Characterization*. 2020 Sep 1;167:110532.

23. Lima-Neto PD, Correia AN, Vaz GL, Casciano PN. Morphological, structural, microhardness and corrosion characterisations of electrodeposited Ni-Mo and Cr coatings. *Journal of the Brazilian Chemical Society*. 2010;21:1968–76.
24. Chen R, Cai Z, Pu J, Lu Z, Chen S, Zheng S, Zeng C. Effects of nitriding on the microstructure and properties of VAlTiCrMo high-entropy alloy coatings by sputtering technique. *Journal of Alloys and Compounds*. 2020 Jun 25;827:153836.
25. Li M, Zhang J, Mi T, Zheng C, Wu D, Chen Z, Yi X. Improvement mechanism of Mo element on cracks, microstructure and properties of laser cladding WC-Ni60AA coating. *Surface and Coatings Technology*. 2024 Aug 15;489:131047.
26. Lala SR, Singh AP, Singh S, Tiwari SC, Srivastava C. Effect of molybdenum partitioning on the evolution of microstructure, grain boundary constitution and corrosion behaviour of electrodeposited zinc-molybdenum coatings. *Materialia*. 2023 Mar 1;27:101709.
27. Pang X, Yao C, Xiong Z, Gong Q, Sun J, Misra RD, Li Z. Comparative study of coatings with different molybdenum equivalent on titanium alloy forged plate for laser cladding: Microstructure and mechanical properties. *Surface and Coatings Technology*. 2022 Sep 25;446:128760.
28. Wang YW, Yang Y, Wang XY, Li W, Gou JF. Microstructure and properties of molybdenum composite coatings by plasma spraying of Mo–SiC and MoO3–Al–SiC composite powders. *Ceramics International*. 2024 Jan 15;50(2):2963–75.
29. Wasekar NP, Verulkar S, Vamsi MV, Sundararajan G. Influence of molybdenum on the mechanical properties, electrochemical corrosion and wear behavior of electrodeposited Ni-Mo alloy. *Surface and Coatings Technology*. 2019 Jul 25;370:298–310.
30. Hwang B, Lee S, Ahn J. Correlation of microstructure and wear resistance of molybdenum blend coatings fabricated by atmospheric plasma spraying. *Materials Science and Engineering: A*. 2004 Feb 5;366(1):152–63.
31. Mitterer C, Mayrhofer PH, Beschliesser M, Losbichler P, Warbichler P, Hofer F, Gibson PN, Gissler W, Hruby H, Musil J, Vlček J. Microstructure and properties of nanocomposite Ti–B–N and Ti–B–C coatings. *Surface and Coatings Technology*. 1999 Nov 1;120:405–11.
32. Mrdak M, Vencl A, Ćosić M. Microstructure and mechanical properties of the Mo-NiCrBSi coating deposited by atmospheric plasma spraying. *FME Transactions*. 2009;37(1):27–32.
33. Overs MP, Harris SJ, Waterhouse RB. Composition, Microstructure and Mechanical Properties of Flame Sprayed Molybdenum Coatings. *Transactions of the IMF*. 1979 Jan 1;57(1):81–5.
34. Pogrebnjak A, Buranych V, Ivashchenko V, Borba-Pogrebnjak S, Maksakova O, Caplovicová M, Goncharov A, Onoprienko A, Skrynskyy P, Sahul M, Konarski P. Tunable TiZrMoC coatings: A comprehensive study of microstructure, mechanical properties, and wear resistance. *Nanomaterials*. 2024 Dec 11;14(24):1986.
35. Pu R, Sun Y, Xu J, Zhou X, Li S, Zhang B, Cai Z, Liu S, Zhao X, Xiao L. Microstructure and properties of Mo-based double-layer MoSi2 thick coating by a new two-step method. *Surface and Coatings Technology*. 2020 Jul 25;394:125840.
36. Zeng Y, Guo Y, Yang Q, Li Z, Wang Y, Wang G, Guo Z, Chen Z. Effect of Ti, Mn and Mo on the microstructure and properties of CoCrFeNi high entropy alloy coatings prepared by laser cladding. *Materials Today Communications*. 2024 Dec 1;41:110509.
37. Yang JF, Yuan ZG, Zhang GG, Wang XP, Fang QF. Manufacture, microstructure and mechanical properties of Mo–W–N nanostructured hard films. *Materials Research Bulletin*. 2009 Oct 1;44(10):1948–53.

38. Ding Z, Fang Z, Wang G, Wu C, Yang S. Microstructure and properties of Mo-Si-B-Y2O3 composite coating on nickel-based alloy by laser cladding. *Optics & Laser Technology*. 2023 Sep 1;164:109473.
39. Choi EY, Kang MC, Kwon DH, Shin DW, Kim KH. Comparative studies on microstructure and mechanical properties of CrN, Cr–C–N and Cr–Mo–N coatings. *Journal of Materials Processing Technology*. 2007 Jun 12;187:566–70.
40. Xiaoli M, Kaiming W, Hanguang F, Jiang J, Yongping L, Dawei Y. Effect of Mo content on microstructure and properties of laser cladding Fe-based alloy coatings. *Surface Review and Letters*. 2018 Apr 1;25(3):1850077.
41. Zhang XH, Pang M, Liu G, Jia L, Chen J. Effect of auxiliary gas flow parameters on microstructure and properties of Mo coatings prepared by high-energy plasma spraying. *Surface and Coatings Technology*. 2020 Mar 15;385:125384.
42. Sampath S, Wayne SF. Microstructure and properties of plasma-sprayed Mo-Mo2C composites. *Journal of Thermal Spray Technology*. 1994 Sep;3(3):282–8.
43. Lin T, Feng M, Lian G, Lu H, Chen C, Huang X. Effects of C content on the microstructure and properties of CoCrFeNiTi0. 5Mo0. 5Cx high-entropy alloy coatings by laser cladding. *Journal of Materials Research and Technology*. 2024 Nov 1;33:1540–57.
44. Wang X, Zhang M, Zou Z, Qu S. Microstructure and properties of laser clad TiC+ NiCrBSi+ rare earth composite coatings. *Surface and Coatings Technology*. 2002 Dec 2;161(2-3):195–9.
45. Iram S, Cai F, Wang J, Zhang J, Liang J, Ahmad F, Zhang S. Effect of addition of Mo or V on the structure and cutting performance of AlCrN-based coatings. *Coatings*. 2020 Mar 23;10(3):298.
46. Erden MA, Yaşar N, Korkmaz ME, Ayvacı B, Nimel Sworna Ross K, Mia M. Investigation of microstructure, mechanical and machinability properties of Mo-added steel produced by powder metallurgy method. *The International Journal of Advanced Manufacturing Technology*. 2021 Jun;114(9):2811–27.
47. Yang T, Yin J, Ying P, Lin C, Zhang P, Wu J, Kovalev A, Huang M, Wang T, Grigoriev AY, Gutsev DM. Influence of molybdenum addition on the structure, mechanical properties, and cutting performance of AlTiN coatings. *Lubricants*. 2024 Dec 3;12(12):429.
48. Lin CM, Kai WY, Su CY, Tsai CN, Chen YC. Microstructure and mechanical properties of Ti-6Al-4V alloy diffused with molybdenum and nickel by double glow plasma surface alloying technique. *Journal of Alloys and Compounds*. 2017 Sep 15;717:197–204.
49. Hou QY, He YZ, Zhang QA, Gao JS. Influence of molybdenum on the microstructure and wear resistance of nickel-based alloy coating obtained by plasma transferred arc process. *Materials & design*. 2007 Jan 1;28(6):1982–7.
50. Danelon MR, Fukumasu NK, Carvalho AA, Rego RR, Machado IF, Souza RM, Tschiptschin AP. Microstructural and mechanical characterization of Nb-doped MoS2 coatings deposited on H13 tool steel using Nb-based interlayers. *Coatings*. 2025 Jan 6;15(1):57.
51. Liang BH, Hsieu FS, Wu FB. Modulation effect on mechanical properties of nanolayered MoN/MoSiN coatings. *Surface and Coatings Technology*. 2022 Apr 25;436:128278.
52. Verma PK, Kumar V, Vasudev H. Bioactivity and corrosion analysis of thermally sprayed hydroxyapatite based coatings. *Journal of Electrochemical Science and Engineering*. 2024;14(4):419–39.
53. Verma PK, Vasudev H, Kumar V. Biocompatibility and corrosion performance of HA-based LVOF sprayed and post-annealed coatings on SS 254 surgical-grade bio-implant material. *Ceramics International*. 2025;51(10):13207–25.

54. Verma PK, Vasudev H, Singh G. Improving the corrosion performance of SS 304 stainless steel using HVOF-sprayed WC-Co-Cr coatings in chloride environments. *Journal of Bio-and Tribo-Corrosion*. 2025;11(4):1–13.
55. Parkash J, Saggu HS, Vasudev H. A short review on the performance of high velocity oxy-fuel coatings in boiler steel applications. *Materials Today: Proceedings*. 2022;50:1442–6.
56. Singh P, Vasudev H, Bansal A. Effect of post-heat treatment on the microstructural, mechanical, and bioactivity behavior of the microwave-assisted alumina-reinforced hydroxyapatite cladding. *Proceedings of the Institution of Mechanical Engineers, Part E: Journal of Process Mechanical Engineering*. 2024;238(3):1060–71.
57. Verma PK, Vasudev H, Singh G, Singh S. A review on the microstructural and biomedical properties of implants manufactured using additive manufacturing. *Journal of The Institution of Engineers (India): Series D*. 2024;106:1587–601.
58. Singh SP, Verma PK, Singh S, Kapoor M, Balu R. Microstructural characterization and electrochemical behaviour of HVOF sprayed WC-Co-Cr coatings on 316L boiler steel stainless steel. *Results in Surfaces and Interfaces*. 2024;16:100248.
59. Verma PK, Minhas AS, Singh P, Kumar S. Slurry erosion behaviour of thermal sprayed Al2O3 and Cr2O3 coatings for turbine steels. *AIP Conference Proceedings*. 2024;2986(1):020012.
60. Dutta V, Thakur L, Singh B, Vasudev H. A study of erosion–corrosion behaviour of friction stir-processed chromium-reinforced NiAl bronze composite. *Materials*. 2022;15(15):5401.
61. Singh M, Vasudev H, Kumar R. Corrosion and tribological behaviour of BN thin films deposited using magnetron sputtering. *International Journal of Surface Engineering and Interdisciplinary Materials Science (IJSEIMS)*. 2021;9(2):24–39.
62. Sharanabasava H, Raviprakash M, Prasad CD, Ramesh MR, Phanibhushana MV, Vasudev H, Kumar S. Microstructure, mechanical and wear properties of SiC and Mo reinforced NiCr microwave cladding. *Advanced Materials and Processing Technologies*. 2023;10(4):3620–33
63. Singh M, Vasudev H, Singh M. Surface protection of SS-316L with boron nitride based thin films using radio frequency magnetron sputtering technique. *Journal of Electrochemical Science and Engineering*. 2022;12(5):851–63.
64. Prashar G, Vasudev H, Thakur L, Bansal A. Performance of thermally sprayed hydroxyapatite coatings for biomedical implants: A comprehensive review. *Surface Review and Letters*. 2023;30(1):2241001.
65. Vasudev H. Wear characteristics of Ni-WC powder deposited by using a microwave route on mild steel: Microwave cladding of Ni-WC. *International Journal of Surface Engineering and Interdisciplinary Materials Science (IJSEIMS)*. 2020;8(1):44–54.
66. Singh J, Gill HS, Vasudev H. Computational fluid dynamics analysis on role of particulate shape and size in erosion of pipe bends. *International Journal on Interactive Design and Manufacturing (IJIDeM)*. 2023;17(5):2631–46.
67. Rao GS, Mukkamala U, Hanumanthappa H, Prasad CD, Vasudev H, Shanmugam B, KishoreKumar KC. Evaluating and optimizing surface roughness using genetic algorithm and artificial neural networks during turning of AISI 52100 steel. *International Journal on Interactive Design and Manufacturing (IJIDeM)*. 2024;18(8):6151–60.
68. Singh J, Singh R, Singh S, Vasudev H, Kumar S. Reducing scrap due to missed operations and machining defects in 90PS pistons. *International Journal on Interactive Design and Manufacturing (IJIDeM)*. 2023;17(5):2527–39.
69. Nagabhushana P, Ramprasad S, Prasad CD, Vasudev H, Prakash C. Numerical investigation on heat transfer of a nano-fluid saturated vertical composite porous channel

packed between two fluid layers. *International Journal on Interactive Design and Manufacturing (IJIDeM)*. 2024;18(5):2927–44.
70. Mahantesh MM, Rao KR, Chandra AP, Vijayakumar MN, Nandini B, Prasad CD, Vasudev H. Design and modeling using finite element analysis for the sitting posture of computer users based on ergonomic perspective. *International Journal on Interactive Design and Manufacturing (IJIDeM)*. 2024;18(8):5875–891.
71. Verma PK, Vasudev H, Kumar V. Dental implant materials and designs and failure during implant placement and loading: A review. In Hitesh Vasudev, Chander Prakash, Mohammad Khalid, Kavindra Kesari (Eds.) *Advanced Manufacturing Processes*, 2025. pp. 138–50, Boca Raton, CRC Press.
72. Verma PK, Vasudev H, Kumar V. Bioactivity and corrosion performance of low-velocity oxy-fuel-sprayed hydroxyapatite-based coatings on a surgical-grade stainless steel. *Transactions of the Indian Institute of Metals*. 2025;78(8):185.
73. Verma PK, Vasudev H, Kumar V. Effect of electro-corrosion behavior of different grades of stainless steel used in bio-implants: A review. In Hitesh Vasudev, Chander Prakash, Mohammad Khalid, Kavindra Kesari (Eds.) *Advanced Manufacturing Processes*. 2025. pp. 269–83, Boca Raton, CRC Press.
74. Verma PK, Vasudev H. Importance of hydroxyapatite coatings in improving new bone osteogenesis: A review. In Hitesh Vasudev, Chander Prakash, Mohammad Khalid, Kavindra Kesari (Eds.) *Advanced Manufacturing Processes*. 2025. pp. 284–99, Boca Raton, CRC Press.
75. Kulkarni SD, Manjunatha, Chandrasekhar U, Manjunath KV, Prasad CD, Vasudev H. Design and optimization of polyvinyl-nitride rubber for tensile strength analysis. *International Journal on Interactive Design and Manufacturing (IJIDeM)*. 2024;18(8):5893–908.
76. Kattimani MA, Venkatesh PR, Masum H, Math MM, Bahadurdesai VN, Mustafkhadri S, Prasad CD, Vasudev H. Design and numerical analysis of tensile deformation and fracture properties of induction hardened inconel 718 superalloy for gas turbine applications. *International Journal on Interactive Design and Manufacturing (IJIDeM)*. 2024;18(8):5939–49.
77. Poojari M, Hanumanthappa H, Prasad CD, Jathanna HM, Ksheerasagar AR, Shetty P, Shanmugam BK, Vasudev H. Computational modelling for the manufacturing of solar-powered multifunctional agricultural robot. *International Journal on Interactive Design and Manufacturing (IJIDeM)*. 2024;18(8):5725–36.
78. Prakash C, Gupta LR, Mehta A, Vasudev H, Tominov R, Korman E, Fedotov A, Smirnov V, Kesari KK. Computing of neuromorphic materials: An emerging approach for bioengineering solutions, *Materials Advances*. 2023;4: 5882–919.

5 Mechanical Behaviour of Cermet Coatings for Wear and Corrosion Applications

Virinder Kumar, Jashanpreet Singh, and Abdul Wahab Hashmi

5.1 INTRODUCTION

Erosion, abrasion, and corrosion are widespread challenges faced across several industries, including hydropower, thermal power, offshore, chemical, food processing, and oil sectors. These problems affect a wide range of components of hydraulic machinery and pipeline circuits (pipelines, elbows, reducers, separators, tees, and seals) [1–3]. Hydraulic machinery and equipment often experience erosion wear, which results in high repair and maintenance expenses. [4]. Corrosion occurs when electrochemical reactions lead to the degradation of a material from its intended surface. The corrosion process involves two basic mechanisms: 'pitting action' and 'crater development' [5]. **Figure** 5.1 illustrates the primary abrasion mechanisms, namely "chip formation" and "detachment", along with the secondary mechanisms, which include "crack formation" and "fracture" [6]. Slurry bubbles cause cavitation erosion. While material toughness, surface roughness, and imperfections play a role, they are not the sole contributors. Instead, erosion is a highly intricate phenomenon that remains not yet fully understood.

Particle properties, target material features, and procedure parameters are among the many aspects that contribute to the intricate process of erosion. The initial behaviour in material hardness determines the material failure mode, according to Finne [7]. In brittle materials, erosion damage typically results from cutting action, while in ductile materials, it is mainly attributed to plastic deformation. Levy [8] discovered that ductile materials were initially more prone to erosion but decreased over time. Bellman and Levy [9] discuss the mechanisms for wearing macroparticles for erosion. The mechanisms of shallow crater formation and platelet development were described as outcomes of macroscopic particle interactions. Erosion mechanisms in ductile solids include deformation, micro-ploughing, micro-cracking, lip development, platelets, and small craters [10]. They further suggested that the fragmentation of brittle materials accelerates particle impingement, leading to surface deterioration. A schematic depiction of erosion processes in ductile and brittle materials is shown in Figure 5.2. The erosion behaviour of ductile and brittle materials differs from their

 DOI: 10.1201/9781003730637-5

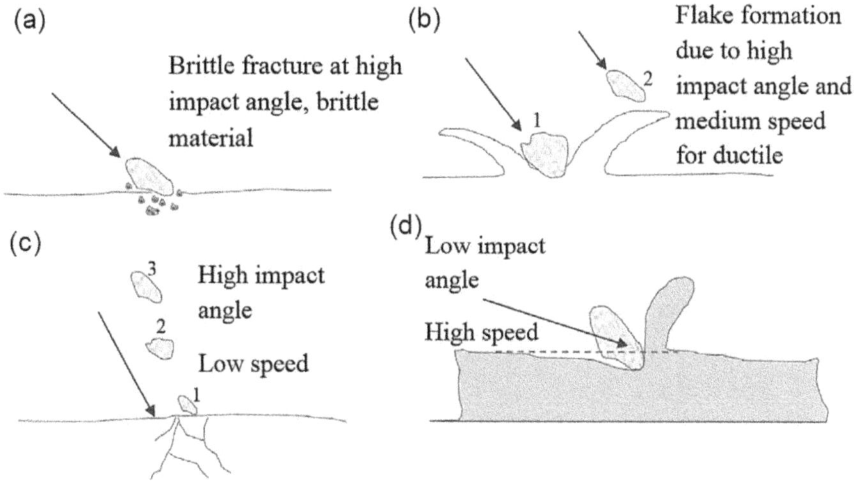

FIGURE 5.1 Various processes of wear: (a) brittle fracture, (b) plastic deformation, (c) fatigue erosion, and (d) abrasive/cutting erosion [6].

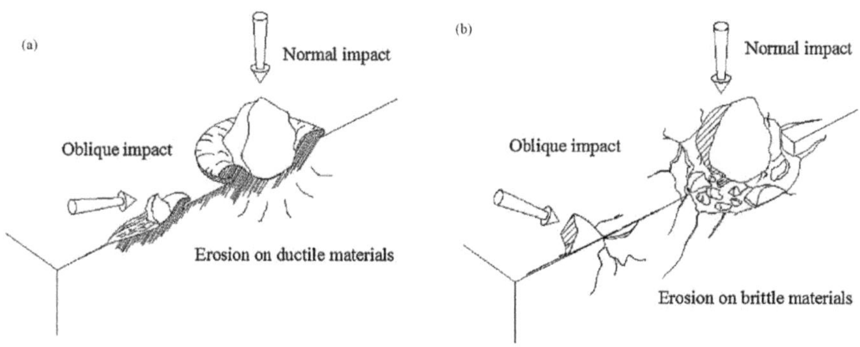

FIGURE 5.2 A schematic depiction of erosion in (a) ductile and (b) brittle materials [10].

basic wear mechanisms. In ductile materials, erosion starts with material deformation, leading to groove formation accompanied by lips. In contrast, brittle materials experience erosion through crack initiation, progressing to material fragmentation as chips. Furthermore, Finne [7] observed that a crater would develop when particles strike the object at low impact angles. Conversely, material loss transpires when particles collide with the target at high impact angles and are shattered by the force of the contact.

It is important to acknowledge that the erosion mechanisms outlined above are not mutually exclusive and may occur in conjunction. The correlation created by Sundararajan et al. [11] to evaluate the erosion efficiency (η) of materials is presented below:

$$\eta = \frac{2VH_T}{M_P V_P^2} \tag{5.1}$$

where MP is the particle mass (kg), VP is the particle velocity (m/s), HT is the hardness of the material (Hv), and V is the volume of removed material (m^3). According to their explanation, a value of $\eta > 1$ implies fragility (brittle mechanism), while a value of $\eta < 1$ suggests plastic erosion deformation (ductile mechanism).

Clark and Wong [12] characterized the deformation (ED) and cutting (EC) processes of erosion wear as follows:

$$E_T = \begin{cases} E_C + E_D \\ \dfrac{\frac{1}{2}M_p\left(V^2{}_N\right)}{\varepsilon} + \dfrac{\frac{1}{2}M_p\left(V^2{}_T \sin 2\alpha\right)}{\phi} \end{cases} \tag{5.2}$$

In this context, the tangential and normal velocity components of the particle are signified by V_T and V_N, respectively. Impingement angle, empirical constant for specific energy necessary in deformation erosion, and the constant for kinetic energy required in cutting erosion are represented by the symbols α, ε, and ϕ, respectively. Grewal et al. [13] presented the technique for estimating the amount of energy needed and consumed for material removal under oblique and normal impact:

$$\xi = \frac{2V\sigma_{cr}\left(\frac{H_T}{K_T}\right)}{M_P V_P^2} \tag{5.3}$$

The critical stress of the material is denoted by σcr, whereas the material toughness (MPa) is represented by K_T (MPa). In the case of brittle metals and alloys, σ_{cr} represents the ultimate tensile strength, whereas in the case of ductile materials, it represents the ultimate shear stress. Ploughing occurs when ξ is less than 1, micro-cutting occurs when ξ is equal to 1, and brittleness occurs when ξ is more than 1.

5.2 CORROSION IN CERMET COATINGS

Corrosion of engineering components subjected to harsh conditions is a serious problem in the mining, oil and gas, energy, marine, and other industries. Mechanical wear, erosion, and corrosion usually work in combination, reducing the life of components and increasing the cost of maintenance. To address these effects, the application of protective surface coating has become a critical measure and cermet coating has become one of the best applications. Ceramics cermets are mixtures of ceramic and metallic phases that are almost as hard, stable to heating and wear, and also have the toughness and ductility of metals [14]. Of them, tungsten carbide (WC)-based cermets, especially WC-Co and WC-Co-Cr, are highly used as they have a good combination of mechanical and chemical characteristics [15]. Chromium enhances the corrosion resistance by creating protective oxide layers and cobalt is used to produce toughness and even distribution of binder.

In addition to WC-based systems, Cr_3C_2-NiCr, TiC-Ni, and Mo_2C-Ni were also found to perform well in harsh operating conditions [16]. Coatings of Cr_3C_2-NiCr are in high demand in high-temperature and oxidizing environments due to the fact that they can be used to form stable chromium oxide scales. TiC-Ni coating is very hard and abrasion resistant and is suitable for cutting tools and sundry aerospace applications. Similarly, Mo_2C-based coating is also characterized by its stability under high temperature and its inhibition of localized corrosion. All these cermet systems may be refined with the addition of secondary reinforcements (e.g. Y_2O_3, ZrO_2, rare-earth oxides) to refine the microstructure, and enhance electrochemical stability [17]. Although cermet coating has better qualities, there is still continued research about its performance under combined erosion-corrosion conditions. Their corrosion resistance in a simulated service environment is commonly tested using electrochemical methods like open-circuit potential (OCP) monitoring, and potentiodynamic polarization, commonly using test media containing salt (saline) or slurry. These techniques serve as useful information on the influence of coating composition, microstructural stability, and reinforcement additions in degradation resistance.

5.3 INFLUENCE OF MATERIAL PROPERTIES ON WEAR

Since the beginning of the twentieth century, the material market has grown rapidly. Many materials can withstand tensile, compressive, impact, and fatigue forces. Researchers and designers have worked extensively to optimize the design improvements, allowing the materials to tolerate these qualities. The phase microstructure and material characteristics, including elasticity, toughness, microhardness, ductility, and elongation, are crucial in surface failure processes. Turbine components are often constructed using high-chrome steels. These include 13Cr4Ni or 16Cr5Ni (martensitic-austenitic steels), 18Cr8Ni (austenitic steel), and 13Cr1Ni (martensitic steel). Currently, most turbines are constructed using 13Cr4Ni or 16Cr5Ni steel [18]. The Swedish grade steel 13Cr6Ni was granted patents in the late 1960s, but its usage was restricted due to patent rights, which supported the fabrication of 17Cr4Ni, 13Cr4Ni, and 16Cr5Ni stainless steels, the latter of which are widely used in the turbine industry. Austenitic steel 18Cr8Ni with a minimum of 0.2% Mo has excellent resistance to corrosion and cavitation, but it has limited resistance to fatigue and creep [18].

Microhardness is a critical feature influencing a material's erosion and wear resistance [19–24]. The hardness ratio of particle hardness to target hardness was reported by researchers [9,25,26]. Oka et al. [27] have developed a model that predicts total erosion wear by using a variety of hardness materials (GPa, H_T) and other method parameters, such as particle diameter (μm, d_p) and impact velocity (m/s, V_p).

$$E_T = k \times V_P^m \times d_P^n \times H_T^{K_1} \times \sin\alpha^{n_1} \left(1 + H_T(1 - \sin\alpha)\right)^{n_2} \tag{5.4}$$

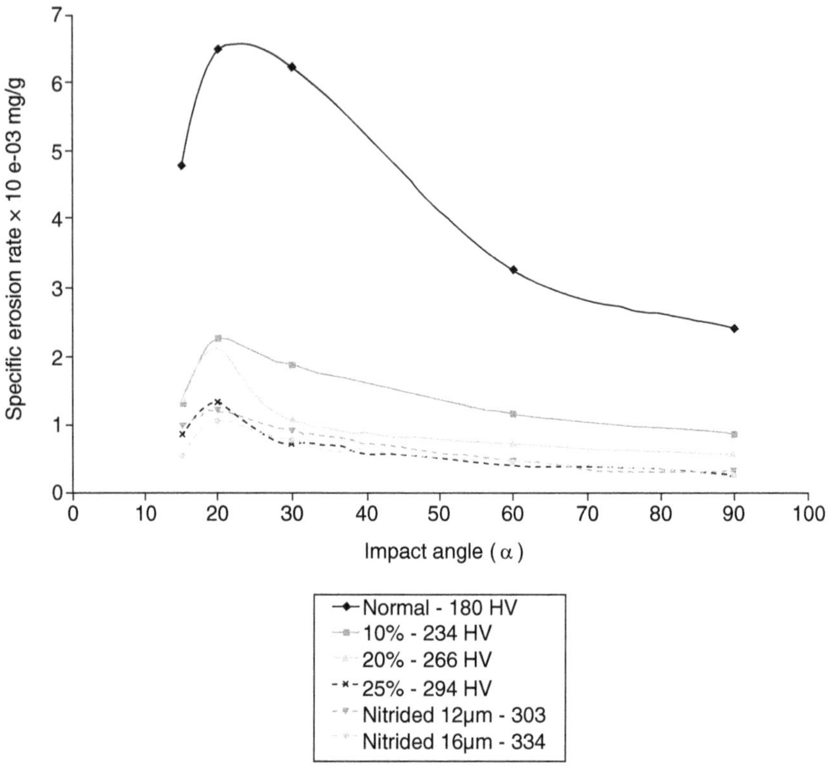

FIGURE 5.3 Material erosion rates under ambient conditions (velocity=32 m/s, particle size=102 μm) [29].

The exponents for particle diameter and speed are represented by m and n, respectively. Experimental and material hardness exponents are denoted by K and K_1, respectively. The angle of impact between the particle and the target determines the constants n_1 and n_2.

Levy and Hickey [28] discovered that harder materials wear faster than softer materials. During impact, the ductility of the material influences plastic deformation through the transfer of particle kinetic energy, which in turn reduces the extent of damage on the target surface. Divakar et al. [29] detailed the erosion patterns of traditional cold-rolled and surface case-hardened SS 316L stainless steel, as illustrated in Figure 5.3.

5.4 APPROACHES TO WEAR MEASUREMENT BASED ON EXPERIMENTAL METHOD

Various testers were employed to forecast turbine wear [3]. Researchers developed pilot plant loops, test rigs, and laboratory-scale testers to evaluate turbine wear. Initial studies used field results to assess erosion wear in working conditions. Still, several limitations exist, such as time duration, material, unpredictable testing variables, and expensive experimentation costs [30–32]. In small-scale laboratory testing, variable parameter control and highly accurate erosion data can be generated at an accelerated rate. The ability to analyze wear material efficiency in the solid flow of particles using a test rig is particularly significant. Since the test wears down the sharp edges of the slurry particles, a major downside of employing a test rig to measure the erosion wear rate is that new slurry is required. Standard tools for gauging wear include a slurry jet tester rig [13], pot tester [33], Coriolis tester [34], whirling arm tester [35], and prototype rigs [36].

The development of the pot tester began around the end of 1985. Schumacher [37] designed a pot tester with three hubs to accommodate 24 specimens, eight in each hub. Although limited in angle and concentration adjustments, this tester paved the way for the development of advanced slurry testers. Early in the 1990s, Clark [38,39] came up with a cylindrical pot tester. Figure 5.4 shows the main parts of this exam Figure 5.4(a). This tester was distinctive because it prevented pressure buildup through ventilation and protected the specimen from corrosion using nylon spacers. Its primary limitation was that specimens had to be cylindrical, and the container could accommodate only two specimens simultaneously. The propeller-type pot tester was also developed by Gupta et al. [40], as shown in Figure 5.4(b). This tester outperformed its predecessors in terms of its capacity to adjust variables such as particle impact angle, solid concentration, and velocity. The components of the pot tester were changed by Gandhi et al. [19], as shown in Figure 5.4(c, d). Figure 5.4(d) shows the fixture that they made. The capacity of this tester to precisely generate impact angles and velocities made it unique. Numerous researchers started making more

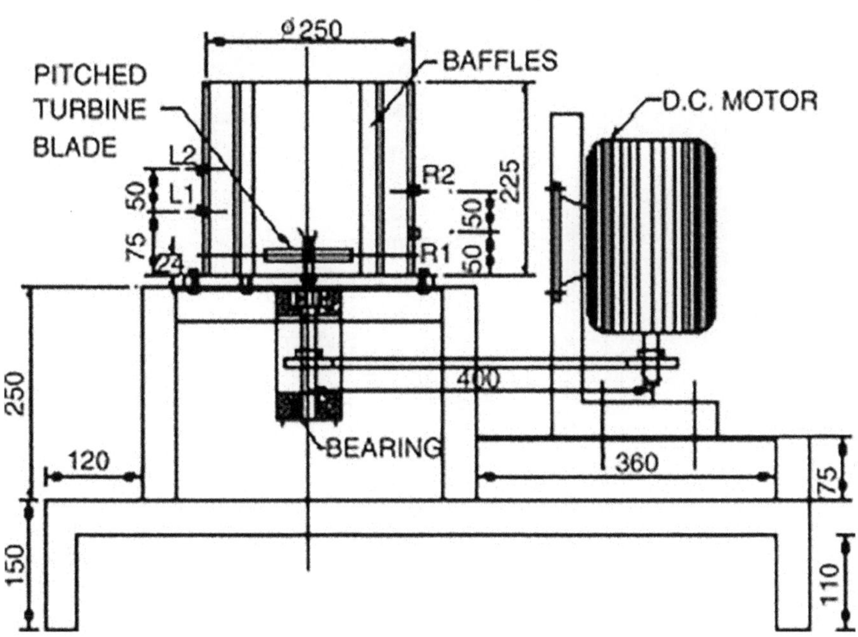

FIGURE 5.4 An illustration of the pot type tester used to test slurry erosion [41].

changes to pot testers around the turn of the millennium. To evaluate erosion at high solid concentrations across several impact angles (0° to 90°), Desale et al. [41] also modified the slurry pot tester. This pot tester is schematically shown in Figure 4(e).

Tuzson created the Coriolis tester in the 1980s [42]. In this tester, a motor rotates the centre of a bowl as slurry particles flow from an overhead tank. The particles then reach a fixture that holds the specimen. This tester operates based on the principle of Coriolis acceleration. This principle states that when solid particles travel at high speeds along a radial path, they generate an outward centrifugal force. The centrifugal force accelerates particles toward the target, promoting particle-target interactions. Coriolis acceleration is the acceleration that causes particles to accelerate through centrifugal force [34,42]. The Coriolis tester's primary benefit was its ability to regulate acceleration and flow velocity. An upgraded version of Tuzson's [42] Coriolis erosion tester was developed by Clark et al. [34].

Various researchers invented the Slurry Jet Erosion (SJE) tester. This tester employs a solid-liquid slurry jet to assess erosion by measuring volume loss. A closed-loop SJE tester is schematically shown in Figure 5.5 [43]. The nozzle jet tester that Matsumura et al. [44] created can accommodate up to four specimens. Reduced power usage, fewer workpieces, reduced erodent amount, accurate impact angle, and easy control over particle concentration were some of the differentiating advantages of this tester. The specimen remained submerged during the trial, which was the biggest drawback of this tester. Additionally, the velocity was not measured.

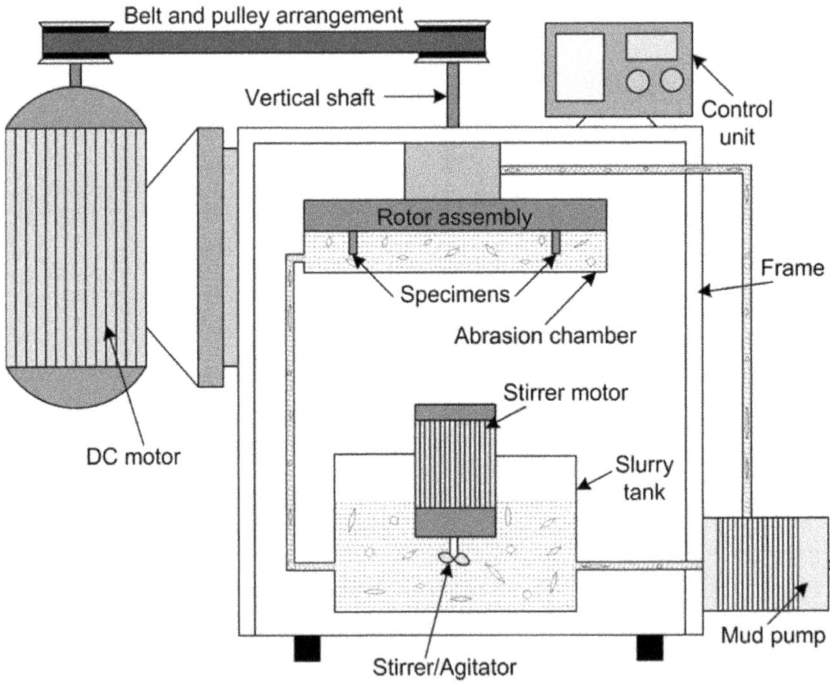

FIGURE 5.5 Schematic diagram of closed-loop erosion test rig [43].

In order to work at high particle concentrations and to allow for adjustment of the impingement angle, Lin and Shao [45] improved upon the conventional SJE tester. The inability to sustain a high velocity was the primary drawback of this testing. An instrument for measuring intrinsic erosion qualities was developed by Iwai et al. [46] to ascertain the impact of variables such as target surface standoff distance, particle size and shape, solid concentration, time duration, and rotating speed. By using an air-operated diaphragm pump to circulate the slurry and monitor erosion at high velocities, Walker and Robbie [26] enhanced the conventional SJE testing. To regulate the rate of flow via the nozzle, they used a diaphragm pump. A number of materials were subjected to erosion tests against slurry at different impact angles using an NRC SJE tester that was developed by Xie et al. [47]. Figure 5.6 shows the NRC slurry jet erosion components.

A hybrid of the slurry pot and the erosion jet tester, the whirling arm erosion tester allows for teamwork. Erosion is measured under certain circumstances using this tester. A motor to adjust the specimen's speed, two arms with holders, a slurry pot, and a vacuum chamber to remove aerodynamic effects make up this tester's four primary components. Determining erosion wear in dry circumstances was the only use of the whirling arm erosion tester in its early years [48,49]. Extreme localized heating and the formation of centrifugal strains were the main issues with this testing. As shown in Figure 5.7, a whirling arm apparatus was developed by Al-Bukhaiti et al. [35] for the purpose of evaluating erosion wear at various impact angles. The main advantage of this tester was its capability to position and lock the specimen holder

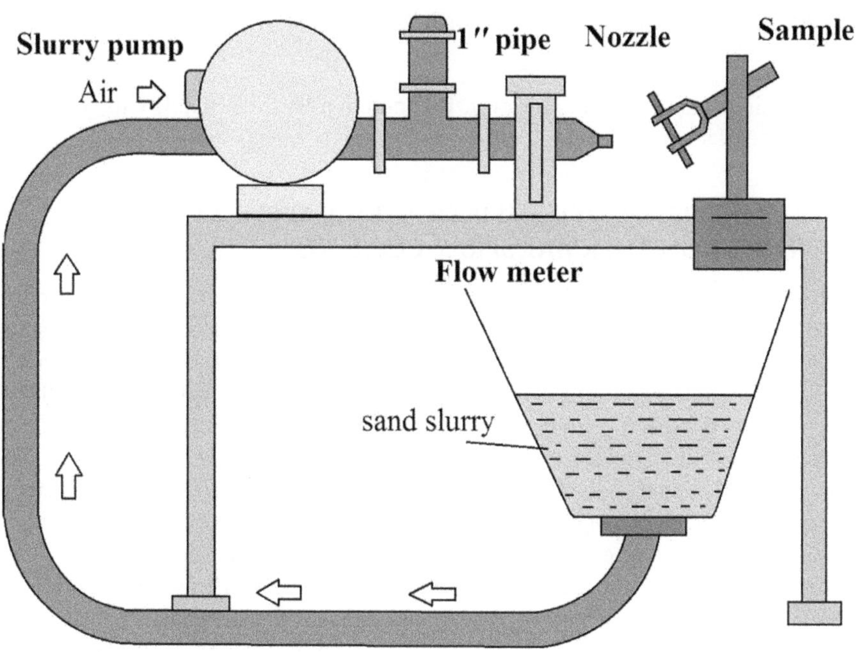

FIGURE 5.6 A slurry jet erosion tester schematic from the NRC [47].

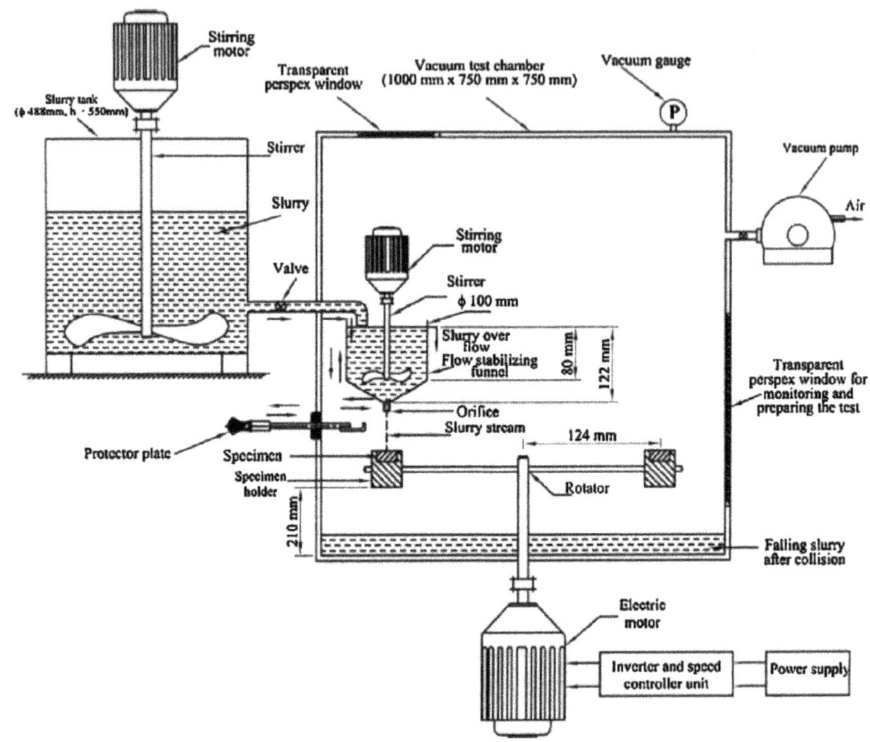

FIGURE 5.7 Schematic diagram of a whirling arm rig [35].

at any desired impact angle. However, its limitation was the inability to generate a controlled erosion model.

5.5 WEAR PROTECTION IN INDUSTRIAL APPLICATIONS THROUGH THERMAL SPRAY COATINGS

In order to safeguard hydraulic and offshore machinery, a variety of surface modification methods are used. Many different coating methods, including thermal spray coating, may be used on both metallic and nonmetallic surfaces. Based on the energy source, it is categorized into electric arcs, plasma arcs, and flame spray methods. Depending on the specific requirement, this process enables coating application using powder, rod, or wire. Fuel gases carry the molten materials to their intended target. The homogeneous lamellar structure is created when molten particles contact the target surface, creating a thick coating layer. A significant advantage of thermal spray coating is its ability to apply diverse materials without degrading them during deposition. Rigid substrates can be used without affecting properties or causing thermal shock. However, a limitation is that the coating is size-restricted and cannot achieve deep penetration.

Thermal spray coatings have a long history of development. From 1882 until 1889, in Zurich, Switzerland, M.U. Schoop worked on an oxyacetylene welding gun for the

Sn (Ti) wire feed [50]. Afterwards, supersonic powder particles were deposited onto the target surface using a welding gun. In 1908, Schoop employed electric arc spray, which allowed the use of Zn and steel via wire-arc metallizing for employing Zn as a corrosion-resistant material on base material [50]. Following World War II, thermal spray coating and plasma arc technologies advanced dramatically. The plasma arc spray technique had a significant advantage over the ignition spray jet because of its greater process temperature. However, advancements in the high-velocity oxy-fuel (HVOF) process continued rapidly until 1995, during which several improvements were introduced, including the High-Velocity Air Fuel (HVAF) and Hypersonic Velocity Impact Fusion (HVIF) processes. In 1992, Browning developed the HVIF method [51]. This process follows the same principle as HVOF, but particles are kept below the melting point to prevent softening. In this process, air replaces oxygen. After Browning Thermal Systems Inc. introduced the HVAF technique in 1993, Irving et al. [52] came up with the idea of using air instead of oxygen in this procedure.

Elasticity, toughness, porosity, bond strength, and thickness are some of the coating qualities that affect waterproofing. Table 5.1 presents the hardness values of several hard coatings. Previous studies indicate that coatings with higher toughness exhibit greater resistance in erosion-prone environments. Thermal spray coatings' characteristics are primarily determined by the coating process used. Table 5.2 provides an overview of coating properties derived from several flame spray processes [53]. Coating thickness and surface roughness also significantly affect adhesion, hardness, and erosion resistance. A coating thickness of around 200 μm is ideal for enhanced slurry erosion resistance [54]. For erosion-resistant coatings, surface roughness should be minimized.

TABLE 5.1
Typical HV0.3 hardness values of major commercial compositions [53]

Property / powder	WC–12Co	WC–17Co	WC–10Co4Cr	WC–(W,Cr)2C–7Ni	Cr_3C_2–25NiCr
Vickers Hardness, HV0.3 (1)	1376	1358	1274	1242	1080
Vickers Hardness, HV0.3 (2)	1250	1200	1350	1300	1150

TABLE 5.2
Properties achieved by different coating processes

S. no.	Coating process	Porosity (%)	Bond strength (MPa)	Thickness (μm)
1	Flame spraying	10–20	15–30	100–2500
2	Electric arc spray	10–20	<20	100–1500
3	APS method	1–7	>30	50–500
4	D-gun spray	0.5–5.0	>70	~300
5	HVOF process	~1%	>40	Not specified

Coating deposition is often accomplished using the best process parameters available in the literature. Other techniques apart from coatings for enhancing erosion resistance of materials in high-temperature oxidation circumstances include cladding, nitriding, carburizing, and laser treatment; nevertheless, they did not exhibit promising performance in offshore environments [58]. When contrasted with thermal spray methods, laser cladding has a number of drawbacks, including immobility, complexity when dealing with bigger components, poor erosion and corrosion resistance, and the difficulty to produce thin coatings. It finds use on carbides such as tungsten carbide, Ni/Co-based alloys (Inconel, Hastelloy, and Stellite), and stainless steels (300 and 400 series) [59–61]. Plasma arc spraying is frequently applied to refurbish components such as piston rods, seals, pump plungers, shafts, and hubs. To make offshore and hydraulic equipment last longer, there are a number of tried-and-true surface coating processes. These include flame spraying, D-gun spraying, high-velocity oxy-fuel, plasma spraying, atmospheric plasma spraying, vacuum plasma spraying, and laser spraying. The nozzle or gun HVOF spray coatings controls localized combustion and increases the energy of the coating powder particles. Typically, HVOF coatings deposit thermal spray powders at hypersonic velocities using fuel gases to provide additional energy. The properties and operational parameters of different thermal spray techniques are summarized in Table 5.3.

5.6 CERMET POWDER CLASSIFICATIONS FOR ENHANCED WEAR RESISTANCE IN THERMAL SPRAY APPLICATIONS

The thermal spray powders such as WC, Co, and Ni-based alloys have been adopted universally to enhance the wear resistance to various engineering materials employed in corrosion and erosion conditions. Usually, the material's hardness and characteristics dictate the coating choice for erosion wear applications. For slurry erosion at low impact angles on ductile materials, the material behaves similarly to abrasion and demands high hardness. In contrast, improving the performance of brittle materials at near-90° impact angles requires a coating with high toughness. A coating with excellent fatigue resistance is essential for cavitation or liquid impact scenarios. Among the most commonly used thermal spray coatings for erosion wear are cermet coatings such as WC-Co, CrC-NiCr (particularly at temperatures above 540 °C), fused self-fluxing alloys, nonferrous alloys like aluminium, bronze, monel, chromium oxide and alumina, as well as several Fe, Ni, Cr and Co-based alloys. Ceramic and cermet materials are strong, capable of withstanding high temperatures, and corrosion-resistant.

5.6.1 WC-based Coatings

Thermal spraying, namely plasma and HVOF, are now widely adopted for the cermet coatings. Tungsten carbide (WC)-based coatings are widely utilized to defend against erosion in various applications. WC-based coatings are renowned for their hardness, making them highly effective against abrasive wear and erosion. Tungsten carbide's high hardness reduces wear on the underlying component, keeping its dimensional

TABLE 5.3
Data was compiled from studies [55–57]Operating parameters and material properties of thermal spray methods

Process no.	Coating method	Process temperature (°C)	Velocity range (m/s)	Bond strength (MPa)	Energy source	Void content (%)	Transport medium	Application
1	Flame spraying (powder)	≈3100	60–70	6–10	Fuel gas (acetylene or propane)	7–12	Same fuel gas	Al, Zn for corrosion protection
2	Flame spraying (wire)	≈3100	120–140	-	Fuel gas (acetylene or propane)	5	Same fuel gas	-
3	Arc spraying	≈7000	100–170	10–20	Electricity	3–15	Compressed air	-
4	Plasma spraying	≈15000	150–600	20–70	Electricity (W-Cathode, Cu- anode)	1–8	Ar, He, N_2, or mixture	Ceramics coating
5	Detonation gun spraying	≈4200	600–800	60	Acetylene + Oxygen	<1	Ar, He, N_2, or mixture and N_2 for purging	-
6	HVOF spraying	≈2750	600–1200	>70	Oxy-propylene, Oxy-hydrogen and kerosene	1-2	Compressed air	Ceramics-metallic or metallic coating

stability and extending its lifespan [62]. WC–10% and Co–4% Cr HVOF coatings, also known as WC-10Co4Cr HVOF coatings, are thermal spray coatings widely used to reduce erosion and corrosion in many industrial applications. There are three widely accepted varieties of tungsten carbide (WC) coatings such as WC-CoCr, WC-170Co, and WC-10Co. WC coatings contain hard carbides such as WC and W_2C, which provide brittleness and hardness to the ductile matrix. According to Singh et al. [63], the SEM image and XRD pattern of the WC-10Co-4Cr coating powder are as presented in Figure 5.8. The coating powder particles can be seen to be embedded in a spherical form in Figure 5.8(a). In Figure 5.8(b) the XRD pattern reveals the existence of crystalline phases including CoC, WC, W2C, and $Cr_{23}C_6$. Both macrocrystalline (WC) and two-phase cast eutectic (WC/W2C) tungsten carbides are the most common types [24,64]. Several mechanical qualities are improved by the W_2C and WC phases in thermal spraying powder, which enhances the exposed surface's erosion performance [64]. WC-based coatings significantly improve resistance to abrasion and erosion at room temperature. Plasma spraying or high-velocity oxy-fuel are used to apply these coatings on different substrates. Plasma arc spraying deposits thick layers of WC at specific locations, while HVOF spraying produces dense, adherent layers of hard carbides tailored to the operating conditions. Combining high particle velocities with a low-temperature flame, HVOF may minimize the breakdown of WC powder. Additionally, incorporating 4–15 wt% of Co and Cr further improves the wear resistance of the coating [24,64]. The erosion performance of WC-Co coatings is much enhanced, according to Jia and Fischer [65], since the carbide grain size is controlled in a ductile cobalt matrix. Because WC coating contains a large proportion (>90%) of fine grains, it enhances the surface qualities of the base material for wear resistance and erosion. WC has a relatively short gap between two fine grains, shielding weak ones. Figure 5.9 illustrates the erosion rates of WC-10Co-4Cr HVOF and WC-12Co coatings at different impact angles and velocities. On the other hand, under 90° impact conditions, WC-12Co HVOF coatings deteriorate at a faster rate than WC-10Co-4Cr HVOF coatings [66]. As demonstrated in the SEM of erosion mechanisms (Figure 5.10), WC particles detach after the initial erosion, which causes binder loss and coating cracking and fracture [66].

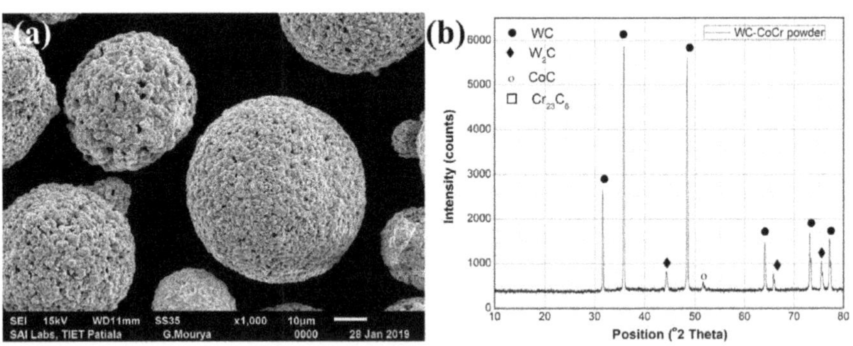

FIGURE 5.8 (a) SEM and (b) XRD images of WC-10Co4Cr powder [63].

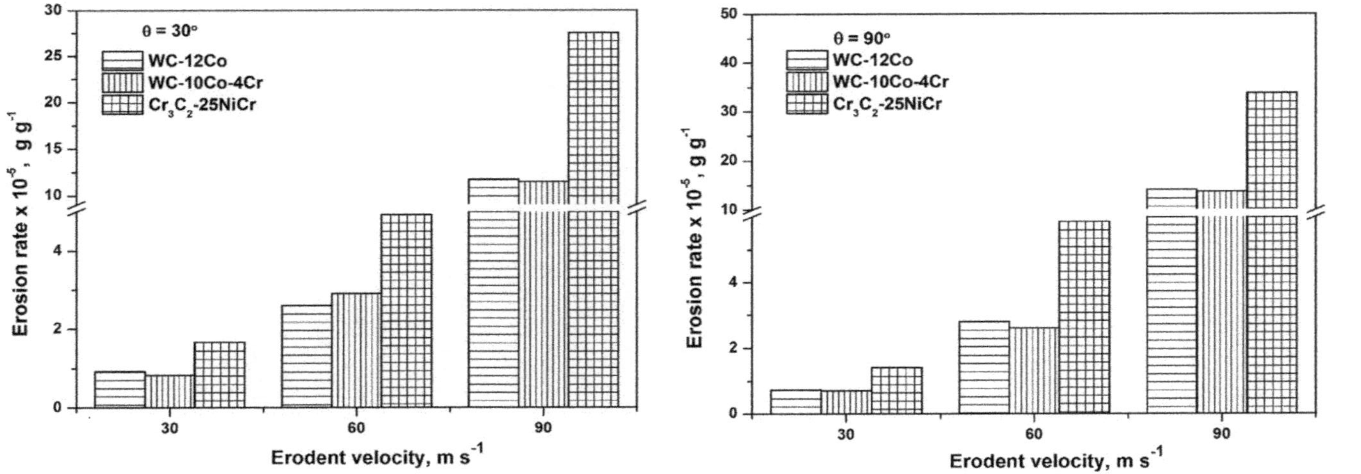

(a) (b)

FIGURE 5.9 Rate of erosion of Cr_3C_2-25NiCr, WC-12Co, and WC-CoCr coatings at (a) 30° and (b) 90°, under the influence of erodent velocity (30, 60, and 90 m/s) [66].

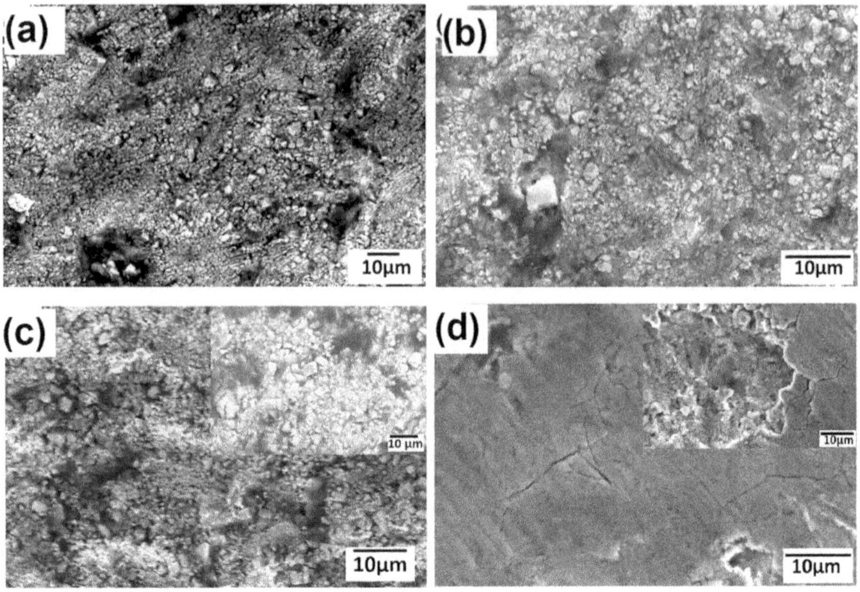

FIGURE 5.10 SEM micrographs of worn regions at 90° impact angle on surface of (a,c) WC-12Co: (a) 30 m/s and (b)90 m/s, (b) WC-10Co-4Cr at 60 m/s and (d) Cr_3C_2–25NiCr at 90 m/s [66].

Sahni and Grewal [67] examined the slurry erosion behaviour of WC-10Co-4Cr-coated CF8M turbine steel under different test conditions. Based on Figure 5.11(a), they reported that the coated specimens had no significant weight loss relative to the uncoated ones at 30° impingement angle, 15 mm stand-off distance, and 10,000 ppm slurry concentration due to the dense and well bonded coating layer. Figure 5.11(b) indicates that the coated sample retained low weight loss (0.01 g) at an increased impingement angle (90°) and stand-off distance (20 mm), whereas the uncoated specimen also exhibited reduced wear, which was explained by the effects of particle orientation and slurry. Figure 5.11(c) shows that the loss of weight of the uncoated specimen was increasing at a steep rate (0.17 g) with the increase in slurry concentration to 20,000 ppm, at 90° impingement angle and 15 mm stand-off distance, but the weight loss of the coated specimen was almost constant (0.01 g). Lastly, Figure 5.11(d) showed that the uncoated specimen lost the most weight (approximately 0.2 g) at a lower stand-off distance of 15 mm, and the coated specimen once again did not exhibit any significant wear. On the whole, the findings proved that the impingement angle, stand-off distance and slurry concentration had a significant impact on erosion in uncoated specimens, whereas WC-10Co-4Cr-coated specimens always had insignificant wear.

Singh et al. [63] investigated the microstructure of erosion processes in WC-10Co-4Cr coating under the impingement of sand particles at 0°. Figure 5.12(a) indicates that impact of particles on the coating surface formed narrow smeared areas on the surface. Figure 5.12(a,b) shows the presence of chips in the craters, and craters formed

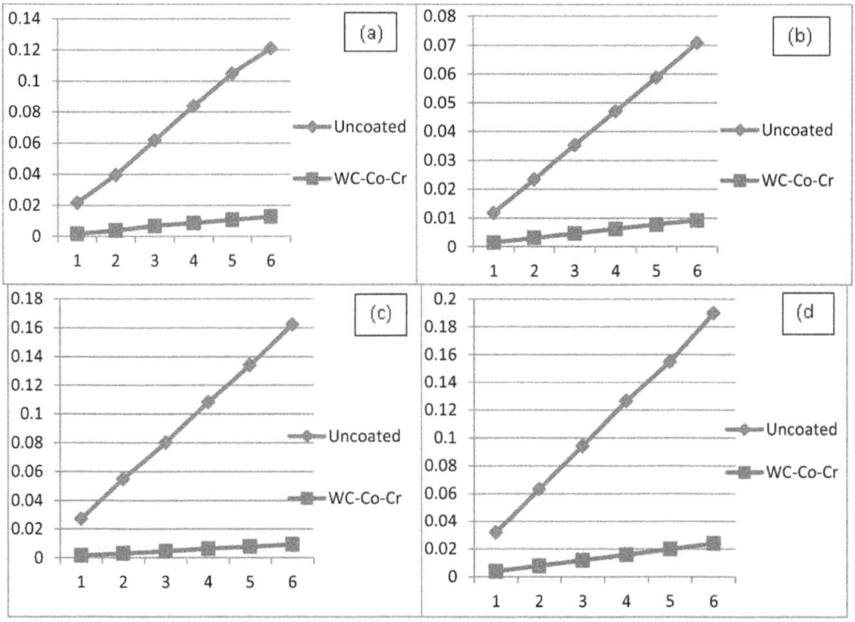

FIGURE 5.11 Wear testing on CF8M steel samples with and without a WC-Co-Cr coating ("Uncoated" word stands for CF8M steel) [67].

on the surface of the coating. Micro-cutting caused by solid particles was also more pronounced at higher magnification (×500) as shown in Figure 5.12(c). Moreover, there were clear images of SEM showing the micro-cutting, lip formation, and intact chips on the eroded WC-10Co-4Cr coating, as shown in Figure 5.12(d).

5.6.2 Comparison of Cermet Coatings with Ceramic Coatings

Ceramic-based coatings have recently been employed for erosion and corrosion applications. Researchers and inventors are extensively utilizing coatings based on TiO_2, Al_2O_3, ZrO_2, Y_2O_3, and Cr_2O_3 [68]. Coatings made of Ni-20Cr_2O_3 are an improvement over Ni chromia coatings. The WC-10Co4Cr HVOF coating exhibited less erosion than the Ni-20Cr_2O_3 HVOF coating (Figure 5.13) [33]. When comparing WC-10Co-4Cr-coated and Al_2O_3-13TiO_2-coated CF8M steel, Goyal et al. [69] discovered that the former had superior erosion resistance (Figure 5.14). But CF8M steel has ductile eroding processes, as shown in Figure 5.15(a). However, the CF8M steel coated with Al_2O_3-13TiO_2 and WC-10Co-4Cr showed brittle erosion behaviour, as shown in Figure 5.15(b,c).

5.7 CORROSION ANALYSIS

WC-10Co4Cr coating is also used for corrosion-related problems. Singh et al. [70] conducted electrochemical corrosion experiments to evaluate the performance of

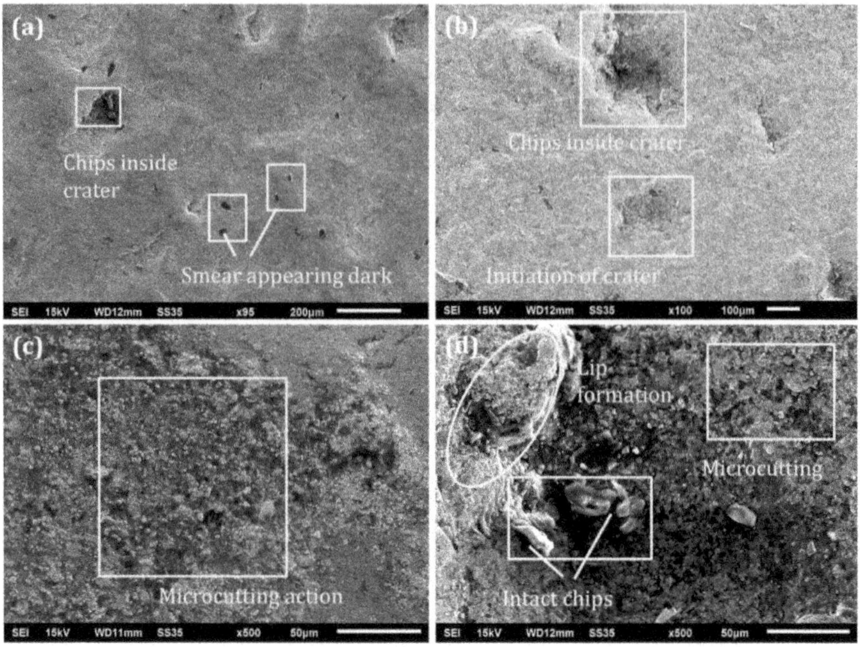

FIGURE 5.12 SEM images of eroded WC-CoCr-sprayed SS316L at impact angle of 0° (N = 1500 rpm, C = 60%, T = 180 min) [63].

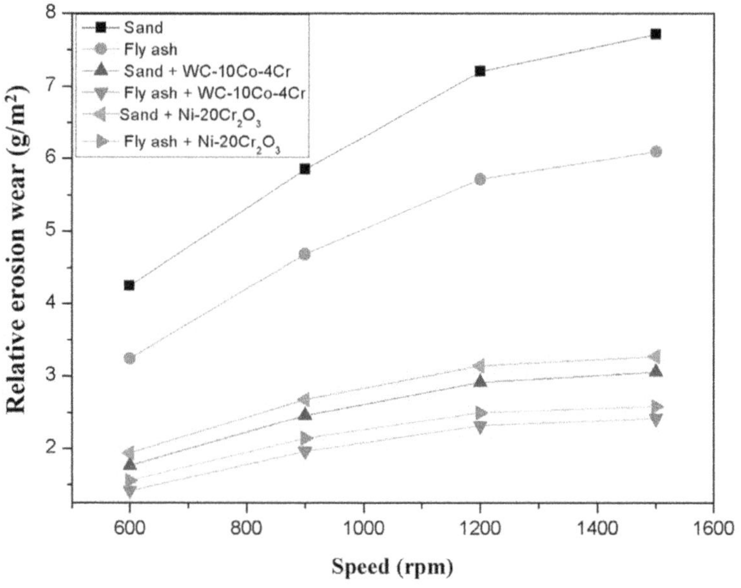

FIGURE 5.13 Erosion wear graph between Ni-20Cr$_2$O$_3$ and WC-10Co4Cr HVOF coating [33].

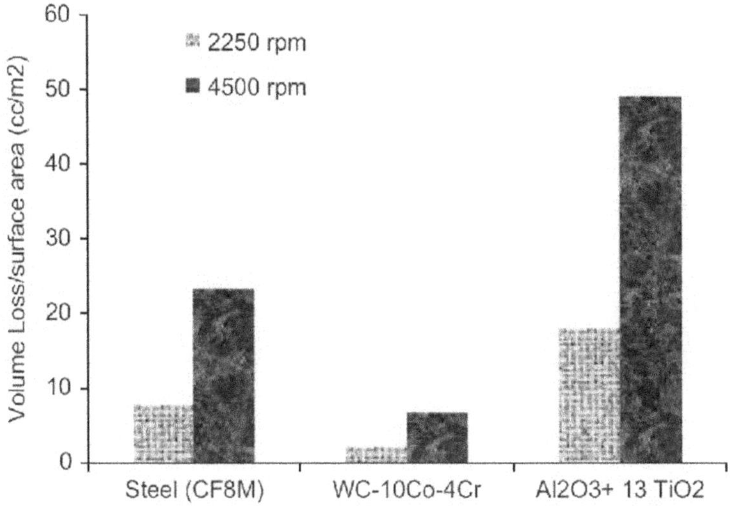

FIGURE 5.14 Volume loss analysis of uncoated CF8M versus WC–CoCr and Al_2O_3–13TiO_2 coatings [69].

WC-Co-Cr coatings under slurry erosion-corrosion conditions. A sand slurry with 30 wt% concentration was prepared by adding a 3.5 wt% NaCl solution, and the coated workpiece samples were exposed to this electrolytic slurry. The corrosion behaviour was assessed by monitoring the open-circuit potential (OCP) and potentiodynamic polarization, with the tests maintained for a fixed duration of 30 minutes at a scanning rate of 0.5 mV/s. During the experiments, the coated surfaces were immersed in the NaCl-containing slurry until the OCP stabilized, after which anodic polarization was performed. To ensure reproducibility, duplicate experiments were carried out. The polarization curves (Figure 5.16) indicated clear differences in the corrosion resistance of the coating compositions. The comparative order of corrosion susceptibility was observed as: WC-10Co-4Cr < WC-10Co-4Cr + 3% Mo_2C < WC-10Co-4Cr + 3% Y_2O_3 > WC-10Co-4Cr + 3% ZrO_2. Among the modified coatings, the incorporation of 3% ZrO_2 showed the highest corrosion resistance compared to Mo_2C and Y_2O_3 additions. This improvement was attributed to the enhanced electronegativity imparted by zirconia, which effectively reduced the corrosion rate of the WC-Co-Cr coating. The observed phenomenon emphasizes the beneficial role of ZrO_2 reinforcement in improving corrosion resistance under aggressive environmental conditions.

The application of thermal spray coatings and surface engineering technologies has proliferated in biomedical, slurry, and high-temperature corrosion systems, particularly when integrated with microwave processing, additive manufacturing, and thin-film coatings via magnetron sputtering [70–84]. Furthermore, coatings offer significant potential for assessing surface roughness, strength (tensile, compressive, and flexural), and efficacy in solar applications; they also facilitate model-based simulations of flow systems such as pipes and their bends, as well as intricate structures like chairs, which also take ergonomic factors into account [85–97].

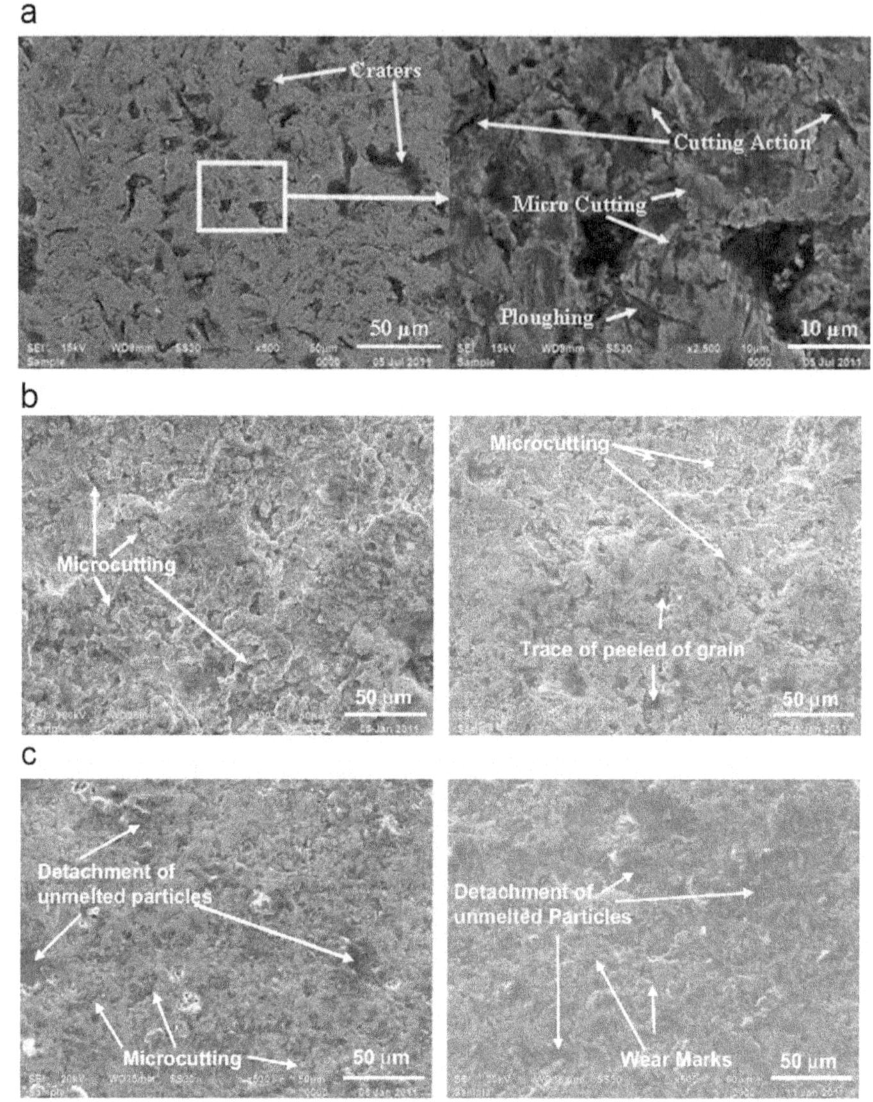

FIGURE 5.15 Scanning electron micrographs of eroded surfaces of (a) CF8M steel, (b) WC-CoCr coating, and (c) Al_2O_3-13TiO_2 coating were obtained under controlled conditions (e.g. 2250 rpm pot tester speed, 10,000 ppm slurry concentration, and 100 mm average particle size) [69].

5.8 CONCLUSIONS

Cermet-based thermal spray coatings are pivotal in mitigating wear and corrosion in critical engineering applications. WC-Co and Cr_3C_2-NiCr coatings deposited by HVOF and related processes demonstrate excellent mechanical properties, including high hardness, toughness, and strong adhesion, which collectively enhance

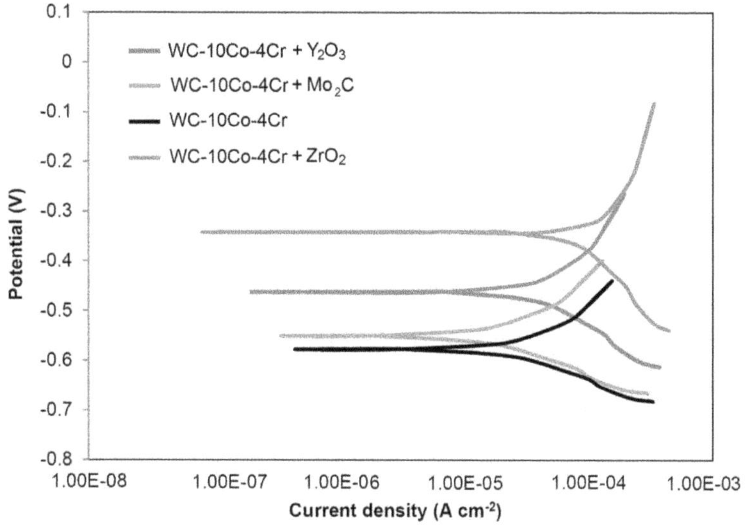

FIGURE 5.16 Potentiograms of polarization curves of different coatings [70].

erosion-corrosion resistance. Comparative analyses with ceramic coatings establish the superior adaptability of cermets under diverse impact angles, temperatures, and environmental conditions. The performance of these coatings strongly depends on process parameters such as spraying velocity, porosity, and coating thickness, with optimized conditions yielding dense, adherent, and durable layers. Advances in thermal spray technology, coupled with the development of nanostructured and composite powders, further improve resistance to erosion and corrosion. Overall, the application of cermet coatings represents a highly effective strategy for extending the operational life of turbines, pipelines, and offshore machinery, thereby reducing maintenance costs and improving system reliability in erosive-corrosive environments.

REFERENCES

[1] Padhy MK, Saini RP. Effect of size and concentration of silt particles on erosion of Pelton turbine buckets. *Energy*. 2009;34:1477–1483. https//doi.org/10.1016/j.energy.2009.06.015.

[2] Neopane HP, Chitrakar S, Dahlhaug OG. Role of turbine testing lab for overcoming the challenges related to hydropower development in Nepal. *IOP Conf Ser Earth Environ Sci.* 2019;240:042012.

[3] Prashar, G., Vasudev, H. and Thakur, L. High-temperature oxidation and erosion resistance of Ni-based thermally-sprayed coatings used in power generation machinery: a review. *Surface review and letters*. 2022;*29*(03):2230003.

[4] Singh J, Kumar S, Gill HS. Review on testing facilities assisting in development of numerical models for erosion calculation in centrifugal slurry pumps. *Int J Interact Des Manuf.* 2023;18:5301–5327. Available from: https://link.springer.com/10.1007/s12008-023-01282-z.

[5] Rajahram SS, Harvey TJ, Wood RJK. Erosion-corrosion resistance of engineering materials in various test conditions. *Wear*. 2009;267:244–254.

[6] Chitrakar S, Neopane HP, Dahlhaug OG. Study of the simultaneous effects of secondary flow and sediment erosion in Francis turbines. *Renew Energy*. 2016;97:881–891. https://doi.org/10.1016/j.renene.2016.06.007.
[7] Finne I. Erosion of surfaces. *Wear*. 1960;3:87–103.
[8] Levy A V. *Solid Particle Erosion and Erosion-Corrosion of Materials*. ASM Int. Ohio: ASM International; 1996.
[9] Bellman R, Levy A. Erosion mechanism in ductile metals. *Wear*. 1981;70:1–27.
[10] Wang YF, Yang ZG. Finite element model of erosive wear on ductile and brittle materials. *Wear*. 2008;265:871–878.
[11] Sundararajan G, Roy M, Venkataraman B. Erosion efficiency-a new parameter to characterize the dominant erosion micromechanism. *Wear*. 1990;140:369–381.
[12] Clark HMI, Wong KK. Impact angle, particle energy and mass loss in erosion by dilute slurries. *Wear*. 1995;186–187:454–464.
[13] Grewal HS, Agrawal A, Singh H. Design and development of high-velocity slurry erosion test rig using CFD. *J Mater Eng Perform*. 2013;22:152–161.
[14] Fedrizzi L, Valentinelli L, Rossi S, et al. Tribocorrosion behaviour of HVOF cermet coatings. *Corros Sci*. 2007;49:2781–2799.
[15] Matikainen V, Rubio Peregrina S, Ojala N, et al. Erosion wear performance of WC-10Co4Cr and Cr3C2-25NiCr coatings sprayed with high-velocity thermal spray processes. *Surf Coatings Technol*. 2019;370:196–212. Available from: https://linkinghub.elsevier.com/retrieve/pii/S025789721930444X.
[16] Vadapalli S, Pathem U, Vuppala VRSN, et al. Corrosion and cavitation erosion properties of sub-micron WC-Co /Cr3C2-NiCr multi-layered coating on aluminium substrates. *J Met Mater Miner*. 2020;30:46–54.
[17] Colen M, Bunshah RF. Synthesis and characterization of yttrium oxide (Y2O3) deposits. *J Vac Sci Technol*. 1975;13:536–539.
[18] Jain A. Silting problems in hydropower projects: Indian scenario. In *Proceedings of the 1st International Conference Silting Probl Hydro Power Plants*. New Delhi, India; 1999. pp. 37–54.
[19] Gandhi BK, Singh SN, Seshadri V. Study of the parametric dependence of erosion wear for the parallel flow of solid-liquid mixtures. *Tribol Int*. 1999;32:275–282.
[20] Gandhi BK, Singh SN, Seshadri V. Performance characteristics of centrifugal slurry pumps. *J Fluid Eng*. 2001;123:271–280.
[21] Desale GR, Gandhi BK, Jain SC. Effect of erodent properties on erosion wear of ductile type materials. *Wear*. 2006;261:914–921.
[22] Clark HM, Llewellyn RJ. Assessment of the erosion resistance of steels used for slurry handling and transport in mineral processing applications. *Wear*. 2001;250–251:32–44.
[23] Clark HMI, Hartwich RB. A re-examination of the "particle size effect" in slurry erosion. *Wear*. 2001;248:147–161.
[24] Stack MM, El-badia TMA. Some comments on mapping the combined effects of slurry concentration, impact velocity and electrochemical potential on the erosion – corrosion of WC / Co – Cr coatings. *Wear*. 2008;264:826–837.
[25] Evans AG, Gulden ME, Rosenblatt M. Impact damage in brittle materials in the elastic-plastic response regime. *Proc R Soc Lond A*. 1978;361:343–365.
[26] Reddy AV, Sundrarajan G. Erosion behaviour of ductile materials with a spherical non-friable erodent. *Wear*. 1986;111:313–323.
[27] Oka YI, Okamura K, Yoshida T. Practical estimation of erosion damage caused by solid particle impact: Part 1: Effects of impact parameters on a predictive equation. *Wear*. 2005;259:95–101.

[28] Levy A, Hickey G. Surface degradation of metals in simulated synthetic fuels plant environments [Internet]. California, USA; 2010. Available from: https://escholarship.org/uc/item/63t511kf.

[29] Divakar M, Agarwal VK, Singh SN. Effect of the material surface hardness on the erosion of AISI316. *Wear*. 2005;259:110–117.

[30] Walker CI, Robbie P. Comparison of some laboratory wear tests and field wear in slurry pumps. *Wear*. 2013;302:1026–1034. https://doi.org/10.1016/j.wear.2012.11.053.

[31] Parent LL, Li DY. Wear of hydrotransport lines in Athabasca oil sands. *Wear*. 2013;301:477–482.

[32] Heilbronner R, Keulen N. Grain size and grain shape analysis of fault rocks. *Tectonophysics*. 2006;427:199–216.

[33] Singh J, Kumar S, Mohapatra SK. Tribological analysis of WC–10Co–4Cr and Ni–20Cr2O3 coating on stainless steel 304. *Wear*. 2017;376–377:1105–1111. https://doi.org/10.1016/j.wear.2017.01.032.

[34] Clark HM, Hawthorne HM, Xie Y. Wear rates and specific energies of some ceramic, cermet and metallic coatings determined in the Coriolis erosion tester. *Wear*. 1999;233–235:319–327.

[35] Al-Bukhaiti MA, Ahmed SM, Badran FMF, et al. Effect of impingement angle on slurry erosion behaviour and mechanisms of 1017 steel and high-chromium white cast iron. *Wear*. 2007;262:1187–1198.

[36] Thapa BS, Dahlhaug OG, Thapa B. Sediment erosion in hydro turbines and its effect on the flow around guide vanes of Francis turbine. *Renew Sustain Energy Rev*. 2015;49:1100–1113.

[37] Schumacher WJ. Ball mill and hub test methods for slurry erosion evaluation of materials. Miller JE, Schmidt JFE, editors. *Slurn Erosion: Uses, Applications and Test Methods*. ASTM Special Technological Publications, Philadelphia, PA, USA; 1987:5–18.

[38] Clark HM. Particle velocity and size effects in laboratory slurry erosion measurements OR … do you know what your particles are doing? *Tribol Int*. 2002;35:617–624.

[39] Clark MCI. Specimen diameter, impact velocity, erosion rate and particle density in a slurry pot erosion tester. *Wear*. 1993;164-162:669–678.

[40] Gupta R, Singh SN, Seshadri V. Study on the uneven wear rate in a slurry pipeline on the basis of mesurements in a pot tester. *Wear*. 1995;184:169–178.

[41] Desale GR, Gandhi BK, Jain SC. Improvement in the design of a pot tester to simulate erosion wear due to solid-liquid mixture. *Wear*. 2005;259:196–202.

[42] Tuzson JJ. Laboratory slurry erosion tests and pump wear rate calculations. *J Fluids Eng*. 2015;106:135–140.

[43] Kumar R, Bhandari S, Goyal A. Performance analysis of CA6NM hydro-turbine steel under accelerated conditions by using an indigenously developed slurry erosion tester. *Trans Indian Inst Met*. 2018;71:753–761.

[44] M. Matsumura YO and MY. Slurry erosion-corrosion of commercially pure iron in fountain-jet testing facility-mechanism of erosion (corrosion under comparable intensity of erosion and corrosion). *Proceedings of the 7th International Conference Erosion by Liquid Solid Impact*, University of Cambridge, Cambridge; 1987. p. 40.

[45] Lin FY, Shao HS. Effect of impact velocity on slurry erosion and a new design of a slurry erosion tester. *Wear*. 1991;143:231–240.

[46] Sun B, Fan J, Wen D, et al. An experimental study of slurry erosion involving tensile stress for pressure pipe manifold. *Tribiology Int*. 2015;82:280–286. https://doi.org/10.1016/j.triboint.2014.07.025.

[47] Xie Y, Jiang JJ, Tufa KY, et al. Wear resistance of materials used for slurry transport. *Wear*. 2015:1–7. https://doi.org/10.1016/j.wear.2015.01.005.

[48] Tilly GP, Sage W. The interaction of particle and material behaviour in erosion processes. *Wear*. 1970;16:447–465.

[49] Madsen BW. A study of parameters using a new constant-wear-rate slurry test. Ludema KC, editor. *Proceedings of the Interncational Conference on Wear of Materials*. New York: American Society of Mechanical Engineers; 1985. pp. 345–854.

[50] Siegmann S, Abert C. 100 years of thermal spray: About the inventor Max Ulrich Schoop. *Surf Coatings Technol*. 2013;220:3–13. https://doi.org/10.1016/j.surfcoat.2012.10.034.

[51] Browning JB. Hypersonic velocity impact fusion spraying. *Proceedings of the International Thermal Spray Conference Expo*. Orlando, FL, USA; 1992. p. 123.

[52] Irving R, Knight R, Smith RW. The HVOF Process-the hottest topic in the thermal spray industry. *Weld J*. 1993 Jul;72(7):25–30.

[53] Berger LM. Application of hardmetals as thermal spray coatings. *Int J Refract Met Hard Mater*. 2015;49:350–364. https://doi.org/10.1016/j.ijrmhm.2014.09.029.

[54] Prashar G, Vasudev H, Thakur L. Performance of different coating materials against slurry erosion failure in hydrodynamic turbines: A review. *Eng Fail Anal*. 2020;115:104622. https://doi.org/10.1016/j.engfailanal.2020.104622.

[55] Goyal DK, Singh H, Kumar H, et al. Erosive wear study of HVOF spray Cr3C2–NiCr coated CA6NM turbine steel. *J Tribol*. 2014;136:1–11. Available from: https://asmedigitalcollection.asme.org/tribology/article/doi/10.1115/1.4027621/374962/Erosive-Wear-Study-of-HVOF-Spray-Cr3C2NiCr-Coated.

[56] Berger L-M, Zieris R, Saaro S. Oxidation of HVOF-sprayed hardmetal coatings. *Proceedings of the International Thermal Spray Conference ITSC*. Basel: DVSVerlag, Düsseldorf; 2005. p. 8.

[57] Oechsle M. Carbide containing spray powders and HVOF-coatings. *Conference Proceedings 7 HVOF-Kolloquium*, 9–10 November 2006. Erding, Germany: emeinschaft Thermisches Spritzen EV, Unterschleißheim; 2006. p. 57–62.

[58] Babu A, Arora HS, Singh H, et al. Microwave synthesized composite claddings with enhanced cavitation erosion resistance. *Wear*. 2019;422–423:242–251. https://doi.org/10.1016/j.wear.2019.01.072.

[59] Singh S, Garg P, Goyal D. Comparison of laser cladding and thermal spraying techniques used for wear control. *Res Cell An Int J Eng Sci*. 2017;25:222–229.

[60] Meghwal A, Anupam A, Murty BS, et al. Thermal spray high-entropy alloy coatings: A review. *J Therm Spray Technol*. 2020;29:857–893. https://doi.org/10.1007/s11666-020-01047-0.

[61] Mostajeran A, Shoja-Razavi R, Hadi M, et al. Evaluation of the mechanical properties of WC-FeAl composite coating fabricated by laser cladding method. *Int J Refract Met Hard Mater*. 2020;88:105199. https://doi.org/10.1016/j.ijrmhm.2020.105199.

[62] Singh J. Tribo-performance analysis of HVOF sprayed 86WC-10Co4Cr & Ni-Cr2O3 on AISI 316L steel using DOE-ANN methodology. *Ind Lubr Tribol*. 2021;73:727–735.

[63] Singh J, Kumar S, Kumar R, et al. Machine learning regression tools for erosion prediction of WC-10Co4Cr thermal spray coating. *Result Surf Interf*. 2023;13:100156. Available from: www.sciencedirect.com/science/article/pii/S2666845923000636.

[64] Thakur L, Arora N, Jayaganthan R, et al. An investigation on erosion behavior of HVOF sprayed WC-CoCr coatings. *Appl Surf Sci*. 2011;258:1225–1234. https://doi.org/10.1016/j.apsusc.2011.09.079.

[65] Jia K, Fischer TE. Sliding wear of conventional and nanostructured cemented carbides. *Wear*. 1997;203–204:310–318.

[66] Vashishtha N, Khatirkar RK, Sapate SG. Tribological behaviour of HVOF sprayed WC-12Co, WC-10Co-4Cr and Cr3C2–25NiCr coatings. *Tribiology Int.* 2017;105:55–68. https://doi.org/10.1016/j.triboint.2016.09.025.
[67] Sahni K, Grewal JS. Slurry erosion behaviour of WC-10Co-4Cr coated CF8M turbine steel. *Mater Today Proc.* 2022;56:2666–2671. Available from: www.sciencedirect.com/science/article/pii/S2214785321060788.
[68] Singh J, Kumar S, Mohapatra SK. Tribological performance of Yttrium (III) and Zirconium (IV) ceramics reinforced WC–10Co4Cr cermet powder HVOF thermally sprayed on X2CrNiMo-17-12-2 steel. *Ceram Int.* 2019;45:23126–23142. https://doi.org/10.1016/j.ceramint.2019.08.007.
[69] Kumar D, Singh H, Kumar H, et al. Slurry erosion behaviour of HVOF sprayed WC–10Co–4Cr and Al2O3+13TiO2 coatings on a turbine steel. *Wear.* 2012;289:46–57. https://doi.org/10.1016/j.wear.2012.04.016.
[70] Singh J, Kumar S, Mohapatra SK. An erosion and corrosion study on thermally sprayed WC-Co-Cr powder synergized with Mo2C/Y2O3/ZrO2 feedstock powders. *Wear.* 2019;438–439. Available from: www.sciencedirect.com/science/article/pii/S0043164819301899.
[71] Dutta V, Thakur L, Singh B, Vasudev H. A study of erosion–corrosion behaviour of friction stir-processed chromium-reinforced NiAl bronze composite. *Materials.* 2022;15(15):5401.
[72] Singh M, Vasudev H, Kumar R. Corrosion and tribological behaviour of BN thin films deposited using magnetron sputtering. *Int J Surf Eng Interdiscip Mater Sci.* 2021;9(2):24–39.
[73] Sharanabasava H, Raviprakash M, Prasad CD, Ramesh MR, Phanibhushana MV, Vasudev H, Kumar S. Microstructure, mechanical and wear properties of SiC and Mo reinforced NiCr microwave cladding. *Adv Mater Process Technol.* 2023;10(4):3620–3633.
[74] Singh M, Vasudev H, Singh M. Surface protection of SS-316L with boron nitride based thin films using radio frequency magnetron sputtering technique. *J Electrochem Sci Eng.* 2022;12(5):851–863.
[75] Vasudev H. Wear characteristics of Ni-WC powder deposited by using a microwave route on mild steel: Microwave cladding of Ni-WC. *Int J Surf Eng Interdiscip Mater Sci.* 2020;8(1):44–54.
[76] Prashar G, Vasudev H, Thakur L, Bansal A. Performance of thermally sprayed hydroxyapatite coatings for biomedical implants: A comprehensive review. *Surf Rev Lett.* 2023;30(1):2241001.
[77] Parkash J, Saggu HS, Vasudev, H. A short review on the performance of high velocity oxy-fuel coatings in boiler steel applications. *Mater Today: Proc.* 2022;50:1442–1446.
[78] Singh P, Vasudev H, Bansal A. Effect of post-heat treatment on the microstructural, mechanical, and bioactivity behavior of the microwave-assisted alumina-reinforced hydroxyapatite cladding. *Proc Inst Mech Eng, Part E: J Process Mech Eng.* 2024;238(3):1060–1071.
[79] Verma PK, Vasudev H, Singh G, Singh S. A review on the microstructural and biomedical properties of implants manufactured using additive manufacturing. *J Inst Eng: Series D.* 2024;106:1587–1607.
[80] Singh SP, Verma PK, Singh S, Kapoor M, Balu R. Microstructural characterization and electrochemical behaviour of HVOF sprayed WC-Co-Cr coatings on 316L boiler steel stainless steel. *Result Surf Interf.* 2024;16:100248.
[81] Verma PK, Minhas AS, Singh P, Kumar S. Slurry erosion behaviour of thermal sprayed Al2O3 and Cr2O3 coatings for turbine steels. *AIP Conf Proc.* 2024;2986(1):020012.

[82] Verma, PK, Kumar V, Vasudev H. Bioactivity and corrosion analysis of thermally sprayed hydroxyapatite based coatings. *J Electrochem Sci Eng.* 2024;14(4):419–439.
[83] Verma PK, Vasudev H, Kumar V. Biocompatibility and corrosion performance of HA-based LVOF sprayed and post-annealed coatings on SS 254 surgical-grade bio-implant material. *Ceram Int.* 2025;51(10):13207–13225.
[84] Verma PK, Vasudev H, Singh G. Improving the corrosion performance of SS 304 stainless steel using HVOF-sprayed WC-Co-Cr coatings in chloride environments. *J Bio- Tribo-Corros.* 2025;11(4):1–13.
[85] Poojari M, Hanumanthappa H, Prasad CD, Jathanna HM, Ksheerasagar AR, Shetty P, Shanmugam BK, Vasudev H. Computational modelling for the manufacturing of solar-powered multifunctional agricultural robot. *Int J Interact Design Manuf.* 2024;18(8):5725–5736.
[86] Prakash C, Gupta LR, Mehta A, Vasudev H, Tominov R, Korman E, Fedotov A, Smirnov V, Kesari KK. Computing of neuromorphic materials: An emerging approach for bioengineering solutions. *Mater Adv.* 2023;4: 5882–5919.
[87] Singh J, Gill HS, Vasudev H. Computational fluid dynamics analysis on role of particulate shape and size in erosion of pipe bends. *Int J Interact Design Manuf.* 2023;17(5):2631–2646.
[88] Rao GS, Mukkamala U, Hanumanthappa H, Prasad CD, Vasudev H, Shanmugam B, KishoreKumar KC. Evaluating and optimizing surface roughness using genetic algorithm and artificial neural networks during turning of AISI 52100 steel. *Int J Interact Design Manuf.* 2024;18(8):6151–6160.
[89] Singh J, Singh R, Singh S, Vasudev H, Kumar S. Reducing scrap due to missed operations and machining defects in 90PS pistons. *Int J Interact Design Manuf.* 2023;17(5):2527–2539.
[90] Nagabhushana P, Ramprasad S, Prasad CD, Vasudev H, Prakash C. Numerical investigation on heat transfer of a nano-fluid saturated vertical composite porous channel packed between two fluid layers. *Int J Interact Design Manuf.* 2024;18(5):2927–2944.
[91] Mahantesh MM, Rao KR, Chandra AP, Vijayakumar MN, Nandini B, Prasad CD, Vasudev H. Design and modeling using finite element analysis for the sitting posture of computer users based on ergonomic perspective. *Int J Interact Design Manuf.* 2024;18(8):5875–5891.
[92] Verma PK, Vasudev H, Kumar V. Dental implant materials and designs and failure during implant placement and loading: A review. In Hitesh Vasudev, Chander Prakash, Mohammad Khalid, Kavindra Kesari (eds.): *Advanced Manufacturing Processes*, 2025, Boca Raton: CRC Press. pp. 138–150.
[93] Verma PK, Vasudev H, Kumar V. Bioactivity and corrosion performance of low-velocity oxy-fuel-sprayed hydroxyapatite-based coatings on a surgical-grade stainless steel. *Trans Ind Inst Metals.* 2025;78(8):185.
[94] Kulkarni SD, Manjunatha, Chandrasekhar U, Manjunath KV, Prasad CD, Vasudev H. Design and optimization of polyvinyl-nitride rubber for tensile strength analysis. *Int J Interact Design Manuf.* 2024;18(8):5893–5908.
[95] Kattimani MA, Venkatesh PR, Masum H, Math MM, Bahadurdesai VN, Mustafkhadri S, Prasad CD, Vasudev H. Design and numerical analysis of tensile deformation and fracture properties of induction hardened inconel 718 superalloy for gas turbine applications. *Int J Interact Design Manuf.* 2024;18(8):5939–5949.
[96] Verma PK, Vasudev H, Kumar V. Effect of electro-corrosion behavior of different grades of stainless steel used in bio-implants: A review. In Hitesh Vasudev, Chander

Prakash, Mohammad Khalid, Kavindra Kesari (eds.): *Advanced Manufacturing Processes,* 2025, Boca Raton: CRC Press. pp. 269–283.

[97] Verma PK, Vasudev H. Importance of hydroxyapatite coatings in improving new bone osteogenesis: A review. In Hitesh Vasudev, Chander Prakash, Mohammad Khalid, Kavindra Kesari (eds.): *Advanced Manufacturing Processes*, 2025, Boca Raton: CRC Press. pp. 284–299.

6 Hot Corrosion Performance of Composite Coatings

Amrinder Mehta, Jashanpreet Singh, Brijesh Prasad, Hitesh Vasudev, and Suresh Gudala

6.1 INTRODUCTION

Together with the advancements in miniaturization, high power, and energy saving technology, the accompanying challenges in heat transfer and heat accumulation have given rise to an urgent need for advanced and efficient thermal management materials. Most metals exhibit a high level of thermal conductivity (TC), making them ideal candidates for heat dissipation materials. However, metals are highly vulnerable to aggressive and corrosive environments, resulting in severe deterioration of the structural and mechanical properties of the metals, which poses a significant economic and security burden [1].

The four-part Figure 6.1 depicts the analyses of a steam turbine blade material and the subsequent stages of sample preparation. In Figure 6.1(a), the fractured steam turbine blade is shown. This illustrates the aftermath of a mechanical failure and the rationale for investigating the blade material. In Figure 6.1(b), an individual bulk sample is shown in preparation for X-ray fluorescence (XRF) analysis, a sample which has also been cut and polished for proper elemental analysis. Figure 6.1(c) shows the die used for sample preparation along with parts for the compressor and shaping that will configure the sample into the needed form for testing. Figure 6.1(d) shows multiple disc-shaped samples, each with a diameter of 13 mm, made from the bulk material, and die so that there is uniformity and reproducibility in the analyses. These pictures collectively demonstrate the process that begins with damaged equipment and proceeds through sample preparation to the production of uniform samples needed for a thorough evaluation of the material and the assurance of its quality [2].

Gas turbines are fundamental engines of innovation in society today, powering aeroplanes and enabling global electricity generation. Their use in the transportation industry is indicative of the turbines' importance to the energy and defence industries, where reliability and performance are paramount [3]. Gas turbine engines are powerful and fuel-efficient, but the engineering develops further to meet the increasing demand for high outputs, fuel efficient engine designs, and eco-friendly alternatives to fossil fuels. The hot section of the turbine operates the hottest, and further engineering developments to improve gas turbine hot section performance are the key to higher gas turbine hot section performance [4].

DOI: 10.1201/9781003730637-6

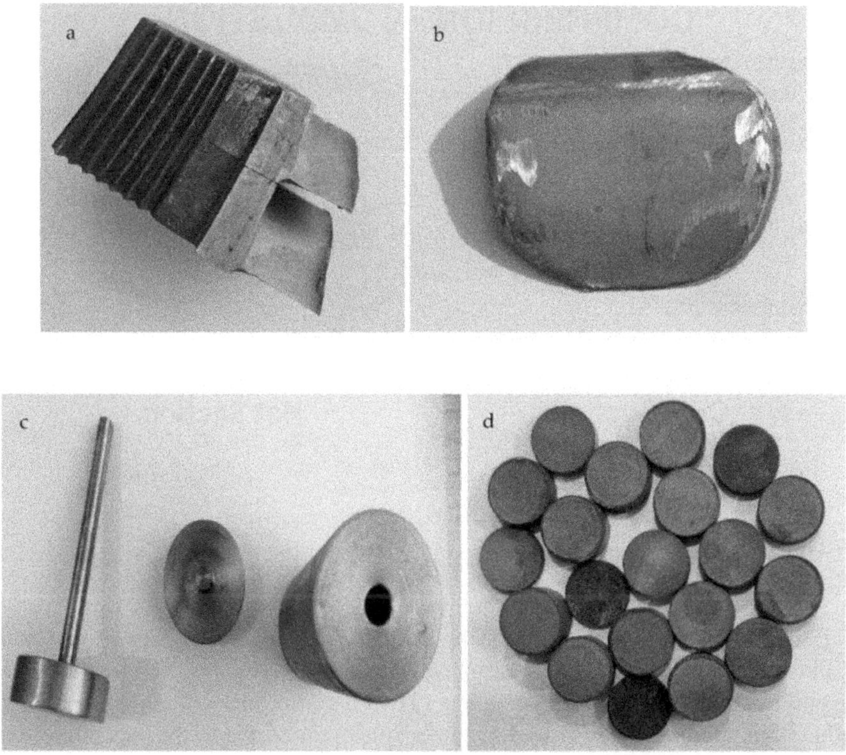

FIGURE 6.1 (a) Broken steam turbine blades, (b) XRF sample, (c) the die utilized in sample preparation, and (d) 13-mm diameter prepared sample [2].

Presented in Figure 6.2 is a cross-sectional schematic representation of a gas turbine engine and its various components and stages of operation. The two sections of the engine are termed the Cold Section and the Hot Section. The Cold Section contains the compression stage where the series of compressor blades compress the incoming air, represented by blue arrows, and thus raises its pressure and temperature. This air then moves into the combustion chambers where it is mixed with fuel and ignited [5].

Both the engine's core power and its overall efficiency depend on the combustion gas temperatures at the engine's critical parts. With the gas-inlet temperature increased, the turbines get more work from each fuel unit and improve work conversion and emission reduction. Higher operational temperatures also reduce harmful NOx emissions, which facilitates compliance with emissions legislation [6]. However, in overcoming these NOx reduction challenges, materials and thermal management must deal with even larger operational extremes. The fuel economy and environmentally safe operational standards can be achieved and more reliably be maintained with advances in protective coating science and out-of-engine cooling methods. For instance, the modern aviation turbine engine's hot-end components can exceed service temperatures of 1500°C [7].

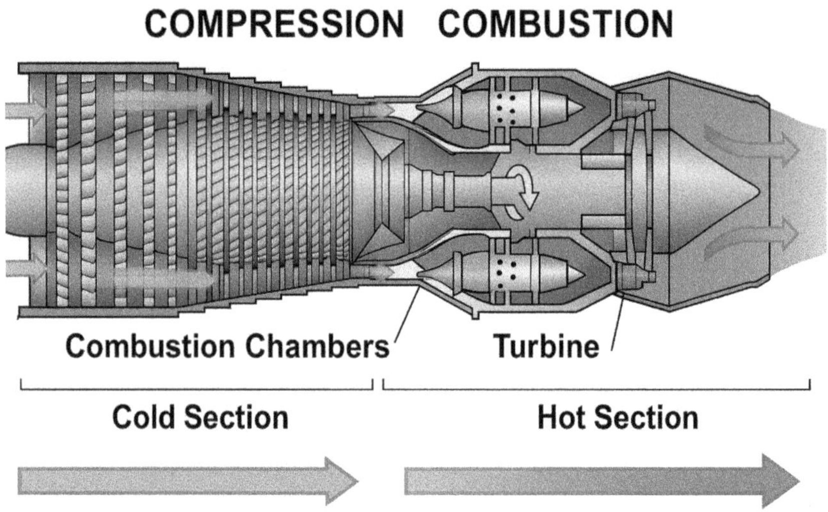

FIGURE 6.2 Gas turbine parts and stages [5].

Progression of gas turbine rev technology for gas turbine engines as presented in Figure 6.3 starts from 1960 and moves up to date. The development of nickel superalloys and thermal barrier coatings (TBCs) is of primary importance. Initially turbine limitations were restricted to lower temperatures due to the superalloys and relied only on convective cooling. Therefore, maximum operating temperatures were restricted. As next-generation superalloys were produced, such as U700, IN738, and IN939, the temperature resistance capabilities improved. The 1970s saw the development of TBCs which increased temperature operating ranges. Further developments in engineering turbine TBC superalloys and advanced coating deposition methods, which include APS since the 1970s, EB-PVD since the 1980s, and PS-PVD since the 2000s, increased turbine operating temperatures to and beyond the superalloy melting temperatures. This has improved turbine efficiency and longevity as engineering solutions and construction of thrust components continued to drive heating and cooling processes to the extremes, melting protective superalloy and thermal protection systems [8].

Structures built for extreme industrial environments require materials with a lower chance of failing. Within the oil and gas sector, plants experience high costs and grave safety risks because of corrosion. Since carbon steel pipes are used for transporting and processing petroleum, they are prone to corrosion because of the harsh environments caused by corrosive gases, liquids, and particulates [9].

The Yttria-Stabilized Zirconia (YSZ) thermal barrier coatings display cross-sectional microstructures and advanced thermal spraying pyrolysis methods, as shown in Figure 6.4. Figure 6.4(e) represents a columnar structure developed with Suspension plasma spraying (SPS) methods. It features vertical, column-like formations, and well-developed feather patterns. These routes provide pathways for strain tolerance and thermal cycling behaviour. Figure 6.4(f) shows a columnar

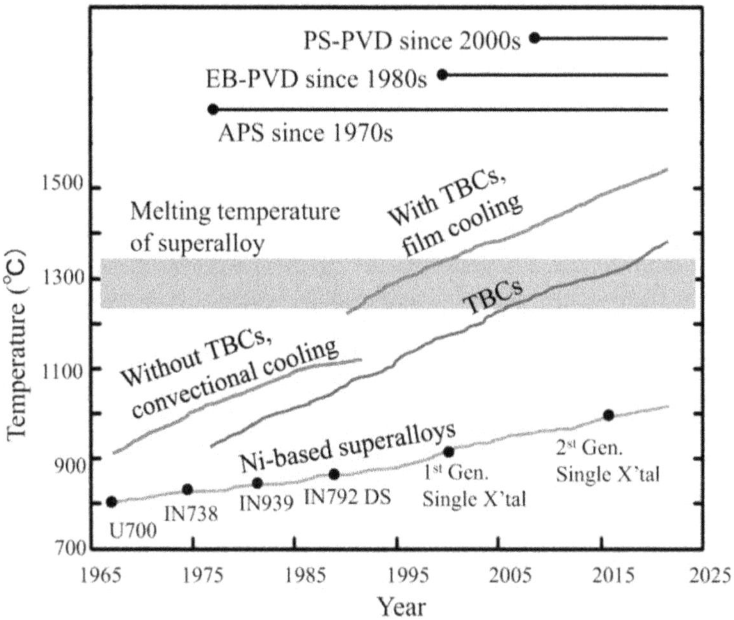

FIGURE 6.3 Modification of gas turbine engine from 1960s to current progression [8].

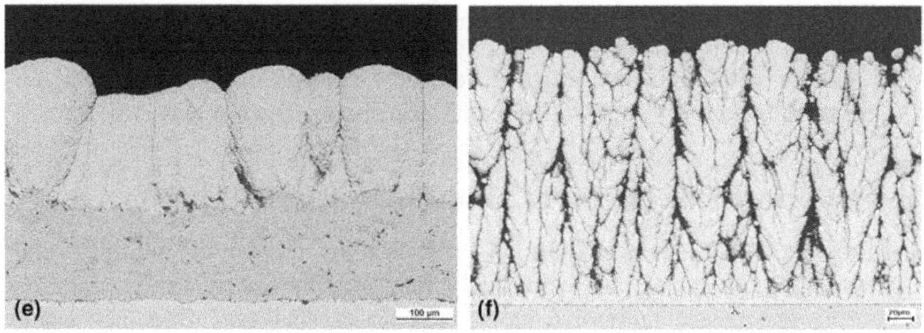

FIGURE 6.4 A variety of YSZ thermal barrier coatings that are thermally sprayed: (a) columnar PS-PVD (columnar PS-porosity level), (b) very porous APS, (c) thick segmented APS, (d) segmented SPS, (e) columnar SPS, and (f) columnar APS [10].

structure created through an innovative APS technique. In this picture, the columns are distinctly separated and organized in a feathered structure. The coating presents a high degree of vertical segmentation which is of utmost importance in helping the coating "diffuse" to achieve thermal expansion mismatches as well as spallation risks. The microstructures are designed to achieve the finest insulation compliance SPS and higher resolvability APS. These are ideal for advanced engine applications and thermal gas protective coatings like gas turbines and aerospace engines [10].

Hot corrosion poses significant challenges in high-temperature industrial settings. Industries such as power generation, aerospace, and shipping face difficulties containing corrosive metals integrated with salts and oxidizing gases. The phenomenon deteriorates critical components like turbine blades, exhaust systems, and boiler tubes, which exacerbates reliability and lifespan issues [11]. While protection may be offered through traditional metallic coatings, they are often ineffective in extreme hot corrosion environments. As a result, high-temperature hot corrosion resilient composite coatings are a viable and promising alternative. The coatings are composed of chemically and mechanically resilient dual repellent barriers of metal and ceramic components. The coatings can be tailored to target specific hot environment corrosive agents such as molten sulfates and chlorides [12]. The performance of such composite coatings under hot corrosion conditions is dependent on the composition and microstructure coating as well as the methods of application. Understanding and evaluating the mechanics of these coatings under conditions that simulate service are necessary to optimize the coatings' protective mechanisms. Hot composite coatings are a fundamental and vital solution for protecting metals and alloys from hot corrosion, particularly in high-temperature environments where aerospace, automotive, power generation, and chemical industries operate [13].

In aerospace applications, corrosive materials including NaCl or KCl, Na_2SO_4/K_2SO_4, and Na_2VO_3 may settle on TiAl alloys. During operation at high temperatures, these alloys undergo hot corrosion with the salts which, in turn, breaches the protective oxide surfaces, and encourages the further deterioration of the alloys [14]. Choosing the right materials, the method of synthesis, filler dispersion, and the environment in which it operates will determine the performance of mixed coatings. This review consolidates contemporary studies on the mechanisms, materials, procedures, and limitations of composite coatings developed for the protection of alloys against hot corrosion, including the protective composite coatings themselves [15].

6.2 MECHANISMS OF HOT CORROSION PROTECTION

Composite coatings enhance substrate protection by employing multiple synergistic strategies. As sturdy physical barriers, they prevent underlying materials from weathering and protect corrosive moisture, oxygen, and salts from penetrating. Composite coatings can also encourage the passive layer formation, which, through chemical reaction, stabilize with the substrate or enclosed elements and form non-reactive surface films. These passive layers slow the corrosion by suppressing the aggressive species interaction with the substrate at reactive layers for corrosion [16]. Designed active fillers in the coating matrix intercept and eliminate corrosive agents. Nanomaterials, including graphene and boron nitride, and metal oxides, enhance these defensive characteristics. Stacked and peculiar nanomaterials (like CNTs, graphene nano-clays) increase protective attributes by further decreasing coating porosity, forming denser and tortuous pathways that slow the route of hazardous materials. Additionally, they can imbue the coating with dead-legged self-healing by repairing microcracks or defects, thus prolonging protective layers. In some coatings, super hydrophobicity is created, and water-beading off surfaces is observed, maximizing the elimination of liquid. This multi-faceted approach is why composite coatings protect substrates in the most extreme and challenging conditions [17].

Figure 6.5 shows how different coatings affect the corrosion behaviour of steel when it is immersed in 3.5 wt% NaCl solution. The coated steel shows the most negative potential compared to bare steel which shows the most positive potential, showing it is the most corrodible. The potential becomes increasingly negative as the coatings become APS-Al_2O_3, high-velocity oxy-fuel (HVOF)-Al_2O_3, and their composites with TiO_2 and MWCNTs. The most positive potential in the combination with TiO_2 and MWCNTs indicates the lowest current densities which means it has the most efficient barrier to corrosion in addition to a reduced corrosion rate. The included TiO_2 and MWCNTs are of great value to the APS and HVOF coatings as they increase the rate at which corrosion is prevented [18].

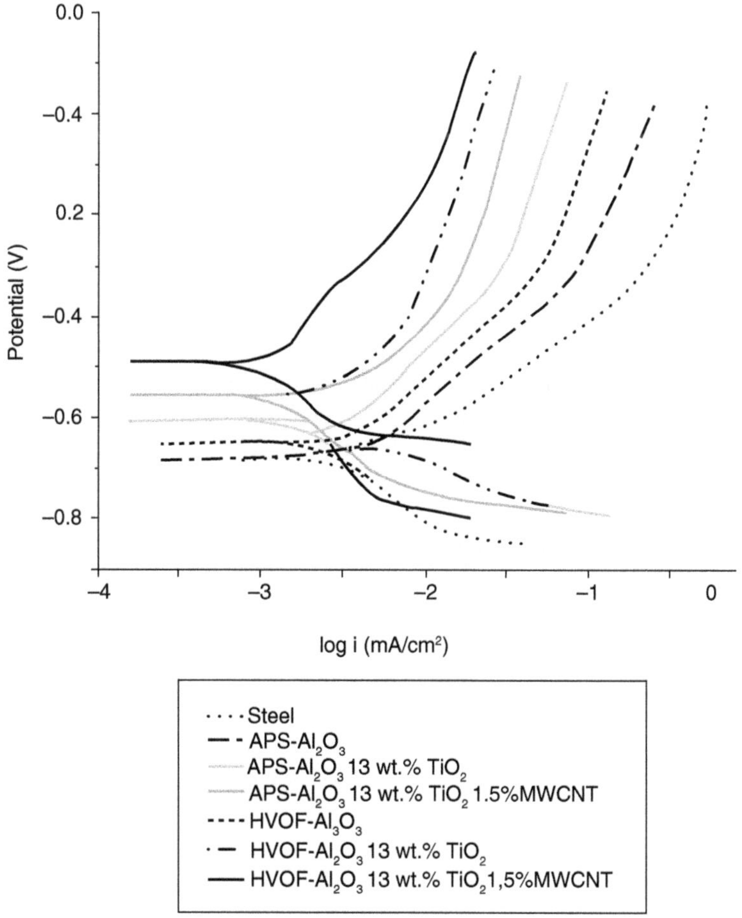

FIGURE 6.5 Potentiodynamic polarization curves of coatings in 3.5 wt% NaCl [18].

The synergy between matrix and filler is crucial for optimal performance. Advancements in protective coatings seek to supplement epoxy and polyurethane polymers with more advanced nanofillers. By introducing nanomaterials that include graphene oxides, boron nitride nanosheets, and various metal oxides, there have been remarkable advancements to coatings with respect to their mechanical, chemical, and thermal attributes [19]. Among these new coatings, graphene coatings hold an advantage because of their substantial thermal stability, chemical inertness, and ability to protect against extreme environmental conditions and aggressive chemicals. The unusual, layered combination of graphite oxides and their derivatives provides extreme impermeability, while the boron nitride nanosheets add additional lubricating and thermal regulating properties [20]. The added metal oxide nanoparticles increase hardness and resistance to oxidation. Unfortunately, the application of these nanocomposite coatings suffers from the inability to achieve a uniform dispersion of the nanofillers throughout the polymer matrix which plays a protective role along with the dispersion efficiency, and dispersion agglomeration controls defects. The ability of graphene to become galvanically corrosive must be controlled due to the conductive properties of graphene, quick micro-cells, and coupling with a metal substrate. In the protective coatings field, there is a reliance on surface modifications, hybrid formulations, and advanced dispersion techniques to achieve appropriate performance and protective coatings [21].

Figure 6.6 presents a schematic of the plasma-spraying process used to deposit yttria-stabilized zirconia (YSZ) coatings, alongside a visual of the finished coating sample. On the left, YSZ powder is fed through a plasma spray gun, which generates a high-temperature plasma flame. The intense heat melts the YSZ particles as they are propelled towards the substrate. As shown in the middle section, these molten particles travel through the plasma jet, impacting and rapidly cooling upon the substrate surface, thus forming a continuous, layered coating. The process enables strong adhesion and controlled thickness of the ceramic layer. The right portion displays the actual sample, a circular coating of bright white YSZ deposited onto a substrate, highlighting the uniformity and quality achievable with plasma spraying. This method is widely used for creating thermal barrier coatings in high-temperature engineering applications due to the excellent thermal and chemical stability of YSZ [22].

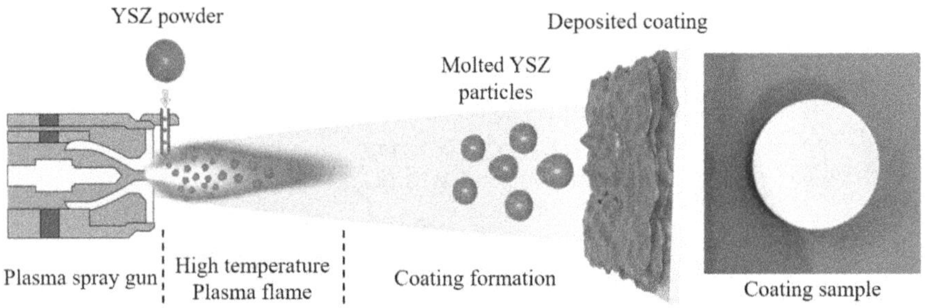

FIGURE 6.6 Schematic diagram of the YSZ coating deposited by a plasma-spraying process and images of the prepared coating samples [22].

There is a considerable amount of research on high-temperature-resistant materials like nickel (Ni) and cobalt (Co) metal matrix composites and zirconate and Cr_3C_2-NiCr ceramic composites. The creation of these materials focuses on trying to merge certain qualities of metals like toughness and ductility with the thermal stability and corrosion resistance of advanced ceramics (which is a positive trait of those composite materials) [23]. Ni and Co composites are mostly of interest for the aerospace science and engineering field, power generation, and a wide range of other industrial applications, because of their mechanical strength at elevated and varying temperatures, and their resistance to oxidation and wear. Performance-enhancing ceramic reinforcements that provide a protective barrier against wear and corrosive agents include chromium carbide (Cr_3C_2) and zirconate phases [24].

6.3 PROCESSING TECHNIQUES AND PERFORMANCE

Advanced methods such as electrodeposition, electroless plating, plasma electrolytic oxidation, and several thermal spray techniques including HVOF spraying are used to create composite coatings. Every composite coating fabrication method provides benefits for customization [25]. The electrodeposition and electroless plating methods are advantageous for the low-temperature coatings needed for the incorporation of nanoscale or thermally sensitive particles. Also, Plasma electrolytic oxidation provides hard, adherent, and uniquely textured oxide layers and increasing wear and corrosion resistance. On the other hand, thermal spray methods such as HVOF are best suited for producing thick, dense coatings on large or complex shaped substrates [26].

Figure 6.7 provides a comparative analysis of the operational flame temperature and particle velocity for different thermal spray systems. The vertical axis indicates the flame temperature which ranges between 0 and 5000°C and beyond. The horizontal axis represents the range of 0 to 1200 m/s for particle velocity. The spaces between the different thermal spray systems and spray techniques are classified and positioned according to the ranges of attainable temperatures and velocities. Atmospheric Plasma Spray (APS) and Vacuum/Low-Pressure Plasma Spray (VPS/LPPS) systems can operate at 12,000°C and achieve a velocity of between 150 and 600 m/s. The HVOF and Detonation Gun (D-Gun) techniques operate at 3000°C, reaching significantly higher velocities of 400 to 1000 m/s. Temperatures of 3000 to 4000°C can be attained in the combustion flame spray and wire arc spray while having significantly lower particle velocities of 150 m/s. With Cold Gas Spray (CGS) the lowest temperatures of 300 to 1000°C are attained at the highest particle velocities of 800 to 1200 m/s. This allows each process's operational window to be defined and subsequently used to customize the desired performance and properties of the coating [27].

Microstructure, porosity, and adhesion depend significantly on process parameters. In electrochemical techniques, current density influences the deposit rate and quality. In spray methods, uniform dispersion and optimal bonding are achieved with balanced particle concentration and spray velocity [28]. Modification of these parameters governs pore size and distribution, the extent of mechanical interlocking at the interface, and the generation of protective barrier phases. The process and parameter

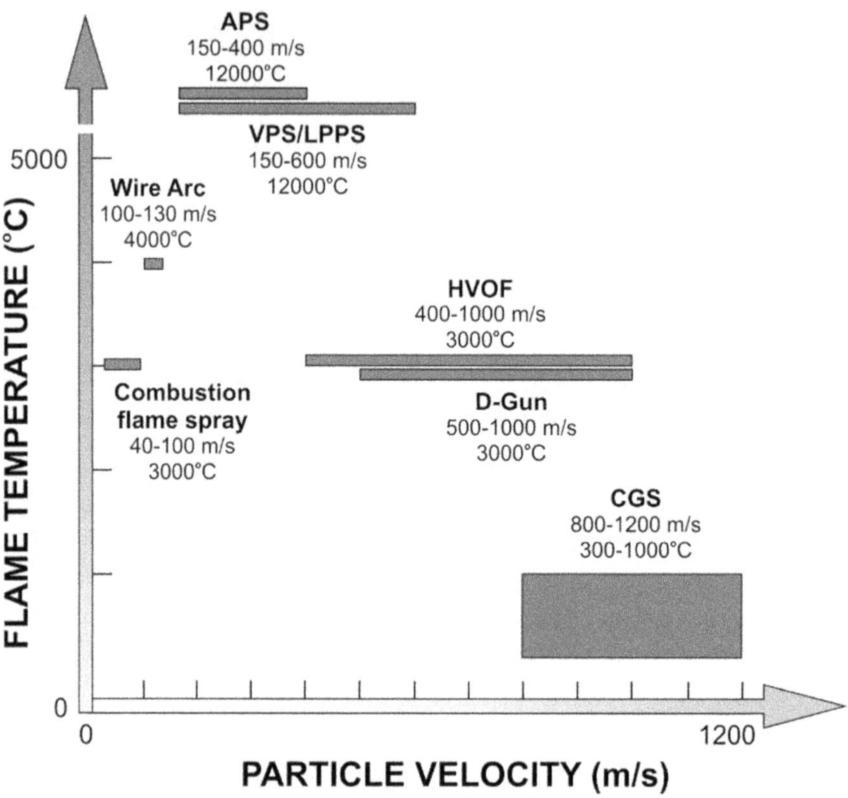

FIGURE 6.7 The operational ranges of flame temperature and particle velocity for a variety of thermal spray systems [27].

selection determines the maximum attainable corrosion resistance and mechanical strength of the protective coatings [29].

Table 6.1 summarizes a comparison of thermal spray methods based on particle concentration, spray velocity, and key findings

Well-dispersed nanofillers and optimized process conditions yield coatings with superior hot corrosion and wear resistance.

6.4 LIMITATIONS AND DEGRADATION MECHANISMS

Composite coatings have improved greatly, but some practical problems still exist, inhibiting widespread use in extreme environments. One major problem is filler agglomeration: more precisely, the clustering of nanoparticles or micro-fillers instead of achieving the desired dispersion within the coating. This clustering or random distribution results in poorly dispersed microstructures, which in turn weakens mechanical strength and compromises barrier properties [38]. Furthermore, during thermal cycling, coating–substrate material differences may generate thermal residual stresses

TABLE 6.1
Illustrates trends and findings

Thermal spray method	Particle concentration	Spray velocity	Key findings
Plasma arc spraying	Optimized via feed rate	High (>400 m/s)	Increased current density and optimized feed rate enhance melting and coating adhesion. Porosity decreases with higher velocity [30].
High-velocity oxy-fuel (HVOF)	Fine particles (5–38 μm)	400–800 m/s	Fine particle concentration improves surface finish; higher velocity forms denser, more corrosion-resistant coatings [31].
Plasma-sprayed	Varied by wire diameter	80–150 m/s (approximate)	Higher current improves energy absorption. Wire parameters control coating thickness and porosity [32].
Detonation gun	Controlled via powder feed	700–1200 m/s	Extremely high velocity produces hard, wear-resistant, and low-porosity coatings [33].
Flame spraying	Coarse particles (20–60 μm)	60–120 m/s	Lower velocity and coarser particles yield more porous coatings; suitable for less demanding wear/corrosion applications [34].
Flame spray	Adjustable	300–600 m/s	Controlled velocity allows softer particles, reducing substrate damage and improving ductility of the coating [35].
Suspension plasma spray	Nanoscale suspension	350–600 m/s	Nanoparticle concentration enhances coating uniformity and surface area. Useful for functional oxide coatings [36].
Cold spray	Fine powder	500–1200 m/s	Particles remain solid; velocity critical for bonding. Yields dense, oxidation-resistant coatings for ductile metals [37].

that, with the potential to crack or delaminate, become critical. Under extreme environments, like rapid temperature swings and high-temperature molten salt exposure, coatings may degrade more quickly [39]. Molten salts aggressively attack coatings, removing weak, protective layers and accelerating the corrosive attacks. In addition, rapid thermal cycling contributes to the problem with repetitive thermal expansion and contraction, which may further compromise coating mechanisms. Addressing the issues of filler agglomeration, poor matrix compatibility, and lack of protective coatings next is crucial to the progression of composite coatings [40].

Microstructural characteristics like grain size and porosity impact corrosion protection; in this regard, smaller grains and lower porosity are preferred. Furthermore, the environmental conditions, particularly high-temperature exposure to corrosive industrial agents, such as sulfur and vanadium, have also been shown to impact the degree to which TBCs are prone to degradation [41]. These oxides can display significant acidity or basicity, which worsens the corrosive properties of the environment.

$$2V + 5/2\ O_2 \rightarrow V_2O_5 \text{ (strongly acidic)} \tag{1}$$

$$S + O_2 \rightarrow SO_2;\ SO_2 + \tfrac{1}{2}\ _{O2} \rightarrow SO_3 \text{ (strongly acidic)} \tag{2}$$

$$2Na + \tfrac{1}{2}\ O_2 \rightarrow Na_2O \text{ (strongly basic)} \tag{3}$$

$$Na_2O \text{ (base)} + SO_3 \text{ (acid)} \rightarrow Na_2SO_4 \text{ (salt)} \tag{4}$$

$$Na_2O \text{ (base)} + V_2O_5 \text{ (acid)} \rightarrow 2NaVO_3 \text{ (salt)} \tag{5}$$

Both vanadium pentoxide (V_2O_5) which is a strong acidic oxide and melts at 690°C and sodium metavanadate ($NaVO_3$), a corrosive salt which melts at 610°C, have been identified as serious threat to TBCs since both sodium and vanadium are present in fuel. It has been shown that the type of material used to manufacture TBCs, dictates the hot corrosion characteristics. Fluoride zirconates as well as other zirconates like $Gd_2Zr_2O_7$, $Sm_2Zr_2O_7$, and $Nd_2Zr_2O_7$ have been researched to varying degrees concerning their corrosion resistance [42].

Figure 6.8 illustrates the stages of hot corrosion on TiAl alloy substrates under the influence of NaCl and Na_2SO_4 salts. In the early stage (Figure 6.8(a)), aggressive ions like O^{2-} and S^{2-} penetrate the surface, attacking the alloy and promoting the formation of initial oxidation products such as TiO_2 and AlO_3^-. At the intermediate stage of corrosion (Figure 6.8(b)), it is the active ingress of these ions alongside the interaction between TiO_2 to form more complex compounds, $Na_2Ti_6O_{13}$, due to the presence of sodium oxide, which acts as a fluxing agent. During the advanced stage (Figure 6.8(c)), the further reaction of TiO_2 is the primary corrosion mechanism whereby the aluminium ion reacts with the chloride ion to produce $AlCl_3$ which further disintegrates and collapses the protective oxide surface. In the culmination of all these processes, corrosive salts illustrate the distinctive reactions and processes that lead to the inevitable corrosive wear of the surface of the TiAl alloy [43].

Failures in coatings may occur due to a set of defined mechanisms like delamination (the protective layer releases/leaves the substrate), cracks developing and propagating through a coating due to imposed coating stress, or the failure of passive, corrosion-resistant layers. Problems with coatings may be aggravated by repeated

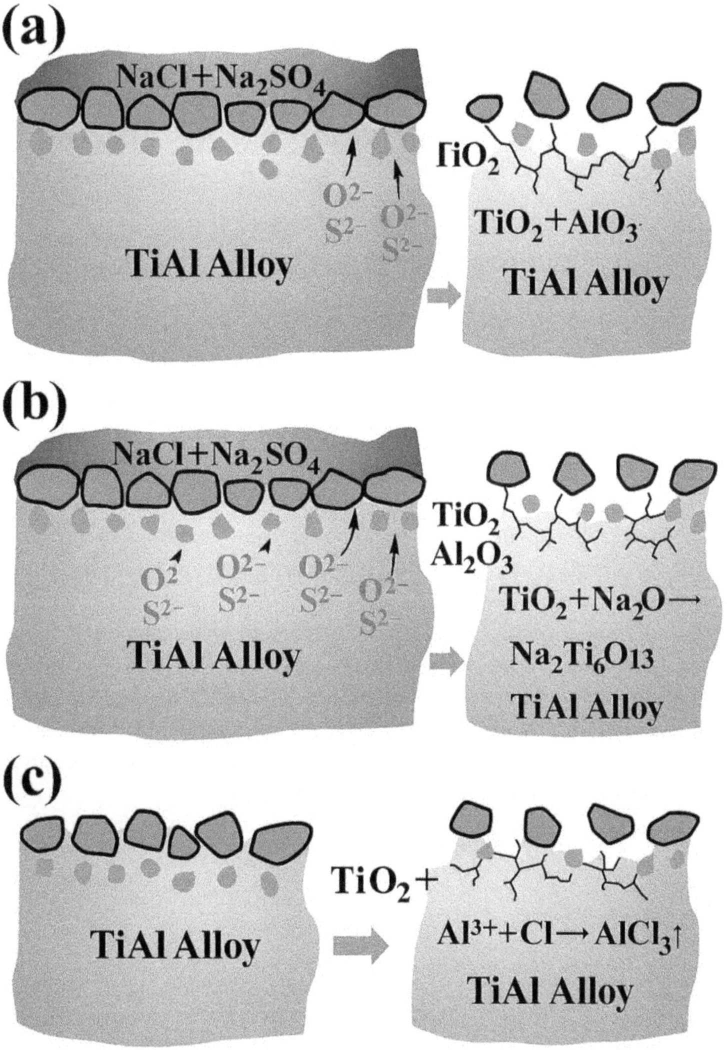

FIGURE 6.8 The schematic diagrams of hot corrosion mechanism of TiAl substrate: (a) early stage, (b) intermediate stage, and (c) late stage [43].

mechanical loadings and variations in ambient temperature which cause cyclic and differential thermal expansion [44]. In the long run, macro and micro cyclic thermally induced stresses that may be imposed on a coating may create microstructural defects that weaken the coating/substrate bond. This results in a protective function of the coating being rapidly lost and the substrate degrading to a level that is highly corrodible and worn [45].

The optical macro-morphologies for three TiAl-based alloys after 100 hours of hot corrosion are shown in Figure 6.9. In panel (a), we see the aluminized alloy which

FIGURE 6.9 Optical images showing macro-morphologies of aluminized (a) aluminium–titanium intermetallic (AA-TiAl) alloy, (b) porous oxidation-assisted aluminium titanium (POAA-TiAl) alloys, and (c) TiAl alloys after hot corrosion for 100 h [46].

exhibits extreme surface spallation with large flakes indicating severe detachment and degradation of the alloy. In panel (b), the AA-TiAl alloy features both spallation and visible cracks. Thus, the AA-TiAl alloy surfaces exhibit structural weakness which results in a corrosion compound being overly susceptible to damage. In contrast, panel (c) shows the POAA-TiAl alloy which has a porous microstructure and relatively good resistance with uniformly distributed pores and very little surface spallation and cracking. These pores attest to a designed microstructure that will alleviate thermal cracking and corrosion resistance. Overall, these images demonstrate the hot corrosion environment destabilizes the interoperability of POAA-TiAl alloys. This shows the POAA-TiAl alloys to be the most stable, hot corrosive environment relative to the AA-TiAl alloy and POAA-TiAl alloys [46].

Addressing these issues requires further innovation in materials design and processing.

6.5 DISCUSSION

Composite coatings, particularly those with nanomaterials, have shown marked improvements in terms of resistance to hot corrosion, mechanical properties, and durability when compared to conventional coatings. The coatings' performances are influenced by the dispersal uniformity of the filler, the interaction between the matrix and filler, and the processing conditions [47]. For instance, nanofillers of graphene and boron nitride provide remarkable barrier properties, although some oversight is required to manage their agglomeration and potential galvanic corrosion. Metallic matrix composites (Ni, Co, Al) and ceramic-based coatings (zirconate, Cr3C2-NiCr) are especially useful for aggressive environments of very high temperatures [48].

Figure 6.10 shows a comparison of ceramic materials used for TBCs, characterized into four classes: t'-tetragonal, c-cubic, p-pyrochlore, and advanced multiphase ceramics. Each of these classes has different rated materials for thermal and other performance characteristics: sinter, CMAS, thermal, phase stability, erosion, strain, and process. The spider chart showing nearly all material characteristics shows the performance levels of each class meeting the rated requirements. Multiphase ceramics

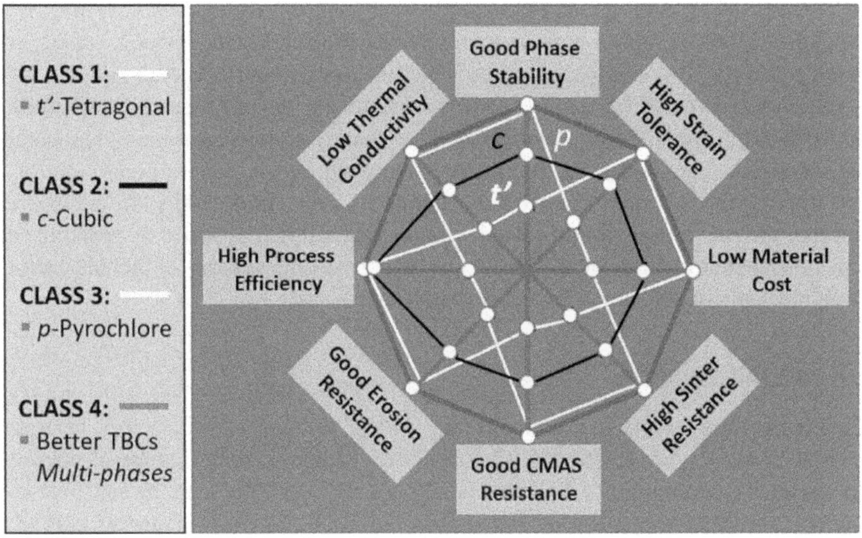

FIGURE 6.10 Comparison of the advancements of novel ceramic materials to fulfil desired characteristics [49].

(Class 4, red line) show high, more balanced fulfilment in most key performance characteristics. The t'-tetragonal (white), c-cubic (black), and p-pyrochlore (yellow) phases do exhibit some performance key attributes but suffer more in the key attributes of thermal conductivity and sinter resistance. This design promoted the advancement of new ceramic materials to focus and accommodate more multi-phase design for high protection and dependability in high-temperature use for the aerospace and energy industries [49].

The effectiveness and longevity of protective coatings remain weak despite advancements in coating technology. Delamination, the separation of the coating from the substrate, and crack propagation have been identified as the primary degradation mechanisms of coatings. Delamination and crack propagation are more pronounced under conditions of constant thermal cycling and repeated mechanical or environmental stresses [50]. Breakdown of passive protective layers, especially under severe corrosive conditions, represents long service periods when coatings can no longer function as an effective barrier. Extreme and aggressive environments, and especially cyclic variations in temperature, can inflict microstructural harm and lead to the irreversible loss of protective characteristics [51]. In developing multifunctional and intelligent coatings, including those enhancing self-healing, superhydrophobicity, or self-cleaning characteristics, coating technology aims to address these service shortcomings. Given the expected enhanced performance and increased coating durability, multifunctional coatings seem to be the future of coating technology. However, practical problems, especially in the large-scale economically viable fabrication of coatings with complex functionalities and maintaining operational stability in coating complex systems, are yet to be solved. Research is vital to resolve these new challenges [52].

Thermal spray coatings and new surface engineering techniques have begun to expand in biomedical and high-temperature corrosion-resistant systems slurry processing sectors. This is especially true when integrated with new technologies such as microwave-assisted fabrication, additive manufacturing, and magnetron sputtering thin-film deposition [53–66]. Furthermore, the coating architectures have great potential in surface topography and robust (tensile, compressive, flexural) and functional performance in solar energy conversion coating systems. Additionally, the engineered surfaces provide a basis for advanced computational fluid dynamics of conduits and pipes with bends and ergonomic structures such as seating systems to include a structural, mechanical, and anthropometric coupling [67–79].

6.6 CONCLUSION

High-temperature protective coatings are critical to achieve satisfactory performance at given service conditions, the capability to resist specific types of environmental attack, compatibility both chemically and mechanically with the substrate and method, and ease of application. The optimization of composite materials is fundamentally predicated on effective selection of constituent parts – metals, ceramics, and polymers – and effective implementation of novel developments in interface engineering. Here, interface engineering refers to the interface that would be developed, that is, made of one constituent of a composite material, polymers, and metals. The optimized materials would thus be predicted to be make the composite materials more durable, particularly the composite materials that would be used in extreme conditions. The chapter has analyzed various aspects of composite coatings materials, methodologies, challenges, and their utility for protecting alloys against hot corrosion in high-temperature terrains. Inhibitive composite coatings serve to augment barrier protection and substrate corrosion protection by physically blocking corrosive media, enhancing passive corrosion protection, and embedding active fillers. The protection properties of coatings can be augmented further by including nanomaterials such as graphene, boron nitride, and various metal oxides. The matrix-filler interface interaction is significant for performance enhancement. Multiple processing techniques, including electrodeposition, plasma electrolytic oxidation, and thermal spray, add composite coatings to substrates, each offering unique benefits. Outstanding issues remain regarding the agglomeration of active fillers, interfacial bonding of coatings and substrates, and dissolution mechanisms in extreme conditions. The long-lasting performance of composite coatings in harsh conditions can be increased by developing multifunctional self-healing coatings and streamlining their production in bulk.

REFERENCES

[1] X. Zhong, Y. Wang, Z. Xu, Y. Zhang, J. Zhang, and X. Cao, "Influence of laser-glazing on hot corrosion resistance of yttria-stabilized zirconia TBC in molten salt mixture of V2O5 and Na2SO4," *Materials Corrosion,* vol. 60, no. 11, pp. 882–888, 2009.

[2] S. Sattar *et al.*, "Corrosion reduction in steam turbine blades using nano-composite coating," *Journal of King Saud University – Science,* vol. 35, no. 8, p. 102861, 2023. www.sciencedirect.com/science/article/pii/S1018364723003233

[3] P.-H. Lee *et al.*, "Thermal cycling behavior and interfacial stability in thick thermal barrier coatings," *Surface and Coatings Technology,* vol. 205, no. 5, pp. 1250–1255, 2010.
[4] M. Nosonovsky and B. Bhushan, "Green tribology: Principles, research areas and challenges," vol. 368, pp. 4677–4694, 2010.
[5] G. Karadimas and K. J. A. S. Salonitis, "Ceramic matrix composites for aero engine applications—A review," *Applied Sciences,* vol. 13, no. 5, p. 3017, 2023.
[6] A. Afrasiabi, M. Saremi, and A. Kobayashi, "A comparative study on hot corrosion resistance of three types of thermal barrier coatings: YSZ, YSZ+Al2O3 and YSZ/Al2O3," *Materials Science and Engineering: A,* vol. 478, no. 1, pp. 264–269, 2008. https://doi.org/10.1016/j.msea.2007.06.001
[7] M. S. Ahmadi, R. Shoja-Razavi, Z. Valefi, and H. Jamali, "Laser surface modification of air plasma-sprayed Al2O3/YSZ multilayer thermal barrier coating to improve hot corrosion resistance in V2O5–Na2SO4 Salts," *Journal of Thermal Spray Technology,* vol. 28, no. 8, pp. 1906–1918, 2019.
[8] X. Zhang *et al.,* "Al2O3-modified PS-PVD 7YSZ thermal barrier coatings for advanced gas-turbine engines," *npj Materials Degradation,* vol. 4, no. 1, p. 31, 2020.
[9] B. Díaz *et al.*, "Failure mechanism of thin Al2O3 coatings grown by atomic layer deposition for corrosion protection of carbon steel," *Electrochimica Acta,* vol. 56, no. 26, pp. 9609–9618, 2011.
[10] R. Vaßen, E. Bakan, D. E. Mack, and O. Guillon, "A perspective on thermally sprayed thermal barrier coatings: Current status and trends," *Journal of Thermal Spray Technology,* vol. 31, no. 4, pp. 685–698, 2022.
[11] K. Derelizade, F. Venturi, R. Wellman, A. Kholobysov, and T. Hussain, "Wear performance of graphene nano platelets incorporated WC-Co coatings deposited by hybrid high velocity oxy fuel thermal spray," *Wear,* vol. 482, p. 203974, 2021.
[12] B. Dermeik and N. Travitzky, "Laminated object manufacturing of ceramic-based materials," *Advanced Engineering Materials,* vol. 22, no. 9, p. 2000256, 2020.
[13] M. Derrick, S. David, M. Frederic, and B. Tajebe, "A review on aerospace-AI, with ethics and implications," *Journal of Civil, Construction and Environmental Engineering,* vol. 10, no. 2, pp. 60–74, 2025. https://doi.org/10.11648/j.jccee.20251002.12
[14] K. M. Doleker, A. C. Karaoglanli, Y. Ozgurluk, and A. Kobayashi, "Performance of single YSZ, Gd2Zr2O7 and double-layered YSZ/Gd2Zr2O7 thermal barrier coatings in isothermal oxidation test conditions," *Vacuum,* vol. 177, p. 109401, 2020.
[15] K. M. Doleker, O. Odabas, Y. Ozgurluk, H. Askerov, and A. C. Karaoglanli, "Effect of high temperature oxidation on Inconel 718 and Inconel 718/YSZ/Gd2Zr2O7," *Materials Research Express,* vol. 6, no. 8, p. 086456, 2019.
[16] K. M. Doleker, Y. Ozgurluk, H. Ahlatci, and A. C. Karaoglanli, "Isothermal oxidation behavior of gadolinium zirconate (Gd2Zr2O7) thermal barrier coatings (TBCs) produced by electron beam physical vapor deposition (EB-PVD) technique," *Open Chemistry,* vol. 16, no. 1, pp. 986–991, 2018.
[17] K. M. Doleker, Y. Ozgurluk, H. Ahlatci, and A. C. Karaoglanli, "Evaluation of oxidation and thermal cyclic behavior of YSZ, Gd2Zr2O7 and YSZ/Gd2Zr2O7 TBCs," *Surface Coatings Technology,* vol. 371, pp. 262–275, 2019.
[18] W. Lan, Z. Chen, X. Zhang, W. Bi, and W. Liu, "Corrosion resistance analysis of Al2O3-TiO2 composite ceramic coatings on carbon steel pipe surfaces," *Alexandria Engineering Journal,* vol. 110, pp. 377–385, 2025. www.sciencedirect.com/science/article/pii/S1110016824011505
[19] H. Bahramnia, H. Mohammadian Semnani, A. Habibolahzadeh, and H. Abdoos, "Epoxy/polyurethane nanocomposite coatings for anti-erosion/wear applications: A review," *Journal of Composite Materials,* vol. 54, no. 22, pp. 3189–3203, 2020.

[20] S. Desai, R. Sonawane, and A. More, "Thermoplastic polyurethane for three-dimensional printing applications: A review," *Polymers for Advanced Technologies,* 2023. https://consensus.app/papers/thermoplastic-polyurethane-for-three%E2%80%90dimensional-more-desai/0516c8e28385557eb034d86654a943f7/
[21] J. C. Ge, J. H. Kim, and N. J. Choi, "Electrospun polyurethane/loess powder hybrids and their absorption of volatile organic compounds," *Advances in Materials Science Engineering,* vol. 2016, no. 1, p. 8521259, 2016.
[22] J. Huang, W. Sun, R. Huang, and W. Ma, "Cracking behavior of atmospheric plasma-sprayed 8YSZ thermal barrier coatings during thermal shock test," *Coatings,* vol. 13, no. 2, p. 243, 2023.
[23] M. Bashirzadeh, F. Azarmi, C. Leither, and G. Karami, "Investigation on relationship between mechanical properties and microstructural characteristics of metal matrix composites fabricated by cold spraying technique," *Applied Surface Science,* vol. 275, pp. 208–216, 2013.
[24] S. Bhatnagar and S. Mullick, "A study on the influence of reinforcement particle size in laser cladding of TiC/Inconel 625 metal matrix composite," *Optics Laser Technology,* vol. 161, p. 109115, 2023.
[25] N. V. Abhijith, D. Kumar, and D. Kalyansundaram, "Development of single-stage TiNbMoMnFe high-entropy alloy coating on 304L stainless steel using HVOF thermal spray," *Journal of Thermal Spray Technology,* vol. 31, no. 4, pp. 1032–1044, 2022. https://doi.org/10.1007/s11666-021-01294-9
[26] N. Abu-Warda, A. López, M. López, and M. Utrilla, "High temperature corrosion and wear behavior of HVOF-sprayed coating of Al2O3-NiAl on AISI 304 stainless steel," *Surface Coatings Technology,* vol. 359, pp. 35–46, 2019.
[27] M. Oksa, E. Turunen, T. Suhonen, T. Varis, and S.-P. Hannula, "Optimization and characterization of high velocity oxy-fuel sprayed coatings: Techniques, materials, and applications," *Coatings,* vol. 1, no. 1, pp. 17–52, 2011.
[28] B. Chen *et al.*, "Carbon nanotube induced microstructural characteristics in powder metallurgy Al matrix composites and their effects on mechanical and conductive properties," *Journal of Alloys and Compounds,* vol. 651, pp. 608–615, 2015.
[29] F. He, Q. Han, and M. J. Jackson, "Nanoparticulate reinforced metal matrix nanocomposites–A review," *International Journal of Nanoparticles,* vol. 1, no. 4, pp. 301–309, 2008.
[30] S. Björklund, S. Goel, and S. Joshi, "Function-dependent coating architectures by hybrid powder-suspension plasma spraying: Injector design, processing and concept validation," *Materials Design,* vol. 142, pp. 56–65, 2018.
[31] M. Bai, L. Reddy, and T. Hussain, "Experimental and thermodynamic investigations on the chlorine-induced corrosion of HVOF thermal sprayed NiAl coatings and 304 stainless steels at 700 C," *Corrosion Science,* vol. 135, pp. 147–157, 2018.
[32] E. Bakan and R. Vaßen, "Ceramic top coats of plasma-sprayed thermal barrier coatings: Materials, processes, and properties," *Journal of Thermal Spray Technology,* vol. 26, no. 6, pp. 992–1010, 2017.
[33] A. Babu, H. S. Arora, and H. S. Grewal, "Microwave-assisted post-processing of detonation gun-sprayed coatings for better slurry and cavitation erosion resistance," *Journal of Thermal Spray Technology,* vol. 28, no. 7, pp. 1565–1578, 2019.
[34] A. Killinger, M. Kuhn, and R. Gadow, "High-velocity suspension flame spraying (HVSFS), a new approach for spraying nanoparticles with hypersonic speed," *Surface Coatings Technology,* vol. 201, no. 5, pp. 1922–1929, 2006.

[35] F. Fanicchia, D. Axinte, J. Kell, R. McIntyre, G. Brewster, and A. Norton, "Combustion flame spray of CoNiCrAlY & YSZ coatings," *Surface Coatings Technology,* vol. 315, pp. 546–557, 2017.
[36] W. Fan and Y. Bai, "Review of suspension and solution precursor plasma sprayed thermal barrier coatings," *Ceramics International,* vol. 42, no. 13, pp. 14299–14312, 2016.
[37] R. H. Aparco, F. Tapia-Tadeo, Y. B. Ascarza, A. L. Ramírez, Y.-L. Huamán-Romaní, and C. C. Otero, "A machine learning approach for analyzing residual stress distribution in cold spray coatings," *Journal of Thermal Spray Technology,* vol. 33, no. 5, pp. 1292–1307, 2024.
[38] Y. Abdesselam, A. Belloufi, I. Rezgui, M. Abdelkrim, T. Catalin, and B. Chiriţă, "Composite coatings based on nickel and Y2O3 nanoparticles: a comprehensive analysis of developments in electrodeposition and functional property optimization," *The International Journal of Advanced Manufacturing Technology,* vol. 137, no. 7, pp. 3273–3332, 2025. https://doi.org/10.1007/s00170-025-15360-x
[39] W. S. AbuShanab, E. B. Moustafa, M. A. Taha, and R. A. Youness, "Synthesis and structural properties characterization of titania/zirconia/calcium silicate nanocomposites for biomedical applications," *Applied Physics A,* vol. 126, pp. 1–12, 2020.
[40] J. P. S. Ahmed, K. Satyasree, R. R. Kumar, O. Meenakshisundaram, and S. Shanmugavel, "A comprehensive review on recent developments of natural fiber composites synthesis, processing, properties, and characterization," *Engineering Research Express,* vol. 5, no. 3, p. 032001, 2023.
[41] I. Gurrappa and A. S. Rao, "Thermal barrier coatings for enhanced efficiency of gas turbine engines," *Surface Coatings Technology,* vol. 201, no. 6, pp. 3016–3029, 2006.
[42] M. J. Khan, G. Moskal, A. Iqbal, M. Mikuśkiewicz, T. Pawlik, and P. Olesik, "Hot corrosion behavior of single-layered Gd2Zr2O7, Sm2Zr2O7, and Nd2Zr2O7 thermal barrier coatings exposed to Na2SO4+ MgSO4 environment," *Coatings,* vol. 13, no. 8, p. 1311, 2023.
[43] Y. Sun *et al.*, "High-temperature hot corrosion resistance of CrAl/NiCoCrAlY/AlSiY gradient composite coating on TiAl alloy," *Coatings*, vol. 14, no. 8. https://doi.org/10.3390/coatings14081067
[44] R. Ajdary *et al.*, "Selective laser sintering of lignin-based composites," *ACS Sustainable Chemistry Engineering,* vol. 9, no. 7, pp. 2727–2735, 2021.
[45] H. I. Akbar, E. Surojo, D. Ariawan, and A. R. Prabowo, "Experimental study of quenching agents on Al6061–Al2O3 composite: Effects of quenching treatment to microstructure and hardness characteristics," *Results in Engineering,* vol. 6, p. 100105, 2020.
[46] Y.-t. Hu *et al.*, "Improving hot corrosion resistance of aluminized TiAl alloy by anodization and pre-oxidation," *Transactions of Nonferrous Metals Society of China,* vol. 31, no. 1, pp. 193–206, 2021. www.sciencedirect.com/science/article/pii/S1003632620654875
[47] L. Al-Naamani, S. Dobretsov, and J. Dutta, "Chitosan-zinc oxide nanoparticle composite coating for active food packaging applications," *Innovative Food Science Emerging Technologies,* vol. 38, pp. 231–237, 2016.
[48] A. Amani, M. Montazer, and M. Mahmoudirad, "Synthesis of applicable hydrogel corn silk/ZnO nanocomposites on polyester fabric with antimicrobial properties and low cytotoxicity," *International Journal of Biological Macromolecules,* vol. 123, pp. 1079–1090, 2019.

[49] X. Ma, K. Rivellini, P. Ruggiero, and G. Wildridge, "Novel thermal barrier coatings with phase composite structures for extreme environment applications: Concept, process, evaluation and performance," *Coatings,* vol. 13, no. 1, p. 210, 2023.
[50] Q. An *et al.,* "Two-scale TiB/Ti64 composite coating fabricated by two-step process," *Journal of Alloys Compounds,* vol. 755, pp. 29–40, 2018.
[51] P. B. Anand *et al.*, "Kenaf fiber and hemp fiber multi-walled carbon nanotube filler-reinforced epoxy-based hybrid composites for biomedical applications: morphological and mechanical characterization," *Journal of Composites Science,* vol. 7, no. 8, p. 324, 2023.
[52] S. Annamalai, B. Anand Ronald, and S. Mohamed Ameer Batcha, "Analysis and prediction of slurry erosion wear response of silicon carbide reinforced Al2124 composite using Taguchi–artificial neural network approach," *Materialwissenschaft und Werkstofftechnik,* vol. 56, no. 3, pp. 399–418, 2025.
[53] M. Singh, H. Vasudev, and R. Kumar, "Corrosion and tribological behaviour of BN thin films deposited using magnetron sputtering," *International Journal of Surface Engineering and Interdisciplinary Materials Science (IJSEIMS*), vol. 9, no. 2, pp. 24–39, 2021.
[54] V. Dutta, L. Thakur, B. Singh, and H. Vasudev, "A study of erosion–corrosion behaviour of friction stir-processed chromium-reinforced NiAl bronze composite," *Materials*, vol. 15, no. 15, p. 5401, 2022.
[55] H. Sharanabasava, M. Raviprakash, C. D. Prasad, M. R. Ramesh, M. V. Phanibhushana, H. Vasudev, and S. Kumar. Microstructure, mechanical and wear properties of SiC and Mo reinforced NiCr microwave cladding. *Advanced Materials and Processing Technologies*, vol. 10. no. 4, pp. 3620–33, 2023.
[56] P. K. Verma, V. Kumar, and H. Vasudev. "Bioactivity and corrosion analysis of thermally sprayed hydroxyapatite based coatings," *Journal of Electrochemical Science and Engineering*, vol. 14, no. 4, pp. 419–439, 2024.
[57] P. K. Verma, H. Vasudev, and V. Kumar, "Biocompatibility and corrosion performance of HA-based LVOF sprayed and post-annealed coatings on SS 254 surgical-grade bio-implant material," *Ceramics International*, vol. 51, no. 10, pp. 13207–13225, 2025.
[58] J. Parkash, H. S. Saggu, and H. Vasudev, "A short review on the performance of high velocity oxy-fuel coatings in boiler steel applications." *Materials Today: Proceedings*, vol. 50, pp. 1442–1446, 2022.
[59] P. Singh, H. Vasudev, and A. Bansal, "Effect of post-heat treatment on the microstructural, mechanical, and bioactivity behavior of the microwave-assisted alumina-reinforced hydroxyapatite cladding," *Proceedings of the Institution of Mechanical Engineers, Part E: Journal of Process Mechanical Engineering*, vol. 238, no. 3, pp. 1060–1071, 2024.
[60] P. K. Verma, H. Vasudev, G., Singh, and S. Singh, "A review on the microstructural and biomedical properties of implants manufactured using additive manufacturing," *Journal of the Institution of Engineers (India): Series D*, vol. 106, pp. 1–15, 2024.
[61] S. P. Singh, P. K. Verma, S. Singh, M. Kapoor, and R. Balu, "Microstructural characterization and electrochemical behaviour of HVOF sprayed WC-Co-Cr coatings on 316L boiler steel stainless steel," *Results in Surfaces and Interfaces*, vol. 16, p. 100248, 2024.
[62] M. Singh, H. Vasudev, and M. Singh, "Surface protection of SS-316L with boron nitride based thin films using radio frequency magnetron sputtering technique," *Journal of Electrochemical Science and Engineering*, vol. 12, no. 5, pp. 851–863, 2022.
[63] G. Prashar, H. Vasudev, L. Thakur, and A. Bansal, "Performance of thermally sprayed hydroxyapatite coatings for biomedical implants: A comprehensive review," *Surface Review and Letters*, vol. 30, vol. 1, p. 2241001, 2023.

[64] H. Vasudev, "Wear characteristics of Ni-WC powder deposited by using a microwave route on mild steel: Microwave cladding of Ni-WC," *International Journal of Surface Engineering and Interdisciplinary Materials Science (IJSEIMS)*, vol. 8, no. 1, pp. 44–54, 2020.

[65] P. K. Verma, A. S. Minhas, P. Singh, and S. Kumar, "Slurry erosion behaviour of thermal sprayed Al2O3 and Cr2O3 coatings for turbine steels," *AIP Conference Proceedings*, vol. 2986, no. 1, p. 020012, 2024.

[66] P. K. Verma, H. Vasudev, and G. Singh, "Improving the corrosion performance of SS 304 stainless steel using HVOF-sprayed WC-Co-Cr coatings in chloride environments," *Journal of Bio-and Tribo-Corrosion*, vol. 11, no. 4, pp. 1–13, 2025.

[67] M. A. Kattimani, P. R. Venkatesh, H. Masum, M. M. Math, V. N. Bahadurdesai, S. Mustafkhadri, C. D. Prasad, and H. Vasudev, "Design and numerical analysis of tensile deformation and fracture properties of induction hardened inconel 718 superalloy for gas turbine applications," *International Journal on Interactive Design and Manufacturing (IJIDeM)*, vol. 18, no. 8, pp. 5939–5949, 2024.

[68] M. Poojari, H. Hanumanthappa, C, D. Prasad, H. N. Jathanna, A. R. Ksheerasagar, P. Shetty, B. K. Shanmugam, and H. Vasudev, "Computational modelling for the manufacturing of solar-powered multifunctional agricultural robot," *International Journal on Interactive Design and Manufacturing (IJIDeM)*, vol. 18, no. 8, pp. 5725–5736, 2024.

[69] C. Prakash, L. R. Gupta, A. Mehta, H. Vasudev, R. Tominov, E. Korman, A. Fedotov, V. Smirnov, and K. K. Kesari, "Computing of neuromorphic materials: An emerging approach for bioengineering solutions," *Materials Advances*, vol. 4, pp. 5882–5919, 2023.

[70] J. Singh, R. Singh, S. Singh, H. Vasudev, and S. Kumar, "Reducing scrap due to missed operations and machining defects in 90PS pistons," *International Journal on Interactive Design and Manufacturing (IJIDeM)*, vol. 17, no. 5, pp. 2527–2539, 2023.

[71] P. Nagabhushana, S. Ramprasad, C. D. Prasad, H. Vasudev, and C. Prakash, "Numerical investigation on heat transfer of a nano-fluid saturated vertical composite porous channel packed between two fluid layers," *International Journal on Interactive Design and Manufacturing (IJIDeM)*, vol. 18, no. 5, pp. 2927–2944, 2024.

[72] M. M. Mahantesh, K. R. Rao, A. P. Chandra, M. N. Vijayakumar, B. Nandini, C. D., Prasad, and H. Vasudev, "Design and modeling using finite element analysis for the sitting posture of computer users based on ergonomic perspective," *International Journal on Interactive Design and Manufacturing (IJIDeM)*, vol. 18, no. 8, pp. 5875–5891, 2024.

[73] P. K. Verma, H. Vasudev, and V. Kumar, "Dental implant materials and designs and failure during implant placement and loading: A review," In Hitesh Vasudev, Chander Prakash, Mohammad Khalid, Kavindra Kesari (eds.): *Advanced Manufacturing Processes*, pp. 138–150, Boca Raton: CRC Press, 2025.

[74] P. K. Verma, H. Vasudev, and V. Kumar, "Bioactivity and corrosion performance of low-velocity oxy-fuel-sprayed hydroxyapatite-based coatings on a surgical-grade stainless steel," *Transactions of the Indian Institute of Metals*, vol. 78, no. 8, p. 185, 2025.

[75] J. Singh, H. S. Gill, and H. Vasudev, "Computational fluid dynamics analysis on role of particulate shape and size in erosion of pipe bends," *International Journal on Interactive Design and Manufacturing (IJIDeM)*, vol. 17, no. 5, pp. 2631–2646, 2023.

[76] G. S. Rao, U. Mukkamala, H. Hanumanthappa, C. D. Prasad, H. Vasudev, B. Shanmugam, and K. C. KishoreKumar, "Evaluating and optimizing surface roughness using genetic algorithm and artificial neural networks during turning of AISI 52100 steel," *International Journal on Interactive Design and Manufacturing (IJIDeM)*, vol. 18, no. 8, pp. 6151–6160, 2024.

[77] P. K. Verma, H. Vasudev, and V. Kumar, "Effect of electro-corrosion behavior of different grades of stainless steel used in bio-implants: A review," In Hitesh Vasudev, Chander Prakash, Mohammad Khalid, Kavindra Kesari (eds.): *Advanced Manufacturing Processes,* pp. 269–283, Boca Raton: CRC Press, 2025.
[78] S. D. Kulkarni, Manjunatha, U. Chandrasekhar, K. V. Manjunath, C. D., Prasad, and H. Vasudev, "Design and optimization of polyvinyl-nitride rubber for tensile strength analysis," *International Journal on Interactive Design and Manufacturing (IJIDeM)*, vol. 18, no. 8, pp. 5893–5908, 2024.
[79] P. K. Verma and H. Vasudev, "Importance of hydroxyapatite coatings in improving new bone osteogenesis: A review," In Hitesh Vasudev, Chander Prakash, Mohammad Khalid, Kavindra Kesari (eds.): *Advanced Manufacturing Processes*, pp. 284–299, Boca Raton: CRC Press, 2025.

7 CNT-Reinforced Thermal Spray Composite Coatings

Amrinder Mehta and Hitesh Vasudev

7.1 INTRODUCTION

Yttria-stabilized zirconia (YSZ) is extensively employed as a thermal barrier coating (TBC) in various applications, including jet engine components, gas turbine engine blades and vanes, and the combustion zone of engines. Nevertheless, its application is restricted to temperatures below 1200°C because of challenges associated with phase transformation and its susceptibility to hot corrosion and oxidation, which degrade the material's performance [1]. Thermal spraying techniques have become highly effective methods for applying protective coatings to surfaces, particularly for preventing corrosion and wear. These coatings have recently gained significant importance. Progress in the production of powders and wires has led to the development of various surface coatings with outstanding wear and corrosion resistance [2]. In 1991, Sumio Iijima's discovery of carbon nanotubes (CNTs) sparked global scientific interest in nanotechnology. A CNT is essentially a sheet of graphite, one atom thick, rolled into a tube with a diameter of one nanometre, and is known as a single-wall carbon nanotube (SWCNT). When additional graphene tubes encircle the core of an SWCNT, it is referred to as a multi-walled carbon nanotube (MWCNT). Owing to their extremely high strength, exceptional elastic modulus, and large aspect ratio, CNTs are ideal nanoreinforcements for metal matrix composites (MMCs) [3]. The incorporation of CNTs as reinforcements is anticipated to enhance the strength and rigidity of metal matrices. Over the last ten years, there has been a growing focus on aluminium matrix composites reinforced with CNTs (CNT/Al) owing to their robust yet lightweight characteristics. This material has emerged as a promising option in the aerospace and automotive sectors. However, a significant challenge is that the increased strength of the composite often leads to a considerable reduction in ductility (toughness). Drawing inspiration from natural nacre, the design of a nanolaminated structure is an effective approach for balancing the trade-off between strength and ductility [4].

Figure 7.1 illustrates the fabrication and advantages of carbon nanotube (CNT)-reinforced WC-Co composite coatings produced via thermal spray techniques. Initially, WC-Co particles and CNTs were sonicated in suspension, which allowed the CNTs to be effectively distributed among the WC-Co grains. Scanning electron microscopy revealed that CNTs were incorporated at the micrometre level

DOI: 10.1201/9781003730637-7

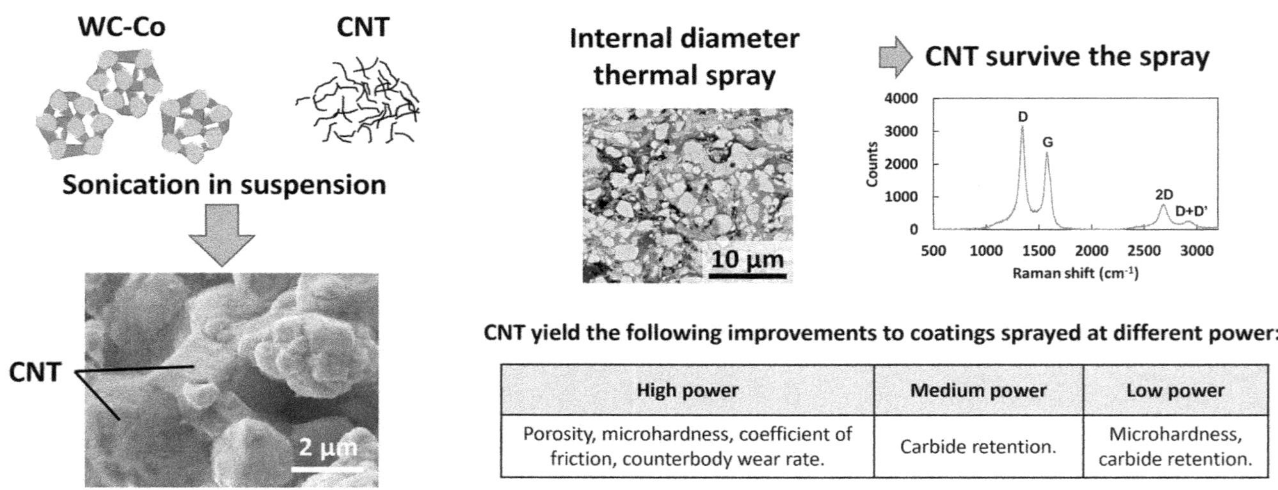

FIGURE 7.1 Carbon nanotube (CNT)-reinforced WC-Co composite coatings [5].

within the composite matrix. The material was subsequently processed using an internal diameter thermal spray, and microstructural analysis verified the persistence of CNTs post-spraying, as evidenced by the distinctive peaks in the Raman spectrum. The table outlines the enhancements in the coating at different thermal spray powers: high power increases porosity and microhardness, reduces friction, and decreases wear rate; medium power preserves carbide retention; and low power supports both microhardness and carbide retention. These findings demonstrate how CNTs strengthen WC-Co composites, improving their mechanical properties and wear resistance, making them appealing options for industrial coatings that require high durability and efficiency [5].

The current study assessed the mobility of grain boundaries in plasma-sprayed aluminium oxide (Al_2O_3) composites reinforced with CNTs. This mobility was determined by examining the grain growth within the spray-dried particles and the thermal history encountered during high-temperature plasma processing [6]. During thermal exposure, CNTs create an interfacial grain boundary layer that restricts the grain growth of the plasma-sprayed coatings. When CNTs are integrated into traditional ceramic or metallic matrices through thermal spray methods like high-velocity oxy-fuel (HVOF), plasma spraying, and detonation gun coating, they can enhance microstructural characteristics, decrease porosity, and strengthen interfacial bonding due to their distinctive mechanical, thermal, and electrical properties. Nonetheless, challenges persist in achieving uniform CNT dispersion, managing agglomeration, and scaling up production while preserving the integrity and performance of the coatings. This review critically explores the synthesis, properties, challenges, and prospects of CNT-reinforced thermal-spray composite coatings, with reference to recent experimental and review studies [7].

7.2 SYNTHESIS AND PROCESSING TECHNIQUES

Carbon nanotubes have gained recognition as promising reinforcement materials and have been successfully integrated into various matrices, such as tungsten carbide-cobalt (WC-Co), zirconia-yttria (ZrO_2-Y2O_3), alumina (Al_2O_3), and chromium carbide-nickel chromium (Cr_3C_2-NiCr), using advanced thermal spray techniques. Techniques such as HVOF [8], plasma spraying (PS) [9], cold spray (CS) [10], and detonation gun (DG) [11], coatings have facilitated the effective dispersion and incorporation of CNTs into these matrices, resulting in composite coatings with enhanced mechanical and functional properties [12]. The inclusion of CNTs in these matrices through thermal spraying results in coatings with improved hardness, wear resistance, toughness, and superior thermal and electrical conductivity. For instance, HVOF spraying is particularly valued for producing dense, well-adhered coatings with high wear and corrosion resistance, making it suitable for demanding industrial applications where durability is critical. Plasma spraying, on the other hand, allows for the deposition of ceramic-based composites such as ZrO_2-Y_2O_3 and Al_2O_3, where the addition of CNTs can further enhance the fracture toughness and thermal stability. The cold spray technique, which operates at lower temperatures, helps preserve the structural integrity of CNTs during deposition, ensuring that their unique properties are retained in the final coating. Detonation gun coating, known for its high particle velocities, facilitates strong bonding and uniform distribution of CNTs within the matrix [13].

Figure 7.2 shows the scanning electron microscopy (SEM) observations of various powder and coated samples involved in advanced protective coatings. Panel (a) shows the Cr3C2–25%NiCr powder with its typical granular and rough surface morphology. Panel (b) presents pure CNTs, where entangled web-like networks of nanotubes are visible, highlighting their unique nanoscale structure. Panel (c) illustrates a composite powder consisting of 7% CNTs incorporated into the Cr3C2–25%NiCr matrix, showing modified surface textures and improved particle connectivity owing to CNT integration. Panel (d) shows the coated sample produced using 7 wt.% CNT, revealing a combination of features: melted particles signifying proper deposition, unmelted particles indicating incomplete melting during spraying, as well as cracks and pores that can affect the mechanical integrity and performance of the coating. Together, these SEM images provide a detailed comparison of the microstructural evolution and coating quality owing to the CNT addition and coating processes in the Cr3C2–NiCr systems [14].

Recent progress in incorporating CNTs into composite coatings has focused on creating innovative setups and powder preparation techniques to improve dispersion and reduce structural damage during thermal spraying [15]. Among these methods, chemical vapor deposition (CVD) has emerged as a leading technique because it

FIGURE 7.2 SEM images of coating powders and coated sample: (a) Cr_3C_2–25%NiCr; (b) Pure CNT; (c) 7% CNT in Cr_3C_2–25%NiCr; (d) 7% wt. of the CNT-coated sample [14].

allows for the precise and uniform application of CNTs onto substrates or within powder matrices. CVD involves the introduction of volatile precursor gases into a reaction chamber, where they decompose or react at high temperatures to form solid CNTs directly on the target surfaces. This method provides excellent control over the purity, alignment, and distribution of CNTs, which is essential for achieving consistent reinforcement and optimal properties in the final composite material [16]. Furthermore, high-shear dispersion techniques have been used to tackle the issue of CNT agglomeration, which can undermine the mechanical and functional advantages of CNTs in the coatings. High-shear mixing applies intense mechanical forces to separate CNT bundles and distribute them evenly within the matrix powders, ensuring that the unique properties of individual nanotubes are preserved and effectively transferred to the composite material. These advanced preparation methods are often combined with carefully controlled thermal spray parameters to further minimize the risk of CNT damage from high temperatures or impact forces during the deposition. By integrating such innovative approaches, researchers and engineers can produce CNT-reinforced coatings with superior mechanical strength, electrical conductivity, and wear resistance while maintaining the structural integrity of the nanotubes [17].

Figure 7.3 presents a detailed energy dispersive spectroscopy (EDS) analysis of the intermetallic compounds formed in a (CNT) Inconel 625 matrix. Panel (a) shows

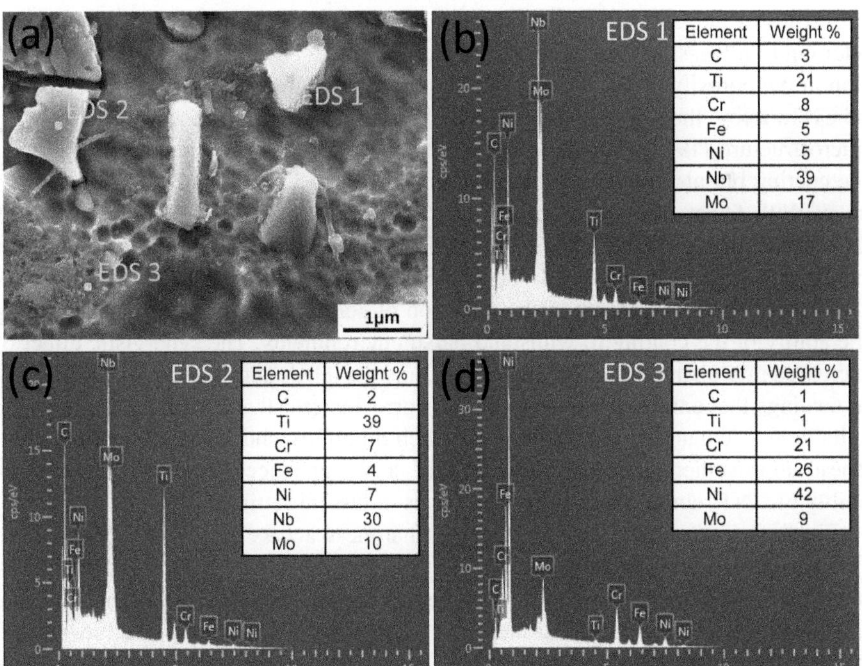

FIGURE 7.3 EDS analysis of intermetallic compounds of TiC-CNTs/Inconel 625. (a) SEM micrograph of the intermetallic compounds of TiC-CNTs/Inconel 625; (b)–(d) EDS spectra of the three points [18].

a scanning electron microscope micrograph with three marked points (EDS 1, EDS 2, and EDS 3) selected for the compositional analysis. Panels (b)–(d) display the EDS spectra and corresponding elemental compositions at these locations. EDS 1 and EDS 2 were rich in titanium and niobium, with significant amounts of chromium, iron, nickel, and molybdenum, indicating the presence of complex intermetallic and carbide phases. EDS 3 is distinguished by a notably high nickel and iron content, with prominent amounts of chromium and molybdenum, suggesting the identification of a matrix region or a nickel-rich intermetallic compound. The overall analysis highlighted substantial variations in elemental distribution across the sample, reflecting the successful integration and microstructural transformations achieved by reinforcing Inconel 625 with TiC and CNTs [18].

7.3 MICROSTRUCTURE AND PROPERTY ENHANCEMENT

Incorporating CNTs into coating materials has been shown to significantly enhance several key properties of the resulting composites. One of the most notable improvements is the increase in microhardness, which is attributed to the exceptional mechanical strength and stiffness of the CNTs. When dispersed within a matrix, CNTs act as effective reinforcement agents, impeding dislocation movement, thereby making the coating harder and more resistant to deformation under applied loads. This enhancement in microhardness is further supported by the elastic recovery properties of the CNTs, which contribute to the overall durability of the coating. In addition to increasing the hardness, the presence of CNTs typically leads to a reduction in the porosity of the coating. The high aspect ratio and excellent dispersibility of CNTs allow them to fill the voids and gaps in the matrix, resulting in a denser microstructure [19]. Lower porosity not only improves the mechanical integrity of the coating but also plays a crucial role in enhancing its wear and corrosion resistance. With fewer pores and defects, the pathways for corrosive agents and wear particles to penetrate the coating are minimized, thereby extending the service life of the material. Furthermore, CNTs themselves possess remarkable chemical stability and can act as barriers to the diffusion of corrosive species, further enhancing the ability of the coating to withstand harsh environments. The synergistic effect of increased microhardness, reduced porosity, and the inherent properties of CNTs leads to coatings that exhibit superior wear and corrosion resistance, making them highly suitable for demanding applications in which durability and longevity are essential. These improvements collectively contribute to the enhanced performance and reliability of CNT-reinforced coatings in various industrial settings [20].

Figure 7.4 shows the variation in microhardness across the cross-section of the ZrO_2-Y_2O_3 coatings reinforced with different concentrations of CNTs. Pure ZrO_2-Y_2O_3 exhibited lower microhardness, whereas the addition of 2 wt. % and 4 wt.% CNT significantly increases hardness, with the 4 wt.% CNT sample achieving the highest values above 1000 Vickers. For all cases, the hardness increased sharply near the interface and remained elevated throughout the coating. This demonstrates that CNT reinforcement effectively enhances the mechanical strength and hardness of ZrO_2-Y_2O_3 ceramic coatings [21].

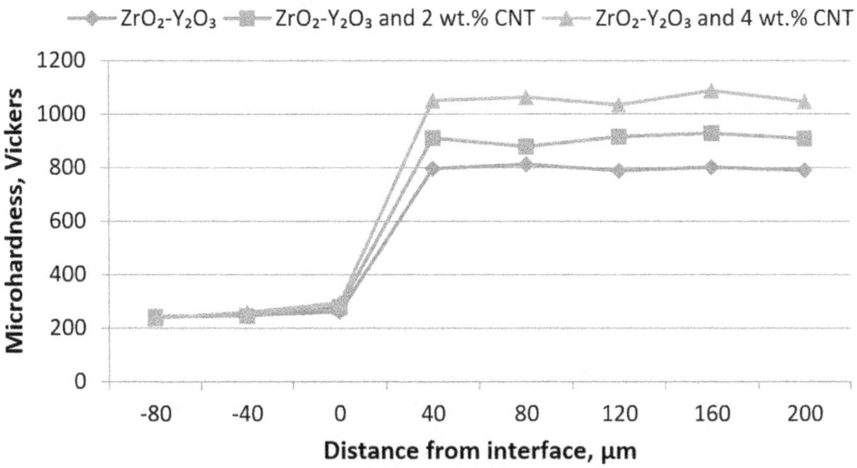

FIGURE 7.4 Microhardness of CNT-ZrO_2-Y_2O_3 reinforced coatings across the cross-section [21].

Increasing the content of CNTs in ZrO_2-Y_2O_3 or YSZ coatings significantly affects the microstructure and performance of these advanced ceramic coatings. As more CNTs are incorporated, the hardness of the coating increases, primarily because the CNTs act as strong reinforcement agents within the ceramic matrix. Their exceptional mechanical properties impede the movement of dislocations and cracks, resulting in a tougher and more resilient surface [22]. Simultaneously, the addition of CNTs reduce the porosity. The high aspect ratio and excellent dispersibility of CNTs allow them to fill microvoids and gaps within the ceramic structure, creating a denser and more uniform coating. This densification not only contributes to the increased hardness but also plays a crucial role in enhancing the resistance of the coating to sliding wear. With fewer pores and a more robust matrix, the coating can better withstand the mechanical stresses and frictional forces encountered during sliding contact, leading to significantly improved wear resistance. Furthermore, the reduction in porosity and the inherent chemical stability of CNTs combine to provide superior corrosion protection [23]. The denser microstructure acts as a more effective barrier against the ingress of corrosive agents, and the CNTs themselves are resistant to chemical attack, further safeguarding the underlying material. Consequently, coatings with higher CNT content exhibit a synergistic improvement in hardness, wear resistance, and corrosion protection, making them highly desirable for applications in harsh environments where durability and longevity are essential. This combination of properties is particularly valuable in industries such as aerospace, energy, and automotive, where advanced coatings are required to protect components from mechanical and chemical degradation [24].

Figure 7.5 illustrates the three distinct fracture mechanisms that occur in hybrid carbon fibre (CF) and CNT-reinforced epoxy composites under tensile loading. Type 1 demonstrates the failure of composites with full CNT reinforcement, where the

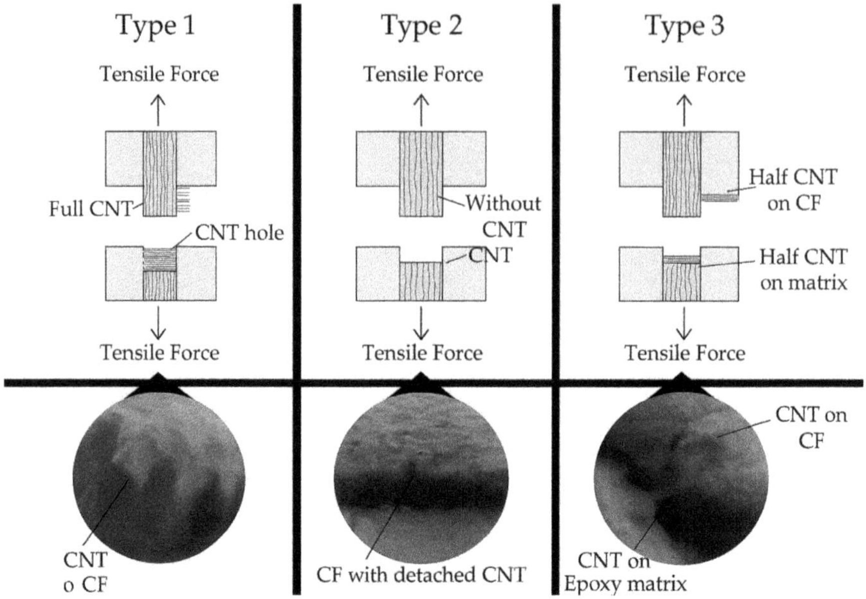

FIGURE 7.5 Schematic of hybrid CF-CNT fracture or pullout from the epoxy matrix [25].

entire CNT-CF structure experiences a tensile fracture with corresponding CNT holes formed in the matrix. Type 2 shows the behaviour of composites without CNT reinforcement, where the carbon fibre experiences detachment from the matrix without CNT involvement, resulting in CF with detached CNTs visible on the fracture surface. Type 3 represents a hybrid configuration with half of the CNTs bonded to the carbon fibre and half to the matrix, creating a more complex failure pattern in which CNTs remain partially attached to both the fibre and matrix during fracture. The corresponding microscopy images below each schematic reveal the actual fracture surfaces, confirming the different failure modes and demonstrating how the CNT distribution affects the mechanical response and energy absorption mechanisms of the composite during tensile loading [25].

Figure 7.6 shows that the atomic force microscopy (AFM) images reveal significant surface topographical changes in epoxy films with varying carbon nanotube (CNT) loadings. Sample (A), EA-0, displayed a relatively smooth and uniform surface characteristic of pure epoxy with minimal height variation, appearing as a dark brown baseline. Sample (B) EPBCNT-4% exhibited moderate surface roughness with scattered bright features indicating CNT aggregation and distribution throughout the matrix. Sample (C), EAJ-0, shows another control sample with a smooth topography similar to EA-0, confirming the baseline epoxy surface characteristics. Sample (D) ECB-CNT-4% demonstrated the most pronounced surface roughness with numerous bright protrusions and valleys, indicating extensive CNT network formation and potential agglomeration. The height scale bars ranging from 0 to 100 nm reveal that CNT incorporation significantly increases the surface roughness compared to pure epoxy systems. These topographical variations directly correlate with the

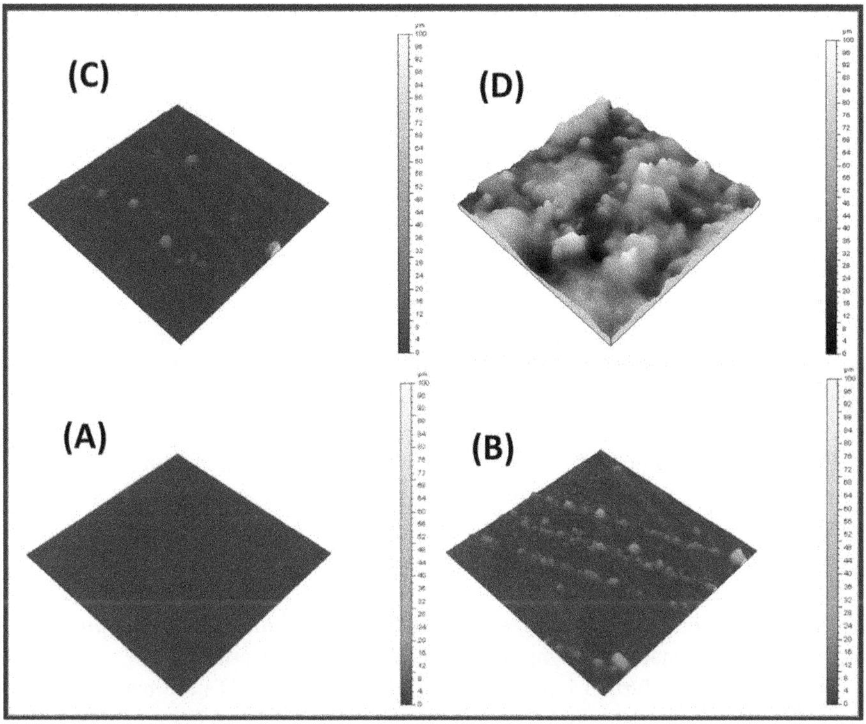

FIGURE 7.6 AFM images of the cured epoxy films with different CNT loadings: (A) EA-0, (B) EPBCNT-4%, (C) EAJ-0, and (D) ECB-CNT-4% composites [26].

CNT dispersion quality and interfacial interactions, which ultimately influence the mechanical and physical properties of the composite materials [26].

Carbon nanotubes are essential for preventing grain growth in composite materials, significantly affecting the microstructure and mechanical properties of both coatings and bulk materials. When materials undergo high-temperature processing, their grains tend to enlarge to minimize the total grain boundary area, thereby reducing the internal energy of the system. However, excessive grain enlargement can result in a coarser microstructure, which typically weakens the mechanical strength and toughness. By incorporating CNTs into the matrix, this process is disrupted, as they act as physical barriers at the grain boundaries, effectively pinning them and preventing excessive grain merging and growth. This process, known as grain boundary pinning, leads to a refined microstructure with smaller and more evenly distributed grains [27]. The reduction in grain size is advantageous because it increases the number of grain boundaries, which hinders dislocation movement and crack propagation. With more grain boundaries, cracks face greater resistance as they attempt to move through the material, making it harder for them to grow and spread. This resistance to crack propagation enhances the fracture toughness and overall durability of the material. Furthermore, CNTs at the grain boundaries can absorb and dissipate the energy associated with crack initiation and growth, further boosting the resistance of the material to mechanical failure. Consequently, CNT-reinforced composites not only show improved hardness and strength due to their refined microstructure but also exhibit superior resistance to cracking and catastrophic failure, making them highly desirable for applications where mechanical reliability is crucial [28].

7.4 CHALLENGES: DISPERSION, AGGLOMERATION, AND SCALE-UP

Ensuring an even distribution of CNTs within coatings is a major challenge, largely because of their strong propensity to clump together. This clumping is caused by the powerful van der Waals forces between the nanotubes, which lead them to form clusters instead of spreading uniformly throughout the matrix. When CNTs aggregate in this manner, their outstanding mechanical, electrical, and chemical properties are not fully harnessed, and the overall performance of the coating can be significantly impaired. Clumped CNTs can create weak spots in the coating, resulting in increased porosity, decreased hardness, and reduced wear and corrosion resistance. Additionally, these clusters can disrupt the uniformity of the microstructure, leading to inconsistent properties across the coated surface [29].

Table 7.1 demonstrates significant progress in CNT-reinforced composites while highlighting the ongoing challenges in maintaining CNT integrity during processing and achieving optimal dispersion throughout the matrix materials.

This challenge is further complicated by the fact that CNTs are naturally insoluble or only very poorly soluble in most common solvents, making it challenging to achieve a stable and uniform dispersion using standard mixing methods. Consequently, specialized techniques such as high-shear mixing, ultrasonication, or the use of surfactants and functionalization strategies are often necessary to overcome these challenges and encourage better separation of individual nanotubes (NTs). However, even with these advanced methods, maintaining a stable dispersion throughout the entire coating process remains challenging, and any re-clumping during deposition can negate the intended improvement. Therefore, the search for reliable and scalable methods to achieve uniform CNT dispersion continues, as it is crucial for unlocking the full potential of CNT-reinforced coatings in demanding applications [40].

Agglomeration, which refers to the clustering or clumping of particles within a material, can significantly impact the microstructure and mechanical properties of engineered materials. When particles agglomerate, they create regions of inhomogeneity within the microstructure, indicating that the distribution of particles is uneven throughout the material. This inhomogeneity leads to localized variations in properties, such as strength, ductility, and toughness, often resulting in weak points where failure is more likely to initiate under stress [41]. The presence of agglomerates disrupts the uniform transfer of load across the material, causing stress concentrations that reduce the overall mechanical performance of the material. Furthermore, these microstructural inconsistencies can hinder the ability of the material to withstand mechanical and environmental challenges, ultimately decreasing its reliability and lifespan. This challenge becomes even more pronounced when scaling up production for industrial applications. Maintaining consistent quality and uniform dispersion of particles across large batches is

TABLE 7.1
Overview of coated CNTs and their use in various composites

Composition	Fabrication method	Coating method	Limitations
Al-CNT nanocomposite	Plasma spray forming	Plasma thermal spray deposition	CNT structural damage during high-temperature processing [30]
Metal/ceramic-CNT coatings	Plasma/HVOF/cold spray	Multi-process thermal spray coating	CNT vaporization in liquid precursor plasma spray [31]
WC-Co-CNT hybrid powder	Sonication + internal diameter thermal spray	HVAF thermal spray with CNT retention	Limited CNT retention at very high spray powers [32]
Al-0.5 wt% CNT composite	Solution mixing vs ball milling	Ethanol-based solution mixing	CNT structural deformation in the ball milling process [33]
Polycarbonate-MWCNT composite	Supercritical fluid treatment (RESS+SAS)	Rapid expansion + antisolvent precipitation	Complex two-step supercritical processing required [34]
Polydopamine/CNT nanocomposite	Layer-by-layer assembly	Polydopamine surface modification	Uncontrollable CNT length and functional groups [35]
AZ31-CNT (0.5-1.5 wt%)	Powder metallurgy + sintering	Compacting and sintering process	CNT agglomeration due to Van der Waals forces [36]
Si_3N_4-MWCNT ceramic composite	Hot isostatic pressing (HIP)	High-pressure sintering technique	Difficulty achieving uniform CNT dispersion [37]
Al_2O_3/ZrO_2/SiC-CNT composites	Heterocoagulation + colloidal processing	Electrostatic heterocoagulation	Process complexity and scalability issues [38]
Nickel-coated CNT reinforcement	Electroless nickel plating + powder metallurgy	Chemical nickel coating on the CNT surface	An additional coating step increases processing cost [39]

difficult because the tendency for particles to agglomerate increases with volume and process complexity. This makes it difficult to replicate the desirable properties observed in small-scale laboratory samples on an industrial scale, where even minor variations in processing conditions can lead to significant differences in microstructure and, consequently, mechanical properties [42]. Consequently, industries face significant obstacles in ensuring that every batch of material meets stringent performance standards, which is critical for applications where safety and durability are paramount. Overcoming agglomeration and achieving homogeneous dispersion thus remain key challenges in the development and large-scale production of advanced materials, directly influencing their commercial viability and application in demanding engineering environments [43].

7.5 APPLICATIONS AND MULTIFUNCTIONALITY

Coatings reinforced with CNTs have attracted considerable interest owing to their outstanding mechanical, thermal, and chemical characteristics, making them ideal for challenging applications. In high-wear environments, such as those involving industrial machinery and automotive parts, CNTs improve the wear resistance and longevity of coatings by offering exceptional hardness and reducing friction, which helps decrease material loss and prolong the lifespan of components [44]. As thermal barriers, CNT-reinforced coatings provide enhanced thermal stability and insulation, which is vital for safeguarding underlying materials in high-temperature environments, such as turbines and engines, where efficiency and durability are critical. Moreover, these coatings act as effective corrosion protection layers, particularly for metals exposed to harsh or corrosive conditions, by creating a strong barrier that restricts the penetration of corrosive agents, thereby maintaining the substrate's integrity. In the biomedical sector, CNT-reinforced coatings are being investigated for implants because of their biocompatibility, mechanical strength, and ability to reduce wear and corrosion of metallic implants. These attributes are especially important in medical devices, such as orthopedic and dental implants, where durability and resistance to bodily fluids are crucial for patient safety and implant performance. Additionally, the unique structure of CNTs can be customized to provide antimicrobial properties, thereby lowering the risk of infection associated with implanted devices [45].

Figure 7.7 illustrates the comprehensive pathway from carbon nanotube synthesis to diverse industrial applications. Single-walled carbon nanotubes (SWCNTs) and multi-walled carbon nanotubes (MWCNTs) undergo functionalization processes involving covalent and non-covalent modifications to enhance their compatibility with various polymer matrices. Figure 7.7 shows multiple polymer systems, including polyacrylate composites, polylactic acid, polystyrene, epoxy, and other advanced materials. Various fabrication techniques, such as melt mixing, solution mixing, sonication, resin transfer moulding, and in situ polymerization, have been employed to create CNT-polymer composites. These processed materials find extensive applications across multiple industries: solar cells benefit from enhanced conductivity, textiles gain improved mechanical properties, electronics utilize superior electrical characteristics, energy storage devices leverage high surface areas, automotive

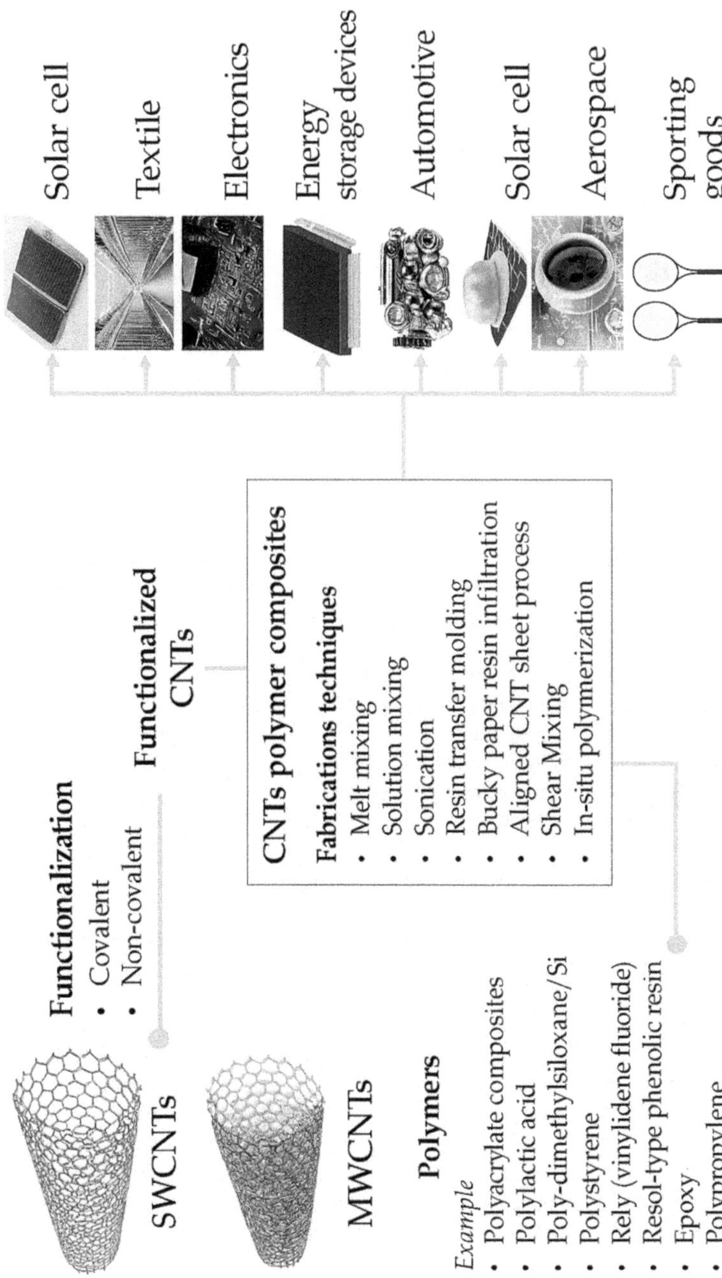

FIGURE 7.7 Single-walled carbon nanotube (SWCNT) and multi-walled carbon nanotube (MWCNT) applications [46].

components achieve lightweight strength, aerospace applications exploit thermal stability, and sporting goods incorporate enhanced durability. This demonstrates the versatility of carbon nanotubes in creating next-generation composite materials for advanced technological applications [46].

As shown in Table 7.2, these studies collectively highlight the critical importance of proper CNT dispersion, interface engineering through coating strategies, and optimized processing parameters to achieve superior composite performance for demanding applications across the aerospace, automotive, defence, and advanced materials sectors.

Their multifunctional properties, such as enhanced electrical and thermal conductivity, open new avenues for smart coatings and integrated sensing applications.

TABLE 7.2
Applications of the CNT optimized processing parameters

Matrix/process	Key findings	Application
High volume CNT/polymer composite	Tensile properties approaching IM7/8552 composites with ~700 MPa/(g/cm³)	Aerospace structures with mass savings and fuel efficiency [47]
AZ31 magnesium/CNT (0.5-1.5 wt%)	0.5 wt% CNT showed optimal corrosion resistance (2.138e-5 g/h)	Lightweight materials for aerospace, defence, and transportation [48]
Al-CNT with h-BN/Ag-Ni coating	Ag/Ni coating enhanced wettability and reduced agglomeration	Advanced ball bearings and high-performance components [49]
CNT/polymer nanocomposites	Synergistic effects in hybrid composites with superior electrical properties	Aerospace, automotive, construction, sensors, and medical devices [50]
CNT/epoxy thermal cycling	Thermal cycling is critical for the performance of aerospace applications performance	Aircraft components under thermal cycling conditions [51]
CNT/polymer functional composites	Enhanced mechanical strength, electrical, and thermal conductivity	Structural health monitoring, wearable technology, and biomedical [52]
CNT-reinforced aluminium coating	Improved corrosion resistance and mechanical strength in coatings	Protective coatings for corrosion resistance applications [44]
Copper-coated CNT/matrix	Cu-coated CNTs improved wettability and interface bonding	Electronic components and conductive material applications [53]
Cobalt-coated CNT/aluminium	Enhanced lightweight properties with good corrosion resistance	Aerospace applications requiring durability and long-term performance [54]
Multiscale CNT/FRP composites	Multiscale reinforcement enhanced electrical and tensile properties	Aerospace wingtips (F-35), rocket motor cases, spacecraft components [55]

7.6 FUTURE DIRECTION

The future direction of CNT-reinforced thermal spray composite coatings is focused on optimizing their performance for advanced industrial applications by enhancing their mechanical, thermal, and functional properties. Research is increasingly aimed at improving the uniform dispersion of CNTs within the coating matrix, as this is crucial for maximizing their reinforcing effects and ensuring consistent properties throughout the coating. Advances in feedstock preparation and spray process parameters are expected to play a significant role in achieving this goal, enabling the development of coatings with superior wear resistance, reduced friction, and enhanced thermal stability [56]. Additionally, there is growing interest in tailoring these coatings for multifunctional purposes, such as integrating self-sensing or self-healing capabilities, which could lead to smart coatings that monitor their own conditions or autonomously repair minor damage. The exploration of new matrix materials and hybrid composites combined with CNTs is anticipated to expand the range of applications, particularly in sectors such as aerospace, energy, and biomedical engineering, where lightweight, high-strength, and corrosion-resistant coatings are in high demand. As thermal spray technology continues to evolve, the integration of CNTs is expected to unlock new possibilities for high-performance, durable, and intelligent surface coatings [57].

Figure 7.8 details the fabrication, morphology, and microstructural characterization of a CNT-reinforced aluminium-silicon (AlSi) composite with a nacre-like, layered architecture. The schematic in panel (a) outlines the process: co-milling Al alloy powder and multi-walled CNTs forms a flaky composite powder, which is fed into a cold spray system to deposit onto a substrate and develop a nanolaminate structure typical of natural nacre. The scanning electron microscopy images in (b) and (c) display the flaky morphology and layered cross-sectional structure of the composite particles. Panel (d) highlights a magnified region, showing the uniform dispersion of CNTs within the matrix. Transmission electron microscopy (TEM) bright-field (BF) imaging in (e) reveals the nanolaminated architecture, with red arrows indicating the presence of CNTs in the nanolaminated architecture. The high-resolution TEM (HRTEM) image in (f) shows individual CNTs embedded within the aluminium matrix, confirming intimate nanoscale integration. The orientation map in (g), obtained by TEM/EBSD, illustrates the Al grain orientation and distribution, supporting the nanostructured design. The histogram in (h) provides the Al grain size statistics, while (i) compares the particle size distribution curves for the composite and pure Al powders, demonstrating the successful modification of the composite particle morphology and size [58].

CNT-reinforced thermal spray coatings offer substantial improvements in mechanical, thermal, and corrosion properties; however, their full potential is limited by persistent challenges in nanoparticle dispersion, agglomeration, and process scalability [59].

The biomedical and high-temperature corrosion-resistant systems slurry processing industries have started to use thermal spray coatings and novel surface engineering techniques. This is particularly true when combined with emerging technologies like additive manufacturing, magnetron sputtering thin-film deposition, and microwave-assisted fabrication [60–73]. Moreover, the coating structures exhibit strong (tensile, compressive, and flexural) and functional performance in solar energy conversion

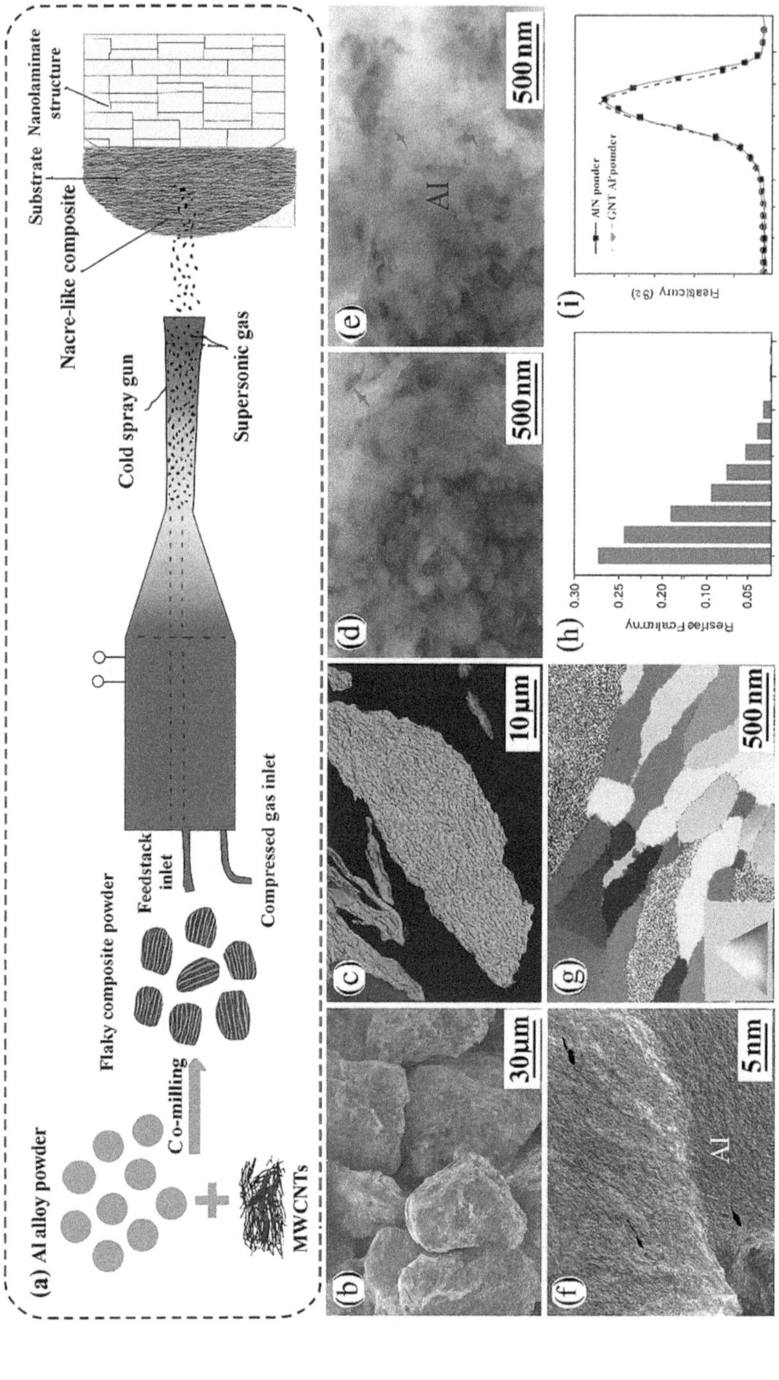

FIGURE 7.8 Fabrication of a CNT/AlSi composite with a self-aligned nacre-like architecture by cold spraying [58].

coating systems, as well as significant potential in surface topography. Furthermore, the designed surfaces serve as a foundation for sophisticated computational fluid dynamics of pipes and conduits with bends and ergonomic seating systems that incorporate anthropometric, mechanical, and structural connection [74–86].

7.7 CONCLUSION

This paper presents a literature review on CNTs. CNT-reinforced thermal spray composite coatings enhance mechanical, thermal, and corrosion properties of industrial components. CNTs have been incorporated into matrices like WC-Co, ZrO_2-Y_2O_3, Al_2O_3, and Cr_3C_2-NiCr using HVOF, plasma spraying, cold spray, and detonation gun coating. Novel fabrication methods that break the limits of current fabrication methods are strongly anticipated. To achieve superior mechanical and physical properties for the composite, the microstructure and interface need to be considered. CNTs affect microstructural properties, such as improving microhardness, wear resistance, and corrosion protection, by reducing porosity, refining the microstructure, and providing nanoscale reinforcement. Challenges persist in achieving uniform CNT dispersion, managing agglomeration, and scaling up production while preserving coating integrity and performance. Future research should focus on innovative powder preparation methods, optimizing spray parameters, and exploring CNTs' multifunctional properties of CNTs for smart coating and sensing applications. The long-term stability, environmental effects, and cost-effectiveness of CNT-reinforced thermal spray coatings require further advancements for successful industrial implementation. Continued research on process optimization, long-term performance, and new applications is essential to fully realize the industrial potential of CNT-reinforced thermal spray coatings. This chapter has reviewed the use of CNTs in thermal spray coatings. These coatings improve the mechanical, thermal, and corrosion properties of industrial components. CNTs have been added to various materials using different spraying techniques. This chapter has highlighted the need for new manufacturing methods to overcome these limitations, and emphasizes the importance of considering the microstructure of the coating and the interaction of CNTs with the base material. CNTs can improve the hardness, wear resistance, and corrosion protection of a material by reducing gaps, refining its structure, and providing reinforcement at a very small scale. However, challenges remain in evenly distributing CNTs, preventing clumping, and scaling up production while maintaining coating quality. Future research should focus on new ways to prepare powders, optimize spraying conditions, and explore the potential of CNTs for smart coatings and sensors. In addition, more work is needed on the long-term stability, environmental effects, and cost-effectiveness of these coatings for successful industrial use. Ongoing research in these areas is crucial to fully realize the potential of CNT-reinforced thermal-spray coatings in the industry.

REFERENCES

[1] Afrasiabi A. Application of plasma sprayed YSZ-AL2O3 composite TBC on Ni-based Super Alloy (IN-738) and evaluation of the microstructure and its thermal stability. 2008.

[2] Ahmadi MS, Shoja-Razavi R, Valefi Z, Jamali H. Laser surface modification of air plasma-sprayed Al_2O_3/YSZ multilayer thermal barrier coating to improve hot corrosion resistance in V_2O_5–Na_2SO_4 salts. *Journal of Thermal Spray Technology*. 2019;28:1906–18.

[3] Iijima S. Carbon nanotubes: Past, present, and future. *Physica B: Condensed Matter*. 2002;323:1–5.

[4] Nurguzhin M, Janikeyev M, Omarbayev M, Yermakhanova A, Meiirbekov M, Zhumakhanov M, et al. Structure and properties of Al–CNT-based composites manufactured by different methods: A brief review. *Materials*. 2025;18:214.

[5] Venturi F, Kamnis S, Hussain T. Internal diameter HVOAF thermal spray of carbon nanotubes reinforced WC-Co composite coatings. *Materials & Design*. 2021;202:109566.

[6] Wei H, Li Z, Xiong D-B, Tan Z, Fan G, Qin Z, et al. Towards strong and stiff carbon nanotube-reinforced high-strength aluminium alloy composites through a microlaminated architecture design. *Scripta Materialia*. 2014;75:30–3.

[7] Mohammed SMAK, Chen DL. Carbon nanotube-reinforced aluminium matrix composites. *Advanced Engineering Materials*. 2020;22:1901176.

[8] Anusha K, Routara BC, Guha S. A review on high-velocity oxy-fuel (HVOF) coating technique. *Journal of The Institution of Engineers: Series D*. 2023;1049(2):1–18.

[9] Ahmadi-Pidani R, Shoja-Razavi R, Mozafarinia R, Jamali H. Evaluation of hot corrosion behavior of plasma sprayed ceria and yttria stabilized zirconia thermal barrier coatings in the presence of Na2SO4+ V2O5 molten salt. *Ceramics International*. 2012;38:6613–20.

[10] Aparco RH, Tapia-Tadeo F, Ascarza YB, Ramírez AL, Huamán-Romaní Y-L, Otero CC. A machine learning approach for analyzing residual stress distribution in cold spray coatings. *Journal of Thermal Spray Technology*. 2024;33:1292–307.

[11] Babu A, Arora HS, Grewal HS. Microwave-assisted post-processing of detonation gun-sprayed coatings for better slurry and cavitation erosion resistance. *Journal of Thermal Spray Technology*. 2019;28:1565–78.

[12] Jiang L, Li Z, Fan G, Cao L, Zhang D. The use of flake powder metallurgy to produce carbon nanotube (CNT)/aluminium composites with a homogenous CNT distribution. *Carbon*. 2012;50:1993–8.

[13] Ozgurluk Y, Gulec A, Ozkan D, Binal G, Karaoglanli AC. Structural characteristics, oxidation performance and failure mechanism of thermal barrier coatings fabricated by atmospheric plasma spraying and detonation gun spraying. *Engineering Failure Analysis*. 2023;152:107499.

[14] Kuruba M, Giridhara G, Jagadeeswaran N. Oxidation performance of CNT-reinforced Cr3C2-NiCr coatings sprayed by HVOF method. *Journal of Mines, Metals and Fuels*. 2022;69:214–23.

[15] Choudhary M, Sharma A, Aravind Raj S, Sultan MTH, Hui D, Shah AUM. Contemporary review on carbon nanotube (CNT) composites and their impact on multifarious applications. *Nanotechnology Reviews*. 2022;11:2632–60.

[16] Chu Z, Xu B, Liang J. Direct application of carbon nanotubes (CNTs) grown by chemical vapor deposition (CVD) for integrated circuits (ICs) interconnection: Challenges and developments. *Nanomaterials*. 2023;13:2791.

[17] Yan Y, Miao J, Yang Z, Xiao F-X, Yang HB, Liu B, et al. Carbon nanotube catalysts: Recent advances in synthesis, characterization and applications. *Chemical Society Reviews*. 2015;44:3295–346.

[18] Zhou S, Xu T, Hu C, Wu H, Liu H, Ma X. Utilizing carbon nanotubes in ceramic particle reinforced MMC coatings deposited by laser cladding with Inconel 625 wire. *Journal of Materials Research and Technology*. 2021;13:2026–42.

[19] Saba F, Nateq B, Sajjadi SA, Zhang F, Heydari S. The enhanced mechanical properties and strain-hardening capability of CNT/Al composites achieved by heterogeneous micro-laminated architecture. *Composites Communications*. 2021;27:100861.

[20] Esawi AMK, Morsi K, Sayed A, Taher M, Lanka S. The influence of carbon nanotube (CNT) morphology and diameter on the processing and properties of CNT-reinforced aluminium composites. *Composites Part A: Applied Science and Manufacturing*. 2011;42:234–43.

[21] Kumar S, Bhatia R, Singh H, Virdi RL. Microstructural and mechanical properties of CNT-reinforced ZrO2-Y2O3 coated boiler tube steel T-91: Original scientific paper. *Journal of Electrochemical Science and Engineering*. 2022;12:877–88.

[22] Xu R, Tan Z, Xiong D, Fan G, Guo Q, Zhang J, et al. Balanced strength and ductility in CNT/Al composites achieved by flake powder metallurgy via shift-speed ball milling. *Composites Part A: Applied Science and Manufacturing*. 2017;96:57–66.

[23] Kim D, Hirayama Y, Liu Z, Takagi K, Kobashi M. Fabrication of Al-CNT composite with high hardness and electrical conductivity by controlling Al4C3 formation. *Journal of Alloys and Compounds*. 2023;942:169102.

[24] Pan Y, Xiao S, Lu X, Zhou C, Li Y, Liu Z, et al. Fabrication, mechanical properties and electrical conductivity of Al2O3 reinforced Cu/CNTs composites. *Journal of Alloys and Compounds*. 2019;782:1015–23.

[25] Zakaria MR, Md Akil H, Omar MF, Abdul Kudus MH, Mohd Sabri FNA, Abdullah MMAB. Enhancement of mechanical and thermal properties of carbon fiber epoxy composite laminates reinforced with carbon nanotubes interlayer using electrospray deposition. *Composites Part C: Open Access*. 2020;3:100075.

[26] Alhendal A. Enhanced thermomechanical properties of epoxy-multiwalled CNT nanocomposites. *RSC Advances*. 2024;14:35360–72.

[27] Chen B, Li S, Imai H, Jia L, Umeda J, Takahashi M, et al. Carbon nanotube induced microstructural characteristics in powder metallurgy Al matrix composites and their effects on mechanical and conductive properties. *Journal of Alloys and Compounds*. 2015;651:608–15.

[28] Peng YZ, You X, Li CJ, Yang CMY, Xu ZY, Lu Q, et al. Strength-ductility synergy in a novel carbon nanotube-high entropy alloy co-reinforced aluminium matrix composite. *Composites Part A: Applied Science and Manufacturing*. 2024;181:108116.

[29] Fan G, Jiang Y, Tan Z, Guo Q, Xiong D-b, Su Y, et al. Enhanced interfacial bonding and mechanical properties in CNT/Al composites fabricated by flake powder metallurgy. *Carbon*. 2018;130:333–9.

[30] Mohammed S, Chen DL, Liu ZY, Ni DR, Wang QZ, Xiao BL, et al. Deformation behavior and strengthening mechanisms in a CNT-reinforced bimodal-grained aluminium matrix nanocomposite. *Materials Science and Engineering: A*. 2021;817:141370.

[31] Kumari JN, Jegadeeswaran N. Development of CNT mixed HVOF coating for water tube boiler material to improve the degradations. *Materials Today: Proceedings*. 2022;54:444–7.

[32] Mohanty D, Kar S, Paul S, Bandyopadhyay PP. Carbon nanotube reinforced HVOF sprayed WC-Co coating. *Materials & Design*. 2018;156:340–50.

[33] Shahrdami L, Sedghi A, Shaeri MH. Microstructure and mechanical properties of Al matrix nanocomposites reinforced by different amounts of CNT and SiCW. *Composites Part B: Engineering*. 2019;175:107081.

[34] Guo J, Liu Y, Prada-Silvy R, Tan Y, Azad S, Krause B, et al. Aspect ratio effects of multi-walled carbon nanotubes on electrical, mechanical, and thermal properties of polycarbonate/MWCNT composites. *Journal of Polymer Science Part B: Polymer Physics*. 2014;52:73–83.

[35] Zou R, Liu F, Hu N, Ning H, Gong Y, Wang S, et al. Graphene/graphitized polydopamine/carbon nanotube all-carbon ternary composite films with improved mechanical properties and through-plane thermal conductivity. *ACS Applied Materials & Interfaces*. 2020;12:57391–400.
[36] Sharma SK, Saxena KK, Salem KH, Mohammed KA, Singh R, Prakash C. Effects of various fabrication techniques on the mechanical characteristics of metal matrix composites: A review. *Advances in Materials and Processing Technologies*. 2024;10:277–94.
[37] Zagyva T, Balázsi K, Balázsi C. Examination of novel electrosprayed biogenic hydroxyapatite coatings on Si3N4 and Si3N4/MWCNT ceramic composite. *Processing and Application of Ceramics*. 2019;13:132–8.
[38] Poyato R, Morales-Rodríguez A, Gutiérrez-Mora F, Muñoz A, Gallardo-López Á. Effect of acid-treatment and colloidal-processing conditions on the room temperature mechanical and electrical properties of 3YTZP/MWNT ceramic nanocomposites. *Ceramics International*. 2017;43:16560–8.
[39] Jagannatham M, Sankaran S. Electroless nickel plating of arc discharge synthesized carbon nanotubes for metal matrix composites. *Applied Surface Science*. 2015;324:475–81.
[40] Zhou W, Zhou Z, Dong M, Fan Y, Shirasu K, Yamamoto G, et al. Enhanced interfacial strength in carbon-nanotubes-reinforced Al matrix composites via an interface substitution strategy. *Composites Part A: Applied Science and Manufacturing*. 2025;195:108955.
[41] Rong X, Chen X, Zhao D, Zhang X, He C, Shi C, et al. High mechanical strengthened CNTs/Al composite concepts with robust interface and intragranular reinforcement achieved via interfacial thermite reaction. *Composites Part A: Applied Science and Manufacturing*. 2023;173:107630.
[42] Yuan C, Tan Z, Fan G, Chen M, Zheng Q, Li Z. Fabrication and mechanical properties of CNT/Al composites via shift-speed ball milling and hot-rolling. *Journal of Materials Research*. 2019;34:2609–19.
[43] Liu L, Li S, Zhang X, Pan D, Gao L, Chen B, et al. Syntheses, microstructure evolution and performance of strength-ductility matched aluminium matrix composites reinforced by nano SiC-cladded CNTs. *Materials Science and Engineering: A*. 2021;824:141784.
[44] Tajzad I, Ghasali E. Production methods of CNT-reinforced Al matrix composites: A review. *Journal of Composites and Compounds*. 2020;2:1–9.
[45] Sharma S, Patyal V, Sudhakara P, Singh J, Petru M, Ilyas RA. Mechanical, morphological, and fracture-deformation behavior of MWCNTs-reinforced (Al–Cu–Mg–T351) alloy cast nanocomposites fabricated by optimized mechanical milling and powder metallurgy techniques. *Nanotechnology Reviews*. 2021;11:65–85.
[46] Mohd NN, Asyraf MRM, Khalina A, Abdullah N, Sabaruddin FA, Kamarudin SH, et al. Fabrication, functionalization, and application of carbon nanotube-reinforced polymer composite: An overview. *Polymers*. 2021;13(7):1047.
[47] Zhang P-X, Xu W-M, Zeng M, Lei Y-S, Yan H. Regulating microstructure, mechanical properties and electrochemical characteristic of 2024-CNTs aluminium composites via decorating nano Ni on the surface of CNTs. *Diamond and Related Materials*. 2022;126:109132.
[48] Han G, Wang Z, Liu K, Li S, Du X, Du W. Synthesis of CNT-reinforced AZ31 magnesium alloy composites with uniformly distributed CNTs. *Materials Science and Engineering: A*. 2015;628:350–7.
[49] Barakat WS, Samir A, Elkady OA, Abu-Okail M, Mohamed AYA, El-Nikhaily A, et al. Improved properties of hybrid Al-CNTs via h-BNs coated with ag and ni for ball bearings. *Scientific Reports*. 2025;15:1251.

[50] Hassanzadeh-Aghdam MK, Mahmoodi MJ, Ansari R. Creep performance of CNT polymer nanocomposites: An emphasis on viscoelastic interphase and CNT agglomeration. *Composites Part B: Engineering*. 2019;168:274–81.

[51] Gardea F, Lagoudas DC. Characterization of electrical and thermal properties of carbon nanotube/epoxy composites. *Composites Part B: Engineering*. 2014;56:611–20.

[52] Kausar A, Rafique I, Muhammad B. Review of applications of polymer/carbon nanotubes and epoxy/CNT composites. *Polymer-Plastics Technology and Engineering*. 2016;55:1167–91.

[53] Nazem Salimi M, Alizadeh Sahraei A, Baniassadi M, Abrinia K, Ehsani M. Synergistic effect of carbon nanotubes and copper particles in an epoxy-based nanocomposite using electroless copper deposited carbon nanotubes: Part I–Mechanical properties. *Journal of Composite Materials*. 2016;50:1909–20.

[54] Iyengar AS, Suresh R. Spark plasma sintering of Al_2O_3 reinforced aluminium alloy metal matrix composites. *Powder Metallurgy and Metal Ceramics*. 2024;62:536–54.

[55] Li J, Zhang Z, Fu J, Liang Z, Ramakrishnan KR. Mechanical properties and structural health monitoring performance of carbon nanotube-modified FRP composites: A review. *Nanotechnology Reviews*. 2021;10:1438–68.

[56] Zhang Z, Huang Y, Zhang P, Wan C. Outstanding EMI shielding properties of Al2O3 reinforced with sandwiched graphene/CNTs. *Journal of the European Ceramic Society*. 2023;43:4082–7.

[57] Salama EI, Morad SS, Esawi AMK. Fabrication and mechanical properties of aluminium-carbon nanotube functionally-graded cylinders. *Materialia*. 2019;7:100351.

[58] Xie X, Chen C, Ji G, Xu R, Tan Z, Xie Y, et al. A novel approach for fabricating a CNT/AlSi composite with the self-aligned nacre-like architecture by cold spraying. *Nano Materials Science*. 2019;1:137–41.

[59] Sang G, Wang C, Zhao Y, He G, Zhang Q, Yang M, et al. Ni@ CNTs/Al2O3 ceramic composites with interfacial solder strengthen the segregated network for high toughness and excellent electromagnetic interference shielding. *ACS Applied Materials & Interfaces*. 2022;14:4443–55.

[60] Dutta V, Thakur L, Singh B, Vasudev H. A study of erosion–corrosion behaviour of friction stir-processed chromium-reinforced NiAl bronze composite. *Materials*. 2022;15(15):5401.

[61] Singh M, Vasudev H, Kumar R. Corrosion and tribological behaviour of BN thin films deposited using magnetron sputtering. *International Journal of Surface Engineering and Interdisciplinary Materials Science (IJSEIMS)*. 2021;9(2):24–39.

[62] Sharanabasava H, Raviprakash M, Prasad CD, Ramesh MR, Phanibhushana MV, Vasudev H, Kumar S. Microstructure, mechanical and wear properties of SiC and Mo reinforced NiCr microwave cladding. *Advanced Materials and Processing Technologies*. 2023;10(4):3620–33.

[63] Singh M, Vasudev H, Singh M. Surface protection of SS-316L with boron nitride based thin films using radio frequency magnetron sputtering technique. *Journal of Electrochemical Science and Engineering*. 2022;12(5):851–63.

[64] Prashar G, Vasudev H, Thakur L, Bansal A. Performance of thermally sprayed hydroxyapatite coatings for biomedical implants: A comprehensive review. *Surface Review and Letters*. 2023;30(1):2241001.

[65] Vasudev H. Wear characteristics of Ni-WC powder deposited by using a microwave route on mild steel: Microwave cladding of Ni-WC. *International Journal of Surface Engineering and Interdisciplinary Materials Science (IJSEIMS)*. 2020;8(1):44–54.

[66] Parkash J, Saggu HS, Vasudev H. A short review on the performance of high velocity oxy-fuel coatings in boiler steel applications. *Materials Today: Proceedings*. 2022;50:1442–6.

[67] Singh P, Vasudev H, Bansal A. Effect of post-heat treatment on the microstructural, mechanical, and bioactivity behavior of the microwave-assisted alumina-reinforced hydroxyapatite cladding. *Proceedings of the Institution of Mechanical Engineers, Part E: Journal of Process Mechanical Engineering*. 2024;238(3):1060–71.

[68] Verma PK, Vasudev H, Singh G, Singh S. A review on the microstructural and biomedical properties of implants manufactured using additive manufacturing. *Journal of the Institution of Engineers (India): Series D*, 2024;106:1–15.

[69] Singh SP, Verma PK, Singh S, Kapoor M, Balu R. Microstructural characterization and electrochemical behaviour of HVOF sprayed WC-Co-Cr coatings on 316L boiler steel stainless steel. *Results in Surfaces and Interfaces*. 2024;16:100248.

[70] Verma PK, Minhas AS, Singh P, Kumar S. Slurry erosion behaviour of thermal sprayed Al2O3 and Cr2O3 coatings for turbine steels. *AIP Conference Proceedings*. 2024;2986(1):020012.

[71] Verma PK, Kumar V, Vasudev H. Bioactivity and corrosion analysis of thermally sprayed hydroxyapatite based coatings. *Journal of Electrochemical Science and Engineering*. 2024;14(4):419–39.

[72] Verma PK, Vasudev H, Kumar V. Biocompatibility and corrosion performance of HA-based LVOF sprayed and post-annealed coatings on SS 254 surgical-grade bio-implant material. *Ceramics International*. 2025;51(10):13207–25.

[73] Verma PK, Vasudev H, Singh G. Improving the corrosion performance of SS 304 stainless steel using HVOF-sprayed WC-Co-Cr coatings in chloride environments. *Journal of Bio-and Tribo-Corrosion*. 2025;11(4):1–13.

[74] Kulkarni SD, Manjunatha, Chandrasekhar U, Manjunath KV, Prasad CD, Vasudev H. Design and optimization of polyvinyl-nitride rubber for tensile strength analysis. *International Journal on Interactive Design and Manufacturing (IJIDeM)*. 2024;18(8):5893–908.

[75] Kattimani MA, Venkatesh PR, Masum H, Math MM, Bahadurdesai VN, Mustafkhadri S, Prasad CD, Vasudev H. Design and numerical analysis of tensile deformation and fracture properties of induction hardened inconel 718 superalloy for gas turbine applications. *International Journal on Interactive Design and Manufacturing (IJIDeM)*. 2024;18(8):5939–49.

[76] Poojari M, Hanumanthappa H, Prasad CD, Jathanna HM, Ksheerasagar AR, Shetty P, Shanmugam BK, Vasudev H. Computational modelling for the manufacturing of solar-powered multifunctional agricultural robot. *International Journal on Interactive Design and Manufacturing (IJIDeM)*. 2024;18(8):5725–36.

[77] Prakash C, Gupta LR, Mehta A, Vasudev H, Tominov R, Korman E, Fedotov A, Smirnov V, Kesari KK. Computing of neuromorphic materials: An emerging approach for bioengineering solutions. *Materials Advances*. 2023;4: 5882–919.

[78] Singh J, Gill HS, Vasudev H. Computational fluid dynamics analysis on role of particulate shape and size in erosion of pipe bends. *International Journal on Interactive Design and Manufacturing (IJIDeM)*. 2023;17(5):2631–46.

[79] Rao GS, Mukkamala U, Hanumanthappa H, Prasad CD, Vasudev H, Shanmugam B, KishoreKumar KC. Evaluating and optimizing surface roughness using genetic algorithm and artificial neural networks during turning of AISI 52100 steel. *International Journal on Interactive Design and Manufacturing (IJIDeM)*. 2024;18(8):6151–60.

[80] Singh J, Singh R, Singh S, Vasudev H, Kumar S. Reducing scrap due to missed operations and machining defects in 90PS pistons. *International Journal on Interactive Design and Manufacturing (IJIDeM)*. 2023;17(5):2527–39.

[81] Nagabhushana P, Ramprasad S, Prasad CD, Vasudev H, Prakash C. Numerical investigation on heat transfer of a nano-fluid saturated vertical composite porous channel

packed between two fluid layers. *International Journal on Interactive Design and Manufacturing (IJIDeM)*. 2024;18(5):2927–44.

[82] Mahantesh MM, Rao KR, Chandra AP, Vijayakumar MN, Nandini B, Prasad CD and Vasudev H. Design and modeling using finite element analysis for the sitting posture of computer users based on ergonomic perspective. *International Journal on Interactive Design and Manufacturing (IJIDeM)*. 2024;18(8):5875–91.

[83] Verma PK, Vasudev H, Kumar V. Dental implant materials and designs and failure during implant placement and loading: A review. In Hitesh Vasudev, Chander Prakash, Mohammad Khalid, Kavindra Kesari (eds.): *Advanced Manufacturing Processes*, 2025. pp.138–50, Boca Raton: CRC Press.

[84] Verma PK, Vasudev H, Kumar V. Bioactivity and corrosion performance of low-velocity oxy-fuel-sprayed hydroxyapatite-based coatings on a surgical-grade stainless steel. *Transactions of the Indian Institute of Metals*. 2025;78(8):185.

[85] Verma PK, Vasudev H, Kumar V. Effect of electro-corrosion behavior of different grades of stainless steel used in bio-implants: A review. In Hitesh Vasudev, Chander Prakash, Mohammad Khalid, Kavindra Kesari (eds.): *Advanced Manufacturing Processes,* 2025. pp. 269–83, Boca Raton: CRC Press.

[86] Verma PK, Vasudev H. Importance of hydroxyapatite coatings in improving new bone osteogenesis: A review. In Hitesh Vasudev, Chander Prakash, Mohammad Khalid, Kavindra Kesari (eds.): *Advanced Manufacturing Processes*, pp. 284–99 Boca Raton: CRC Press.

8 Optimizing Thermal Spray Parameter using Machine Learning

Umesh Ghorpade, Shubhangi Suryawanshi, and Satyajeet Gaikwad

8.1 INTRODUCTION

One of the biggest obstacles to creating high-performance coatings for a range of industrial uses is still optimizing thermal spray settings. Although specifications supplied by the original equipment manufacturer (OEM) give a solid starting point, achieving the best results frequently necessitates a great deal of trial and error. Although thermal spray methods are widely used, there is currently no systematic, physics-based justification in the literature to reduce these repetitive efforts when optimizing coating recipes. [1]

In order to close this gap, the current work suggest a methodical way to assist practitioners in creating a more logical and effective process parameter selection methodology. This chapter seeks to simplify the optimization process by taking into account the complex relationships among material characteristics, coating quality, and machine capabilities [2, 3]. The need to speed up coating development cycles is growing in modern industrial environments, which are frequently referred to as "responsive" since they must quickly adapt to shifting supply and demand. The process is made more difficult by a number of issues, including the need for repeatable results, the trade-off between cost and performance, and the variability of feedback and equipment [4, 5, 6]. These industrial imperatives motivate this chapter's introduction of a framework that incorporates contemporary machine learning methods to tackle the intricate problems of process parameters creation. Utilizing data-driven models, the suggested methodology facilitates quicker and more precise parameter optimization, which eventually aids in the creation of premium coating with less expense and development time. Many thermal spray technologies, such as wire arc, flame spray, high velocity oxy-fuel (HVOF), and air plasma spray (APS), can use this architecture [7, 8, 9].

8.2 PROCESS AND EQUIPMENT SELECTION BASED ON MATERIALS

8.2.1 Materials

Understanding the materials to be coated thoroughly is the first step in choosing the best thermal spraying method. Establishing a clear connection between the ideal

DOI: 10.1201/9781003730637-8

spraying technique and the intrinsic physical qualities of the coating material, such as its melting point, shape, and desirable microstructure characteristics, is essential before developing process parameters. Instead of depending on trial-and-errors techniques, this methodical connection allows for a logical approach to process selection. It is common practice to use decision-support flowcharts to aid in this decision-making process (Figure 8.1). These frameworks, while not all-inclusive, offer a methodical manner to match coating materials with certain production procedures. Beyond material properties, they serve as a starting point that can be broadened by adding more factors. Practical considerations that might greatly impact the ultimate process selection include, for example, the masking requirements to prevent overspray and the expected post-coating surface finish [10, 11].

An objective viewpoint must be used while developing process parameters. The most effective or long-lasting remedies might not necessarily be found in conventional methods. Therefore, attaining optimal performance outcomes requires being receptive to different material-process combinations. Environmental sustainability is now a key consideration when choosing process and equipment in modern manufacturing settings. Thermal spray process design must place a grater emphasis on environmentally friendly methods as firms throughout the world work to attain a "net zero" carbon footprint. In order to guarantee that the chosen procedure satisfies both technical specifications and global sustainability goals, it is necessary to strike a balance between coating performance, economic feasibility, and decreased environmental effect.

8.2.2 Inert Gases vs. Combustibles: An Educated Choice

In thermal spray processes, the choice of gases is crucial for determining coating quality, process effectiveness, and long-term sustainability. High-velocity oxy-fuel and flame spraying methods commonly use combustible gases including hydrogen, oxygen, acetylene, propylene, and propane, whereas inert gases like nitrogen, argon, and helium are typically thought to be cleaner and more stable options for plasma spraying (Figure 8.2).

Scientifically speaking, the dissociation and ionization process in atmospheric pressure plasma jets must be taken into account while selecting the gases. These days, plasma spraying is mostly used for four gases: hydrogen and nitrogen (diatomic), argon and helium (monoatomic). Prior to ionization, diatomic gases needed more energy to dissociate into their monoatomic form. For instance, hydrogen, which has a single outer electron, needs a lot of energy to dissociate but, once ionized, releases a significant amount of thermal energy, which makes it appropriate for melting ceramics and tribological alloys. In a similar manner, nitrogen dissociates and then ionizes, generating a high plasma enthalpy that works well for spraying materials with high melting temperatures [12].

The utilization of nitrogen and hydrogen is not without restrictions, though. Both gases can encourage the oxidation of metallic and alloy powders during air plasma spraying, resulting in coating with reduced structural integrity. Despite being expensive, inert gases like argon and helium reduce oxidation because of their chemical

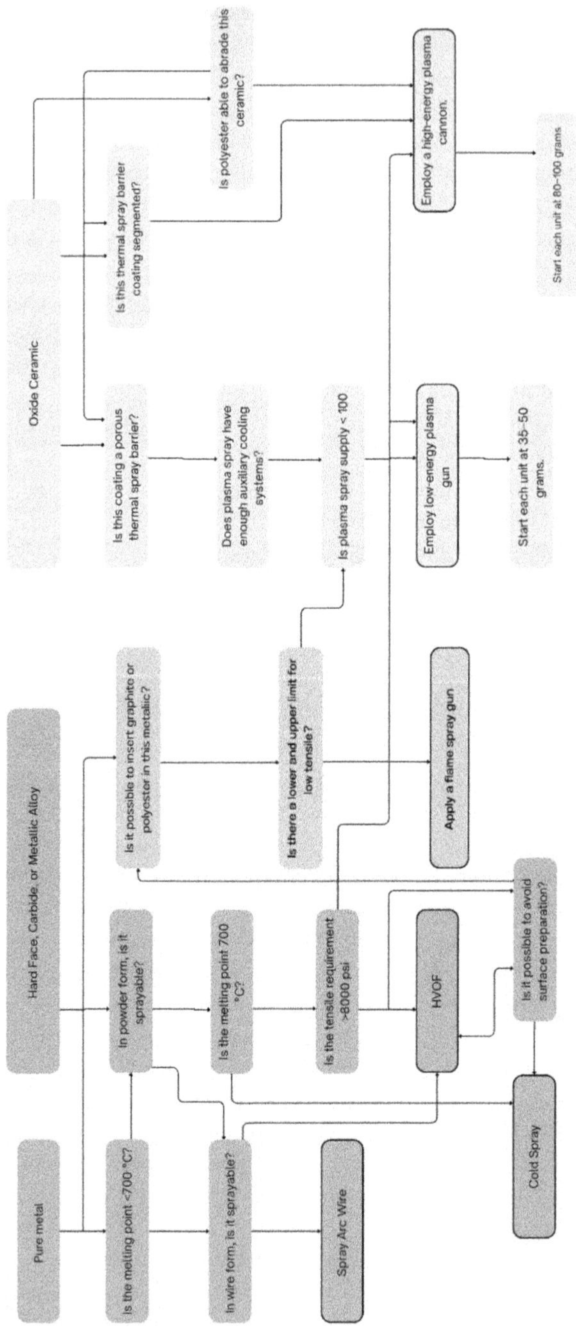

FIGURE 8.1 Flow diagram for selection of the best spraying technique.

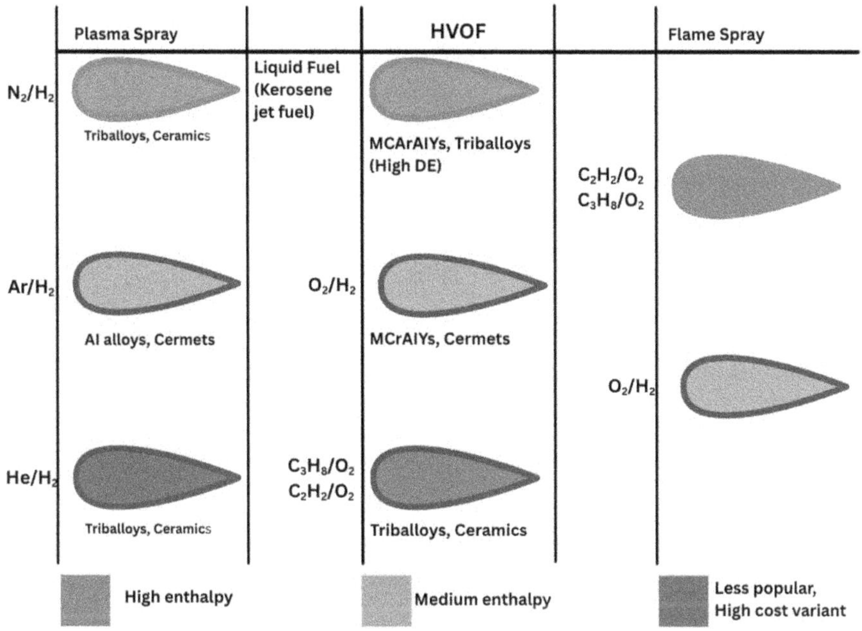

FIGURE 8.2 Suitable materials that depend on thermal spray plumes that exhibit high, medium, and low enthalpy producing gases.

stability. Ionization of inert gases requires more energy because of their fully occupied outer electron shells, but once it is accomplished, the ensuing plasma provides enough heat to efficiently melt metals and alloys. This benefit helps to create coatings that function better and have less oxide. A logical foundation for choosing the best gas for certain material requirements is provided by a comparison of the enthalpy values for various gases (Figure 8.3). In order to balance performance and process cost, practitioners can make well-informed decisions about inert and combustible gases by comparing plasma enthalpy with the melting behaviour of target materials [13].

These process and thermodynamics factors can be important components of predictive modelling in machine learning-driven optimization. Machine learning can direct process engineers toward gas selections that maximize efficiency, decrease oxidation, and guarantee environmental sustainability by training algorithms on datasets that include materials parameters, gas enthalpy values, and coating performance measures. In thermal spraying, this fusion of data-driven intelligence and physics-based knowledge provides a technique to attain both technical excellence and a lower environmental impact.

8.2.3 Choices Within an Air Plasma Spray Process

Since there is a large selection of commercially available equipment with power ratings that typically range from 20 to 200 kW, choosing the right plasma spray cannon is an essential part of process design. Although it provides flexibility, this

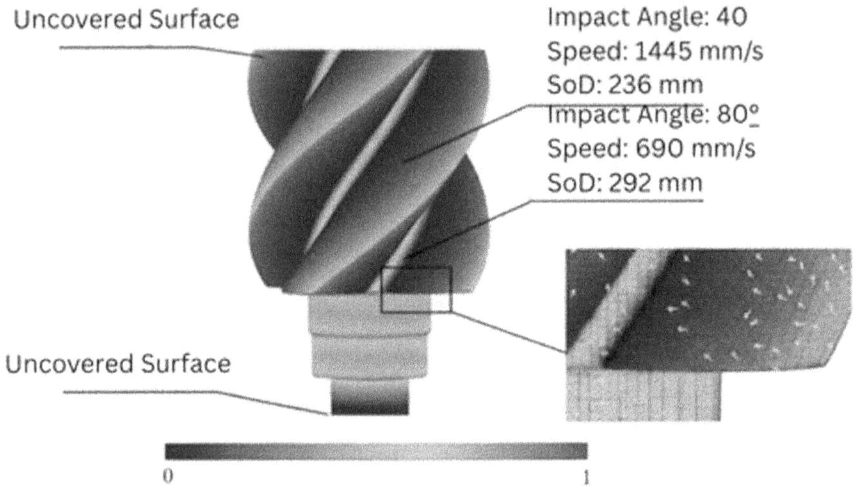

FIGURE 8.3 Coating thickness evolution simulation on a complex industrial component: HVOF thermal spraying for oil and gas mud rotor.

variation frequently leaves practitioners unsure about the best arrangement. An evaluation of the coating portfolio and possible future explanation plans should be the first step in a methodical decision-making process [14, 15]. For instance, tribological alloys, Al-Si polyester composites, and thermal barrier coatings (TBCs) are examples of porous coatings that may typically be deposited using plasma spray guns with a power range of 60–80 kW. On the other hand, it makes more sense to purchase a 200 kW power supply right away if the planned applications involve sophisticated or segmented TBCs. These higher-capacity systems offer the adaptability to manage both dense and porous coatings without requiring future, expensive retrofits [16].

The spray gun's power rating has a direct impact on the amount of the capital needed and the infrastructure needed. Equally powerful electrical transformers and associated infrastructure are also required for high-capacity plasma spray weapons. Moreover, long-term hardware reliability and performance depend on effective temperature management systems. For dense, segmented coatings, high-energy guns may need a cooling water flow of 55–60 L/min, while guns in the 60–100 kW range usually need a flow of 35–45 L/min. Thus, a closed-loop chiller with a heat exchanger prolongs component life and allows for steady performance across a range of gun capacities [17, 18]. The incorporation of optimization based on machine learning presents new possibilities for well-informed equipment selection. Predictive models can include operational parameters as a feature, such as coating type, gun power, cooling needs, and infrastructure compatibility. Manufactures may choose plasma spray systems that not only satisfy current needs but are also scalable for future applications thanks to the decision support tools these data-driven approaches offers. By doing this, machine learning ensures the best results in plasma spraying by balancing performance, cost-effectiveness, and sustainability.

8.2.4 Physical Properties and Parameter Correlations

A thorough grasp of physical characteristics of feedstock powders is essential for choosing the right carrier gases, auxiliary equipment, spray guns, and thermal spray procedures. Important powder characteristics, including melting point, apparent density, particle size distribution, morphology, and chemical compositions, are crucial in determining process parameters and, eventually, coating quality.

A material's compatibility with various spray procedures is significantly influenced by its thermal properties. For instance, plasma spray is rarely used for low-melting-point metals like zinc or aluminium, whereas ceramics are typically inappropriate for HVOF spraying because of their high melting points. For high enthalpy needs, plasma spray systems work especially well, allowing refractory materials like tungsten carbide (WC) and zirconium oxide to be deposited. On the other hand, particularly in carbide-based chemistries, HVOF systems are more efficient in creating thick coating with better surface finishes [19].

The energy of the plasma jet is strongly influenced by the composition of the gas. For example, high hydrogen flow rates minimize the presence of unmelted inclusions by providing enough enthalpy to allow for the complete melting of a feedback particle. On the other hand, in order to effectively melt low-density metals like titanium and aluminium, greater carrier gas flow is necessary to combat plume turbulence. High-density powders, like molybdenum or WC, on the other hand, are easier to entrain into the plasma jet and require less assistance from carrier gases [20].

Another important element that directly affects feeder performance is the density of the powder. Making the distinction between apparent density and tap density is crucial for determining the capacity of the powder feeder and guaranteeing continuous production runs. The morphology of the powder also influences flowability; spherical powder flows better through feed hoses than irregularly shaped particles. Particular care must be taken with nanopowders since, before spraying, morphological control and drying are required to reduce agglomeration and guarantee uniform feeding [21].

The choice of carrier gas for powders is determined by their chemical reactivity. Using inert gases like argon or helium instead of hydrogen is advantageous for materials that are prone to oxidation. Brittle microstructures and a change in coating hardness, occasionally above acceptable upper limits, might result from oxidation during spraying. To maintain coating integrity, gas selection must therefore strike a compromise between thermodynamics and chemical factors.

The compatibility of substrate and coating is equally crucial. Performance and adhesion may be harmed by a discrepancy in thermal expansion coefficients. For example, the high enthalpy produced by plasma gases makes it difficult to deposit Ni-Cr alloys onto aluminium substrates using plasma spraying, which increases the risk of substrate damage. Alternative methods, including wire arc spraying, are better in these situations because they provide coatings that are structurally strong while maintaining lower substrate temperatures [22].

A comprehensive strategy that incorporates material characteristics, carrier gas composition enthalpy generation, and injector configurations with knowledge of periodic trends in sprayability is necessary to provide consistent and high-performance coatings. A viable approach to predictive optimization is to incorporate these connections into machine learning systems. Selection of data-driven parameters that increase reproducibility, boost coating quality, and guarantee scalability in industrial applications can be accomplished by training models on material-process relationships and coating results [23, 24].

8.2.5 Process Parameter Development

The thermal choice of power source and spray gun is only one aspect of optimizing thermal spray procedures; other crucial aspects must be carefully adjusted. The hopper pressure, powder feed rate, carrier gas flow, spray distance, robot traverse speed, nozzle design, electrode type, and gas composition are some of these. Usually, comprehensive spray iterations, repeatability checks, and validation through stringent material testing feedback are required to achieve the ideal balance among these variables.

Structured approaches like Design of Experiments (DOE), which includes Taguchi designs, are frequently used to reduce reliance on trial-and-error experimentation. DOE greatly reduces the number of experimental trials needed while facilitating methodical study of the process window. Additionally, particle diagnostic tools, such as substrate temperature measurement and particle velocity tracking, function as efficient screening techniques that make it possible to create more targeted and effective experimental programs [25, 26]. DOE frequently starts the process of developing process parameters, which are then backed up by strong statistical analysis. While sophisticated data-driven methods like random forest regression (RFR) offer deeper insights into the intricate relationship between process variables, traditional techniques like analysis of variance (ANOVA) enable the discovery of statistically significant parameters [27, 28].

Usually, trial runs on flat substrates are carried out to create baseline conditions before developing coating parameters. These characteristics can be applied to more intricate geometries, including round or cylindrical components, after they have been tuned. This methodical comprehension of parameter-property correlation serves as the foundation for methodical optimization techniques and makes it possible to include machine learning techniques to forecast and improve coating results.

8.3 COATING DEPOSITION MODELLING

One of the most important tools for improving thermal and cold spray processes is the creation of precise and adaptable coating deposition models. The transfer of novel spray coating materials from lab experiments to industrial use has been slow, despite extensive research. The problem is brought on by differences in application techniques, kinematics of deposition, intricate component geometries, and the lack of data on quoted component performance under actual operating circumstances. Although academic testing offers important insights into characteristics like wear rate, porosity, and hardness, it is still difficult to translate these findings to industrial practice [2, 3]

Predictive and simulation-based methods are necessary in high-value industries like aerospace, automotive, and defence to cut down on design iterations and qualification expenses. Therefore, uniform properties, minimizing coating thickness variations, and optimizing robotic spray paths under complex geometries and shadowing conditions all depend on accurate deposition models. From a computational perspective, these models need to strike a balance between efficiency and accuracy, providing predictions in real time with the least amount of numerical error [5, 7].

8.4 SIMULATION APPROACHES

Forecasts of coating deposition parameters, including coating thickness distribution, have been developed via a number of computational methods. For complex axisymmetric geometries, level set-best numerical approaches have shown encouraging accuracy. In order to optimize this gun kinematics nozzle positioning, and spray trajectories for both planar and free-form surfaces, robot simulation platforms like robotstudio™ have been used extensively. In order to improve spray patterns by taking into consideration heat and mass transfer during deposition, optimization algorithms, such as nonlinear conjugate gradient methods, have been used [9, 10].

Geometric and mathematical modelling has been applied to the cold spray process to solve issues like thickness prediction and shadow effects. Advanced models incorporate robot trajectory planning for better process control and use evolving 3D geometries to simulate coating build up morphology. Other methods, like MATLAB™ best routines, map coating properties (like thickness, porosity, and microhardness) onto triangulated parts geometries using lookup tables generated from experimental calibration data. Although these techniques offer quick, error-free computations, they mainly rely on manual optimization and have limited automation [11–13, 29].

8.5 MACHINE LEARNING PROSPECTS

In order to handle the inherent non-linearities and data-driven requirements of coating deposition modelling, machine learning has become a potent tool. Kohonen developed Self-Organizing Maps (SOM), which provide unsupervised learning for high-dimensional spray data. By dynamically adjusting node positions, spray angles, and velocities to reduce thickness variation, SOMs can set up ideal robotic spray trajectories for thermal and cold spray applications. SOM-based trajectory planning produces more uniform coatings than traditional expert-driven methods, according to experimental validation, particularly for complex geometries like mud rotors used in oil and gas applications [30, 31].

Cold spray additive manufacturing (CSAM) has also seen success with artificial neural networks (ANNs). When it comes to predicting coating track geometry, especially when spray angle varies, ANN-based data-driven models perform better than conventional mathematical techniques (such as Gaussian or parabolic functions). However, the requirements for sizable datasets, the expense of data generation, and the existence of non-linear physical phenomena that are not adequately represented by current models limit the accuracy of ANN models. To increase data efficiency and predictive reliability, hybrid approaches that combine ANN with mathematical regression models have been proposed. These combinatorial techniques improve model generalization over a wider range of process conditions while lowering experimental demands [32].

8.5.1 Process Optimization, Monitoring, and Control

Thermal spray (TS) and cold spray technologies have become competitive substitutes for hard chrome plating (HCP) in a variety of applications due to more stringent regulations. However, to guarantee consistent coating quality and high process efficiency, this transition calls for more advanced techniques for process monitoring and control.

Particles' adherence to the substrate in low-temperature TS processes, like HVOF and high-velocity air-fuel (HVAF) is primarily controlled by kinetic energy rather than total melting. This adds complexity even though it makes coating with better mechanical performance possible. Significant changes in particle velocity and temperature at the point of impact are frequently caused by a wide particle size distribution and intrinsic variations in flame intensity [33]. Larger particles usually have insufficient kinetic energy when they arrive at the substrate in a solid state, which causes elastic rebound. The smaller particles, on the other hand, are frequently completely melted. which can increase porosity and reduce compressive residual stresses required for thick and long-lasting coatings. The significance of sophisticated process control in maximizing coating integrity and deposition efficiency is highlighted by the delicate balance [34]. Despite offering reproducible input parameters, including feed rates, pressures, gun movement, and gas flow rates, programmable logic controllers (PLC) and mass flow controllers have limitations when it comes to identifying hardware-related problems. Undetected fluctuations caused by nozzle wear, unstable powder feeding, gas leaks, and line erosion can seriously impair coating quality and possibly lead to catastrophic component failures. Therefore, in order to meet increasing standards for industrial quality, it is essential to continuously monitor the characteristics of the process and record any fluctuations.

Initially, thermal and optical diagnostic tools were used to monitor thermal spray processes. Particle temperature was recorded by fast parameters, particle trajectories were monitored by CCD cameras, particle velocity was measured using laser Doppler anemometry (LDA), and size and velocity was measured simultaneously using phase Doppler techniques [35]. Commercial diagnostic systems, such as the DPV-2000 (Tecnar), Spray watch (Oseir), and accuspray (Tecnar), were developed as a result of these fundamental methods. These devices allowed for real-time monitoring of in-flight practical parameters like velocity, temperature, size distribution, and morphology. Moreover, closed-loop control techniques were put forth to create a direct connection between coating characteristics and diagnostic measurements, opening the door to intelligent process control.

Another technique that has shown a lot of promise is acoustic emission (AE). Particle impact energy, crack initiation, and coating delamination can all be tracked by contact sensors affixed to substrates. This capability is further expanded by non-contact laser AE techniques, although they increase system complexity and necessitate exact calibration. Despite being underutilised in thermal spraying, airborne AE monitoring has shown great promise for adoption in TS processes due to its successful application in related fields like welding and machining [36].

Significant restrictions still exist in spite of these advancements. Small and medium-sized business (SMEs) find it challenging to justify the high cost of commercial diagnostic systems. Additionally, they are difficult to incorporate into torch-mounted setups for dynamic real-time monitoring and are best suited for static monitoring. Their industrial applicability is further limited by other problems like dust accumulation, high boost temperatures, and limited torch movement when spraying complex geometry. Compact reliable and reasonably priced sensing technologies that can record process-relevant data and instantly correlate it to coating properties are therefore the way of the future.

8.5.2 Machine Learning Approaches

The nonlinear and multivariate nature of thermal spray processes make machine learning (ML) a potent substitute for traditional process optimization techniques. Large datasets with nonlinear dependencies are difficult to handle for traditional statistical methods like DOE, response surface methodology (RSM), and ANOVA. In contrast, ML can process a variety of data streams quickly and predict process outcomes [37]. In order to correlate in-flight particle properties with coating performance metrics like adhesion strength, bond strength, and wear resistance, ANNs have been widely used. For example, coating adhesion strength has been accurately predicted by ANN models trained using data from the DPV-2000 optical sensing system. In a similar vein, wear rate prediction using optimized ANN frameworks has been developed, allowing for better coating lifetime estimation and faster testing.

These capabilities are expanded by convolutional neural networks (CNN), which process unstructured input like audio signals and images. According to recent research, real-time coating quality monitoring with inexpensive microphones is possible when the airborne acoustic signal is processed using ANN and CNN architectures. In addition to thermal spray, CNN-based models have demonstrated cutting-edge performance in welding and additive manufacturing (AM), where they have been applied to identify different types of defects, detect plume fluctuations, and evaluate the quality of the process. Their potential applicability to TS processes is highlighted by the cross-disciplinary success [38].

In the future real-time monitoring and adaptive control may be made possible by combining ML-driven techniques with smaller sensor systems. New approaches include intelligent airborne acoustic monitoring for economical, in-process quality assurance, hybrid machine learning frameworks that combine ANNs, CNNs, and physics-informed models for coating property prediction, and computer vision combined with CNNs for plume stability assessment. With these developments, thermal spraying could become a data-driven, self-optimizing manufacturing process that can satisfy demanding industrial requirements.

8.6 OVERVIEW OF MACHINE LEARNING APPROACHES

The optimization, monitoring, and control of thermal spray processes have greatly benefited from the use of ML approaches because of their capacity to handle intricate and nonlinear interactions. ML models may create correlations between process inputs and outputs by utilizing simulated or experimental datasets, which increases

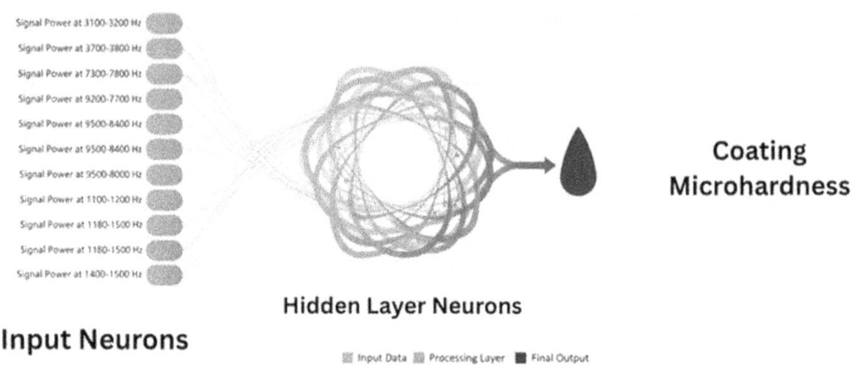

FIGURE 8.4 Neural network architecture used to predict the coating microhardness as a function of frequency power inputs.

the accuracy of target variable predictions. This predictive capacity helps to achieve better coating quality and enables adaptive process control (Figure 8.4) [38, 39].

8.6.1 Artificial Neural Networks (ANNs)

One of the most popular machine learning approaches in thermal spray research is the use of artificial neural networks. ANNS work using two basic methods, forward propagation and backpropagation, and are based on the idea of biological brain networks.

- Forward propagation involves the application of weights, biases, and nonlinear activation functions such as the sigmoid, Tanh, or structured input data, thereby enabling the network to capture complex patterns.
- Backpropagation serves to optimize the weights of the model iterative algorithms including Gradient Descent, Adam optimizer, and RMSProp, ensuring improved accuracy over successive training cycles.

In thermal spray applications, ANN been effectively trained using diagnostic data from instruments like the DPV-2000 to create associations between coating adhesion strength and in-flight practical characteristics. An optimized ANN model has also been used to forecast coating loss, material wear, and overall process stability. The viability of employing airborne acoustic signals in conjunction with ANN inference as an affordable substitute for intricate optical diagnostic system for tracking HVOF processes was demonstrated by an intriguing development [40].

8.6.2 Convolutional Neural Networks (CNNs)

Convolution neural networks process unstructured datasets like text, audio, and images, extending the capabilities of ANNs. They can extract and classify hierarchical features straight from raw process data thanks to their architecture, which consists of convolutional, pooling, activation (like ReLU), and fully connected layers. CNN models trained with process simulation datasets have been demonstrated to optimize plasma spray parameters, thus allowing for the prediction of coating properties without requiring extensive experimental trials. This makes CNNs promising tools for process monitoring and control in the field of thermal spraying.

- Using just a basic microphone configuration, CNN-based frameworks applied to airborne acoustic datasets showed the capacity to identify plume fluctuations and categorize coating quality in real time.
- CNNs applicability in two time scale processes is further supported by insights from related fields like welding and additive manufacturing (AM) which highlight their capacity to identify flaws, track plume stability, and forecast product quality.

8.6.3 Hybrid and Emerging Approaches

Hybrid frameworks that combine ANNs, CNNs, and physics-based models have been proposed to improve predictive robustness. For instance, real-time control of particle characteristics during spraying is made possible by ANNs combined with diagnostic sensors.

- Inspired by advancements in AM, CNNs applied to plume and spatter images show promise for assessing plume stability in thermal spray.
- To create small, affordable monitoring devices appropriate for industrial deployment, data fusion techniques that combine visual and acoustic inputs with machine learning models are basing investigated.

Compared to other manufacturing domains, ML applications in thermal spray are still in their infancy; preliminary research shows that they have the potential revolutionize the industry. In particular, ML techniques have shown that they can connect adhesion strength, wear resistance, and coating durability to in-flight diagnostics.

- They make predictive monitoring possible with reasonably priced sensor systems, like cameras and microphones.
- To guarantee constant quality, optimize process variables such as feedback delivery, gas flow rates, and standoff distances [41, 42].

8.6.4 ANNs′ Function in Surface Coatings

Artificial neural networks (ANNs) have become effective tools for surface coating process optimization, allowing for major advancements in quality control, performance prediction, formulation design, and process monitoring. ANNs can create correlations between process parameters and coating properties by learning from past experimental or sensor datasets. This reduces the need for trial-and-error experiments and increases process efficiency overall [43, 44]. Figure 8.5 shows an ANN's typical architecture, with the arrows representing the inward and outward flow of neurons.

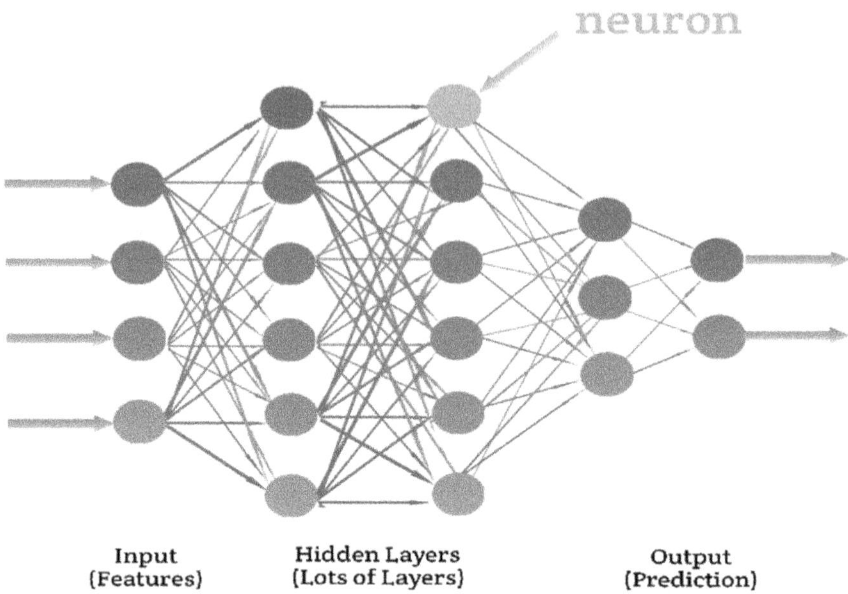

FIGURE 8.5 An ANN's typical architecture, with the arrows representing the inward and outward flow of neurons.

Artificial neural networks offer real-time monitoring and adapt to control coating processes by analyzing a variety of input data, including temperature, viscosity, coating thickness, optical images, or spectral signals. By anticipating possible flows, non-Comfortis (Contextual Interpretation), or early failure under service conditions, they further improve quality assurance. Higher coating reliability and less testing are guaranteed by the predictive capability [45].

ANNs have achieved very high accuracy (R2 > 0.99 in some cases) in predicting coating thickness in thermal spray hot-dip galvanizing, and the electrodeposition process. Often used in conjunction with Genetic Algorithms (GA) to further improve predictive accuracy and lesson experimental workload, optimized ANN models (Figure 8.6) can model coating trends based on time, feed rate, air flow, and voltage [46, 47].

ANN frameworks (Figure 8.7) have been created for industrial tool coatings, WC-CoCr coatings, NiTiN nano-coatings, and TiO_2 thin films, with up to 98% prediction accuracy reported [13]. These models show that hardness and process variables like duty ratio, current density, pulse frequency, and particle concentration are strongly correlated [48, 29].

With prediction errors as low as 0.3 nm, ANN models have been used for plasma nitride steels, TiAIN/MO multilayer coatings, and PTA hardfacing. Hybrid methods, like ANN in conjunction with Taguchi optimization, have additionally shown effective roughness prediction while reducing the quantity of experimental traits trials [49].

The effects of bath temperature, saccharin concentration, and current density on nanocrystalline coatings have all been simulated using ANNs. According to published

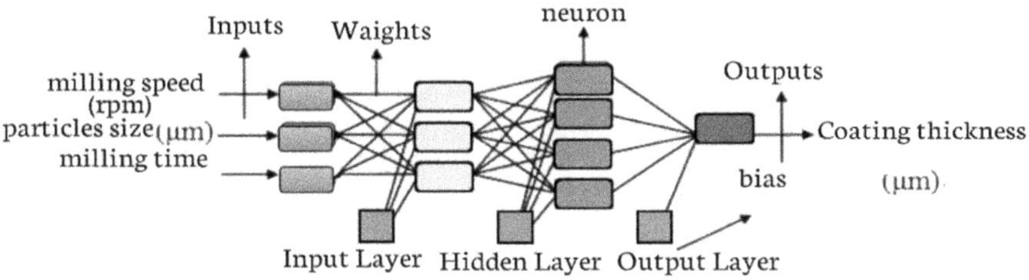

FIGURE 8.6 The coating thickness prediction model's ANN architecture.

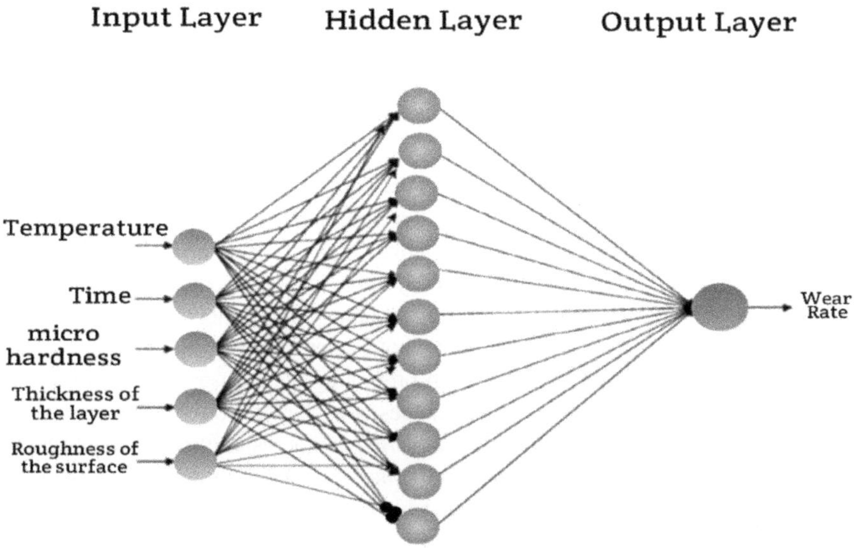

FIGURE 8.7 ANN schematic models for predicting the wear rate of 316 L stainless steel.

research, ANNs can predict grain size with an accuracy of over 96% making them a dependable tool for better coating design and longer tool life [45]. Gas ratios, current, carrier gas flow, and powder feed rates are some of the input variables that ANNs have used to predict porosity in APS alumna-titania coatings, HVOF coatings, and plasma spray deposits. The coating density and overall performance are improved by these reactive models, which help to maintain low porosity levels [50].

Multilayer nitride films, SiC-filled PEEK coatings, Mo coatings, and TiN/NbN multilayers have all been effectively modelled using ANNs. In order to enable predictive wear testing, the results validate that ANNs can precisely correlate process conditions, such as applied load, sliding velocity, and deposition rate, with tribological outcomes [51]. ANNs have demonstrated high reliability in predicting oxidation kinetics in PVD/HIPIMS, boride, and nanocrystalline coatings. With prediction accuracy of up to 99% achieved by coupling ANNs with electrochemical test data, accelerated evolution techniques for coating durability and long-term protection are now available [52, 53].

The biomedical and high-temperature corrosion-resistant systems slurry processing sectors have begun to employ thermal spray coatings and innovative surface engineering techniques. This is especially true when combined with developing technologies such as additive manufacturing, magnetron sputtering thin-film deposition, and microwave-assisted fabrication [54–67]. Furthermore, the coating structures demonstrate high strength (tensile, compressive, and flexural) and functional performance in solar energy conversion coating systems, as well as substantial potential in surface topography. Furthermore, the developed surfaces provide a platform for advanced computational fluid dynamics of pipes and conduits with bends, as well as ergonomic seating systems that include anthropometric, mechanical, and structural connections [68–80].

8.7 SUMMARY

Reliable coatings for industrial applications depend on the precise coordination of materials, process gases, spray equipment, and robotic handling in thermal spray processes. Despite their usefulness, conventional approaches frequently fail to handle the nonlinear interactions between several process parameters. A potent substitute is offered by the incorporation of machine learning, which offers predictive insights linking coating performance, particle diagnostics, plume stability, and feedstock characteristics. Thermal spray operations can attain increased sustainability, shorter development cycles, and increased repeatability by integrating robotics, sensors, and real-time monitoring into ML-driven frameworks. In addition to promoting the industrial transition to net-zero carbon manufacturing, this strategy creates new chances for adaptive control and high-value applications in the biomedical, energy, and aerospace industries. Accurate, effective, and future-ready next-generation thermal spray systems are made possible by the combination of data-driven intelligence and physics-based knowledge.

REFERENCES

1. Herman, H., Sampath, S., McCune, R.: Thermal spray: Current status and future trends. *MRS. Bull.* 25, 17–25 (2000).
2. Viswanathan, V., Katiyar, N. K., Goel, G., Matthews, A., Goel, S.: Role of thermal spray in combating climate change. *Emergent. Mater.* 4, 1515–1529 (2021).
3. Chen, C., Liao, H., Montavon, G., Deng, S.: Nozzle mounting method optimization based on robot kinematic analysis, *J. Therm. Spray Technol.* 25(6), 1138–48 (2016). https://doi.org/10.1007/s11666-016-0429-1
4. Faisal, N.H., Prathuru, A., Ahmed, R., Rajendran, V., Hossain, M., Venkatachalapathy, V., Katiyar, N.K., Li, J., Liu, Y., Cai, Q., et al. Application of thermal spray coatings in electrolysers for hydrogen production: Advances, challenges, and opportunities. *ChemNanoMat.* 8, e202200384 (2022).
5. ABB. *RobotStudio. robotstudio.* Available online: https://new.abb.com/products/robotics/ (accessed on 31 October 2021).
6. Katiyar, N.K., Goel, G., Goel, S.: Emergence of machine learning in the development of high entropy alloy and their prospects in advanced engineering applications. *Emergent. Mater.* 4, 1635–1648 (2021).

7. Shi, Z.J., Guo, J.: A new algorithm of nonlinear conjugate gradient method with strong convergence. *Comput. Appl. Math.* 27, 93–106 (2008). https://www.scielo.br/j/cam/a/vB59TGtwGZnmx9djLf8qFKq/?lang=en
8. Vardelle, M., Vardelle, A., Fauchais, P.: Spray parameters and particle behavior relationships during plasma spraying. *J. Therm. Spray Technol.* 2, 79–91 (1993).
9. Fang, D., Deng, S., Liao, H., Coddet, C.: The effect of robot kinematics on the coating thickness uniformity, *J. Therm. Spray Technol.* 19, 796–804 (2010), https://doi.org/10.1007/s11666-010-9470-7
10. Devaraj, S., Anand, B., Gibbons, M., McDonald, A., Chandra, S.: Thermal spray deposition of aluminum and zinc coatings on thermoplastics. *Surf. Coat. Technol.* 399, 126114 (2020).
11. Feuerstein, A., Knapp, J., Taylor, T., Ashary, A., Bolcavage, A., Hitchman, N.: Technical and economic aspects of current thermal barrier coating systems for gas turbine engines by thermal spray and EBPVD: A review. *J. Therm. Spray Technol.* 17, 199–213 (2008).
12. Prashar, G. and Vasudev, H. High temperature erosion behavior of plasma sprayed Al2O3 coating on AISI-304 stainless steel. *World Journal of Engineering*. 18(5), 760-766 (2021).
13. Reyes-Mojena, M., Ángel, M. S.-O., Carvajal-Fals, H., Sagaró-Zamora, R., and Camello-Lima, C. R.: A comparative study on slurry erosion behavior of HVOF sprayed coatings. *Dyna 84*, 202, 239–246 (2017).
14. Lima, R.S.: Porous APS YSZ TBC manufactured at high powder feed rate (100 g/min) and deposition efficiency (70%): Microstructure, bond strength and thermal gradients. *J. Therm. Spray Technol.* 31, 396–414 (2022).
15. Sahraoui, T., Feraoun, H.I., Fenineche, N., Montavon, G., Aourag, H., Coddet, C.: HVOF-sprayed Tribaloy©-400: Microstructure and first principal calculations. *Mater. Lett.* 58, 2433–2436 (2004).
16. Öksüz, M., Yıldırım, H., Erturan, S.: Microstructure and wear properties of plasma-sprayed aluminum–silicon–polyester coatings. *J. Appl. Polym. Sci.* 100, 3609–3614 (2006).
17. Karger, M., Vaßen, R., Stöver, D.: Atmospheric plasma sprayed thermal barrier coatings with high segmentation crack densities: Spraying process, microstructure and thermal cycling behavior. *Surf. Coat. Technol.* 206, 16–23 (2011).
18. Bolelli, G., Lusvarghi, L., Manfredini, T., Pighetti Mantini, F., Polini, R., Turunen, E., Varis, T., Hannula, S.-P.: Comparison between plasma-and HVOF-sprayed ceramic coatings. Part I: Microstructure and mechanical properties. *Int. J. Surf. Sci. Eng.* 1, 38–61 (2007).
19. Jandin, G., Liao, H., Feng, Z.Q., Coddet, C.: Correlations between operating conditions, microstructure and mechanical properties of twin wire arc sprayed steel coatings. *Mater. Sci. Eng. A.* 349, 298–305 (2003).
20. Zhang, Z., Zhang, D., Xie, Y.: Experimental study on water droplet erosion resistance of coatings (Ni60 and WC-17Co) sprayed by APS and HVOF. *Wear.* 432–433, 202950 (2019).
21. Viswanathan, V., Rea, K.E., Vaidya, A., Seal, S.: Role of spray drying of nanoagglomerates in morphology evolution in nanostructured APS coatings. *J. Am. Ceram. Soc.* 91, 379–386 (2008).
22. Brossard, S., Munroe, P.R., Tran, A.T.T., Hyland, M.M.: Effects of substrate roughness on splat formation for Ni-Cr particles plasma sprayed onto aluminum substrates. *J. Therm. Spray Technol.* 19, 1131–1141 (2010).

23. Rahman, M., Saifur, M.R., Hasan, K.A., Rikta, A.: A novel graphene coated surface plasmon resonance biosensor with tungsten disulfide (WS2) for sensing DNA hybridization. *Opt. Mater*. 75, 567–573 (2018).
24. Mauer, G., Vaßen, R., Stöver, D.: Comparison and applications of DPV-2000 and accuraspray-g3 diagnostic systems. *J. Therm. Spray Technol*. 16, 414–424 (2007).
25. Zimmermann, S., Vogli, E., Kauffeldt, M., Abdulgader, M., Krebs, B., Rüther, B., Landes, K., Schein, J., Tillmann, W.: Supervision and measuring of particle parameters during the wire-arc spraying process with the diagnostic systems accuraspray-g3 and LDA (laser-Doppler-anemometry). *J. Therm. Spray Technol*. 19, 745–755 (2010).
26. Mahendru, P., Tembely, M., Dolatabadi, A.: Artificial intelligence models for analyzing thermally sprayed functional coatings. *J. Therm. Spray Technol*. 32(2), 388–400 (2023).
27. Prashar, G. and Vasudev, H. Thermal Spray Nanostructured Coatings Integrated with Artificial Intelligence: A Sustainable Approach in Industry for Corrosion Applications. *Current Nanoscience*. 20 (2024).
28. Kohonen, T.: Self-organized formation of topologically correct feature maps. *Biol. Cybern*. 43(1), 59–69 (1982). https://doi.org/10.1007/BF00337288
29. Kazemi, M., Madandoust, R., Chastre, C., Esfahani, M.R., Courard, L.: Numerical study on the flexural behaviour of normal-and high-strength concrete beams reinforced with GFRP bar, using different amounts of transverse reinforcement. *Structures*. 34, 3113–3124 (2021).
30. Kubota, N., Nojima, Y., Kojima, F., Fukuda, T., Shibata, S.: Intelligent control of selforganizing manufacturing system with local learning mechanism. In: IECON Proc. (Industrial Electron. Conf (1999). https://doi.org/10.1109/iecon.1999.816540
31. Lobo, V.J.A.S.: Application of self-organizing maps to the maritime environment. In: Popovich, V.V., Claramunt, C., Schrenk, M., Korolenko, K.V. (eds) *Lecture Notes Geoinformation and Cartography*. Cham: Springer, pp. 19–36 (2009). https://doi.org/10.1007/978-3-642-00304-2
32. Sartwell, B.D., Bretz, P.E.: HVOF thermal spray coatings replace hard chrome. *Adv. Mater. Process*. 156, 25–33 (1999).
33. Bodger, B.E., McGrann, R.T.R., Somerville, D.A.: The evaluation of tungsten carbide thermal spray coatings as replacements for electrodeposited chrome plating on aircraft landing gear. *Plat. Surf. Finish*. 84, 28–23 (1997).
34. Kordijazi, A., Zhao, T., Zhang, J., Alrfou, K., Rohatgi, P.: A review of application of machine learning in design, synthesis, and characterization of metal matrix composites: current status and emerging applications. *JOM*. 73, 2060–2074 (2021). https://doi.org/10.1007/s11837-021-04701-2.
35. Holm, E.A., Cohn, R., Gao, N., Kitahara, A.R., Matson, T.P., Lei, B., Yarasi, S.R.: Overview: Computer vision and machine learning for microstructural characterization and analysis. *Metall. Mater. Trans. A Phys. Metall. Mater. Sci*. 51, 5985–5999 (2020). https://doi.org/10.1007/s11661-020-06008-4
36. Wang, Z.L., Ogawa, T., Adachi, Y.: Machine-learning-based image similarity analysis for use in materials characterization. *Adv. Theory Simul*. 3(3), 1900237 (2020), https://doi.org/10.1002/adts.201900237
37. Wang, Z., Bovik, A.C., Sheikh, H.R., Simoncelli, E.P.: Image quality assessment: From error visibility to structural similarity. *IEEE Trans. Image Process*. 13(4), 600–612 (2004). https://ieeexplore.ieee.org/document/1284395
38. Yegnanarayana, B.: Artificial neural networks for pattern recognition. *Sadhana* (1994). https://doi.org/10.1007/BF02811896
39. Teuwen, J., Moriakov, N.: Convolutional neural networks. In S. Kevin Zhou, Daniel Rueckert, and Gabor Fichtinge (eds.): *Handbook of Medical Image Computing*

and Computer Assisted Intervention. (2019). https://doi.org/10.1016/B978-0-12-816176

40. Garland, A.P., White, B.C., Jared, B.H., Heiden, M., Donahue, E., Boyce, B.L.: Deep convolutional neural networks as a rapid screening tool for complex additively manufactured structures. *Addit. Manuf. 35*, 101217 (2020). https://doi.org/10.1016/j.addma.2020.101217
41. Mia, M., Khan, M.A., Dhar, N. R.: Study of surface roughness and cutting forces using ANN, RSM, and ANOVA in turning of Ti-6Al-4V under cryogenic jets applied at flank and rake faces of coated WC tool. *Int. J. Adv. Manuf. Technol.* 93, 975–991 (2017).
42. Chowdhury, A., Kautz, E., Yener, B., Lewis, D.: Image driven machine learning methods for microstructure recognition, *Compute. Mater. Sci.* 123, 176–187 (2016). https://doi.org/10.1016/j.commatsci.2016.05.0
43. Baştürk, E., Alver, A.: Modeling azo dye removal by sono-fenton processes using response surface methodology and artificial neural network approaches. *J. Environ. Manage.* 248, 109300 (2019).
44. Nasirzadehroshenin, F., Sadeghzadeh, M., Khadang, A., Maddah, H., Ahmadi, M.H., Sakhaeinia, H., Chen, L.: Modeling of heat transfer performance of carbon nanotube nanofluid in a tube with fixed wall temperature by using ANN–GA. *Eur. Phys. J. Plus.* 135(2), 217 (2020).
45. Zhao, J., Jiang, Z., Wang, Z., Sang, S., Dobrzański, L.A., Yang, M., Ma, X., Wang, Y.: An analysis of micro deep drawing of ferritic stainless steel 430 using crystal plasticity finite element method. *J. Mater. Res. Technol.* 20, 2247–2261 (2022).
46. Kumaran, G., Sivaraj, R., Ramachandra Prasad, V., Anwar Beg, O., Leung, H.-H., Kamalov, F.: Numerical study of axisymmetric magneto-gyrotactic bioconvection in non-fourier tangent hyperbolic nano-functional reactive coating flow of a cylindrical body in porous media. *Eur. Phys. J. Plus.* 136(11), 1107 (2021).
47. Das, A., Chawla, V., Matos, R.S., Da Fonseca Filho, H. D., Yadav, R. P., Ţălu, Ş., Kumar, S. Surface microtexture and wettability analysis of quasi two-dimensional (Ti, Al) N thin films using fractal geometry. *Surf. Coat. Technol.* 421, 127420 (2021).
48. Wu, G., Chen, S.: Direct numerical simulation of a moving droplet impacting a moving particle: Effects of particle–droplet density ratio, bond number, and Reynolds number. *Eur. Phys. J. Plus.* 138(8), 723 (2023).
49. McPherson, R.: A review of microstructure and properties of plasma sprayed ceramic coatings. *Surf. Coat. Technol.* 39–40, 173–181.
50. Ferguen, N., Mebdoua-Lahmar, Y., Lahmar, H., Leclerc, W., Guessasma, M.: DEM model for simulation of crack propagation in plasma-sprayed alumina coatings. *Surf. Coat. Technol.* 371, 287–297 (2019).
51. Yadav, S., Kumar, V., Misra, J.P., Singh, R.K., Upadhyay, V.: Surface modification by electrical discharge machining: A systematic literature reviewbibliometric analysis. *Proc. Inst. Mech. Eng. Part E: J. Process Mech. Eng.* 239(5), 3045–3061 (2023).
52. Borgaonkar, A.V., Syed, I.: Effect of coatings on rolling contact fatigue and tribological parameters of rolling/sliding contacts under dry/lubricated conditions: A review. *Sādhanā.* 45, 1–16 (2020).
53. Asadollahi, A., Rashidi, S., Mohamad, A.A.: Removal of the liquid from a micro-object and controlling the surface wettability by using a rotating shell-Numerical simulation by lattice–boltzmann method. *J. Mol. Liq.* 272, 645–655 (2018),
54. Dutta, V., Thakur, L., Singh, B., Vasudev, H.: A study of erosion–corrosion behaviour of friction stir-processed chromium-reinforced NiAl bronze composite. *Materials.* 15(15), 5401 (2022).
55. Singh, M., Vasudev, H., Kumar, R.: Corrosion and tribological behaviour of BN thin films deposited using magnetron sputtering. *Int. J. Surf. Eng. Interdiscip. Mater. Sci. (IJSEIMS)*, 9(2), 24–39 (2021).

56. Sharanabasava, H., Raviprakash, M., Prasad, C.D., Ramesh, M.R., Phanibhushana, M.V., Vasudev, H., Kumar, S.: Microstructure, mechanical and wear properties of SiC and Mo reinforced NiCr microwave cladding. *Adv. Mater. Process. Technol.* 10(4), 36220–33 (2023).
57. Singh, M., Vasudev, H., Singh, M.: Surface protection of SS-316L with boron nitride based thin films using radio frequency magnetron sputtering technique. *J. Electrochem. Sci. Eng.* 12(5), 851–863 (2022).
58. Prashar, G., Vasudev, H., Thakur, L., Bansal, A.: Performance of thermally sprayed hydroxyapatite coatings for biomedical implants: A comprehensive review. *Surf. Rev. Lett.* 30(1), 2241001 (2023).
59. Vasudev, H.: Wear characteristics of Ni-WC powder deposited by using a microwave route on mild steel: microwave cladding of Ni-WC. *Int. J. Surf. Eng. Interdiscip. Mater. Sci. (IJSEIMS*), 8(1), 44–54 (2020).
60. Parkash, J., Saggu, H.S., Vasudev, H.: A short review on the performance of high velocity oxy-fuel coatings in boiler steel applications. *Mater. Today: Proc.* 50, 1442–1446 (2022).
61. Singh, P., Vasudev, H., Bansal, A.: Effect of post-heat treatment on the microstructural, mechanical, and bioactivity behavior of the microwave-assisted alumina-reinforced hydroxyapatite cladding. *Proc. Inst. Mech. Eng., Part E: J. Process Mech. Eng.* 238(3), 1060–1071 (2024).
62. Verma, P.K., Vasudev, H., Singh, G., Singh, S.: A review on the microstructural and biomedical properties of implants manufactured using additive manufacturing. *J. Inst. Eng. (India): Ser. D.* 106, 1587–1601 (2024).
63. Singh, S.P., Verma, P.K., Singh, S., Kapoor, M., Balu, R.: Microstructural characterization and electrochemical behaviour of HVOF sprayed WC-Co-Cr coatings on 316L boiler steel stainless steel. *Result Surf. Interf.* 16, 100248 (2024).
64. Verma, P.K., Minhas, A.S., Singh, P., Kumar, S.: Slurry erosion behaviour of thermal sprayed Al2O3 and Cr2O3 coatings for turbine steels. *AIP Conf. Proc.* 2986(1), 020012 (2024).
65. Verma, P.K., Kumar, V., Vasudev, H.: Bioactivity and corrosion analysis of thermally sprayed hydroxyapatite based coatings. *J. Electrochem. Sci. Eng.* 14(4), 419–439 (2024).
66. Verma, P.K., Vasudev, H., Kumar, V.: Biocompatibility and corrosion performance of HA-based LVOF sprayed and post-annealed coatings on SS 254 surgical-grade bio-implant material. *Ceram. Int.* 51(10), 13207–13225 (2025).
67. Verma, P.K., Vasudev, H., Singh, G.: Improving the corrosion performance of SS 304 stainless steel using HVOF-sprayed WC-Co-Cr coatings in chloride environments. *J. Bio- Tribo-Corrosion.* 11(4), 1–13 (2025).
68. Prakash, C., Gupta, L.R., Mehta, A., Vasudev, H., Tominov, R., Korman, E., Fedotov, A., Smirnov, V., Kesari, K.K.: Computing of neuromorphic materials: An emerging approach for bioengineering solutions. *Mater. Adv.* 4, 5882–5919 (2023).
69. Singh, J., Gill, H.S., Vasudev, H.: Computational fluid dynamics analysis on role of particulate shape and size in erosion of pipe bends. *Int. J. Interact. Design Manuf. (IJIDeM*). 17(5), 2631–2646 (2023).
70. Verma, P.K., Vasudev, H. and Kumar, V.: Effect of electro-corrosion behavior of different grades of stainless steel used in bio-implants: A review. In Hitesh Vasudev, Chander Prakash, Mohammad Khalid, Kavindra Kesari (eds.): *Advanced Manufacturing Processes,* pp.269–283 (2025), Boca Raton: CRC Press.
71. Verma, P.K., Vasudev, H.: Importance of hydroxyapatite coatings in improving new bone osteogenesis: A review. InHitesh Vasudev, Chander Prakash, Mohammad

Khalid, Kavindra Kesari (eds.): *Advanced Manufacturing Processes*, pp.284–299 (2025), Boca Raton: CRC Press.

72. Rao, G.S., Mukkamala, U., Hanumanthappa, H., Prasad, C.D., Vasudev, H., Shanmugam, B., KishoreKumar, K.C.: Evaluating and optimizing surface roughness using genetic algorithm and artificial neural networks during turning of AISI 52100 steel. *Int. J. Interact. Design Manuf. (IJIDeM)*. 18(8), 6151–6160 (2024).
73. Singh, J., Singh, R., Singh, S., Vasudev, H., Kumar, S.: Reducing scrap due to missed operations and machining defects in 90PS pistons. *Int. J. Interact. Design Manuf. (IJIDeM)*. 17(5), 2527–2539 (2023).
74. Nagabhushana, P., Ramprasad, S., Prasad, C.D., Vasudev, H., Prakash, C.: Numerical investigation on heat transfer of a nano-fluid saturated vertical composite porous channel packed between two fluid layers. *Int. J. Interact. Design Manuf. (IJIDeM)*. 18(5), 2927–2944 (2024).
75. Mahantesh, M.M., Rao, K.R., Chandra, A.P., Vijayakumar, M.N., Nandini, B., Prasad, C.D., Vasudev, H.: Design and modeling using finite element analysis for the sitting posture of computer users based on ergonomic perspective. *Int. J. Interact. Design Manuf. (IJIDeM)*. 18(8), 5875–5891 (2024).
76. Kulkarni, S.D., Manjunatha, Chandrasekhar, U., Manjunath, K.V., Prasad, C.D., Vasudev, H.: Design and optimization of polyvinyl-nitride rubber for tensile strength analysis. *Int. J. Interact. Design Manuf. (IJIDeM)*. 18(8), 5893–5908 (2024).
77. Kattimani, M.A., Venkatesh, P.R., Masum, H., Math, M.M., Bahadurdesai, V.N., Mustafkhadri, S., Prasad, C.D., Vasudev, H.: Design and numerical analysis of tensile deformation and fracture properties of induction hardened inconel 718 superalloy for gas turbine applications. *Int. J. Interact. Design Manuf. (IJIDeM)*. 18(8), 5939–5949 (2024).
78. Poojari, M., Hanumanthappa, H., Prasad, C.D., Jathanna, H.M., Ksheerasagar, A.R., Shetty, P., Shanmugam, B.K., Vasudev, H.: Computational modelling for the manufacturing of solar-powered multifunctional agricultural robot. *Int. J. Interact. Design Manuf. (IJIDeM)*. 18(8), 5725–5736 (2024).
79. Verma, P.K., Vasudev, H., Kumar, V.: 2025. Dental implant materials and designs and failure during implant placement and loading: A review. In Hitesh Vasudev, Chander Prakash, Mohammad Khalid, Kavindra Kesari (eds.): *Advanced Manufacturing Processes*, pp.138–150 (2025), Boca Raton: CRC Press.
80. Verma, P.K., Vasudev, H., Kumar, V.: Bioactivity and corrosion performance of low-velocity oxy-fuel-sprayed hydroxyapatite-based coatings on a surgical-grade stainless steel. *Trans. Indian Inst. Metals*. 78(8), 185 (2025).

9 Microstructure and Wear of HVOF-Sprayed High-Entropy Alloy Coatings

Praveen Kumar Verma and Hitesh Vasudev

9.1 INTRODUCTION

Surface deterioration caused by corrosion and wear is the most important problem in contemporary engineering practice, especially in industries like aerospace, energy, oil and gas, and marine engineering. In comparison to terrestrial drilling, deep-sea drilling at sea is more challenging and dangerous. Drill pipes and drill bits are permanently subjected to aggressive environments in which, apart from corrosive media like chloride ions in marine water and drilling fluids, they are exposed to extensive wear through repeated contact with rock cuttings. Materials with excellent corrosion and wear resistance are required due to the severe nature of the service conditions [1]. Economically and technically speaking, it is crucial to coat drilling equipment with a wear-resistant material so that it lasts longer, works better, and requires less maintenance. Improved mechanical and functional materials are now within reach, thanks to the advent of high-entropy alloys (HEAs). The opposite is true of HEAs, which typically contain five or more main ingredients in nearly equal proportions. They are distinct in four key respects: high-entropy effect, lattice distortion, slow diffusion, and cocktail effect. As a result of these events, HEAs have useful properties, such as stabilizing solid solutions, decreasing the rate of grain formation, improving the thermal stability and strength of microstructures, and many more [2]. Figure 9.1 shows the configurational entropy of low-, medium-, and high-entropy alloys and how they are different from each other.

The microstructural diversity of HEAs is another reason why they have better characteristics. Most HEAs easily form simple FCC or BCC solid solutions because the atomic radii of their parts, such as Co, Cr, Fe, and Ni, are comparable. Of them, the CoCrFeNi and CoCrFeMnNi systems have received a lot of interest because they are structurally stable and have a balanced strength and ductility. Heat treatment improves the microstructural evolution of these alloys even more by making the phases more homogeneous and the grains finer. Research indicates that enough annealing might lead to the precipitation of strengthening phases, such as sigma or BCC second phases, which enhance yield strength and wear resistance [4].

There has also been a lot of research on how heat treatment affects the wear properties of HEAs. In some systems, heat treatment makes wear resistance stronger by changing the microstructures, while in others, it might make the material softer or change how oxides develop, which makes wear resistance weaker. For instance, research on FeCoCrNiMnAl

 DOI: 10.1201/9781003730637-9

Low Entropy Alloy	Medium Entropy Alloy	High Entropy Alloy
$\Delta S_{Conf} \leq 1.0R$	$1.0R \leq \Delta S_{Conf} \leq 1.5R$	$\Delta S_{Conf} \geq 1.5R$
One or two dominant elements. Formation of solid solution and/or compounds	Two to four dominant elements	Five or more dominant elements. Single phase solid solution phase formations (FCC/BCC or both)

FIGURE 9.1 Differences between low-, medium-, and high-entropy alloys based on their configurational entropy [3].

coatings produced via laser cladding indicated decreased wear resistance upon heat treatment at 600°C, while other HEA systems, i.e. AlCoCrFeNiSi, showed dramatically lowered friction coefficients upon high-temperature annealing. Likewise, research on HEA performance in corrosive settings, hydrogen peroxide, or lubricated circumstances underscores the complex interplay between microstructure, environment, and wear mechanisms. People are interested in thermal spray technologies because they can use HEA's new features in surface engineering. Plasma spray coating has shown itself to be a flexible and effective way to make thick, protective layers of advanced metals and composites. Plasma spraying is the use of a high-temperature plasma jet to melt feedstock particles, which are then sped up to the substrate to make a coating by quickly solidifying [5].

Some early research on using thermal spray to process HEAs and CCAs has shown promise. Atmospheric plasma spray (APS) has been used to treat alloys like AlCoCrFeNi and AlCoCrFeNiTi from gas-atomized powders. The resulting coatings are very uniform, durable, and resistant to wear [6]. Compositional design is executed meticulously to attain phase stability and customization of mechanical response. For example, equimolar AlCoCrFeNiTi alloys have been shown to have dendritic microstructures with two BCC phases. These alloys have high compressive strength and strong wear resistance. When compared to other traditional steels like SUJ2 and SKH51, HEAs are superior at resisting wear. This shows that they could be used to replace or strengthen existing materials in high-end applications. Because of these things, plasma spray coatings of HEAs are a particularly interesting area of research and industry [7]. Studying plasma-sprayed HEA coatings is both scientifically and technologically relevant. It fills the gap between alloy design and surface engineering, giving the industry what it needs: high-performance materials that can handle harsh conditions [8].

9.2 HIGH-ENTROPY ALLOYS (HEAS) FOR COATING APPLICATIONS

Coatings are essential for prolonging the life of components exposed to harsh environments like aerospace turbines, oil and gas drilling equipment, and marine structures.

In contrast to traditional alloys, HEAs are engineered through the mixing of more than one principal element, often five or more in near-equimolar ratios, resulting in uncomplicated solid solution microstructures stabilized by high configurational entropy. The novel properties of HEAs, such as solid-solution strengthening, lattice distortion, slow diffusion, and the cocktail effect, make them especially suitable for coating purposes where both mechanical and chemical stability are desirable [9].

From a typological perspective, HEAs can be generally categorized into refractory, lightweight, and corrosion- or wear-resistant systems based on their intended applications. Refractory HEAs, which tend to contain Nb, Mo, Ta, Ti, and W, are intended for high-temperature coatings and show exquisite thermal stability and hardness. For example, Nb–Mo–Ti–Ta–Cr-based HEA coatings deposited via laser cladding on 316L steel showed a microstructure consisting of BCC solid solutions with secondary Laves and HCP phases. The increment of Nb content grain-refined the grains and increased hardness to ~1066 HV, and annealing increased mechanical strength and wear resistance by precipitation hardening and stabilization of the microstructure. Lightweight HEAs, however, contain low-density constituents like Al and Ti to lower density without sacrificing strength. One example is the AlCoCrFeNiTi coating fabricated by laser cladding, which not only showed excellent metallurgical bonding with the steel substrate but also showed superior hardness and wear resistance in comparison with the substrate alloy. Lastly, wear- and corrosion-resistant HEAs like AlCoCrFeNi, CoCrFeNi, and CoCrFeMnNi are especially well suited for thermal spray coatings. Atmospheric plasma-sprayed AlCoCrFeNi coatings deposited on stainless steel exhibited enhanced microhardness after annealing and excellent wear resistance at high temperatures, whereas FeCoCrNi coatings deposited at optimum spray powers exhibited low wear rates and constant friction coefficients under sliding tests [10, 11]. Figure 9.2 depicts a schematic of material property categories for assessing mechanical performance in extreme environmental applications.

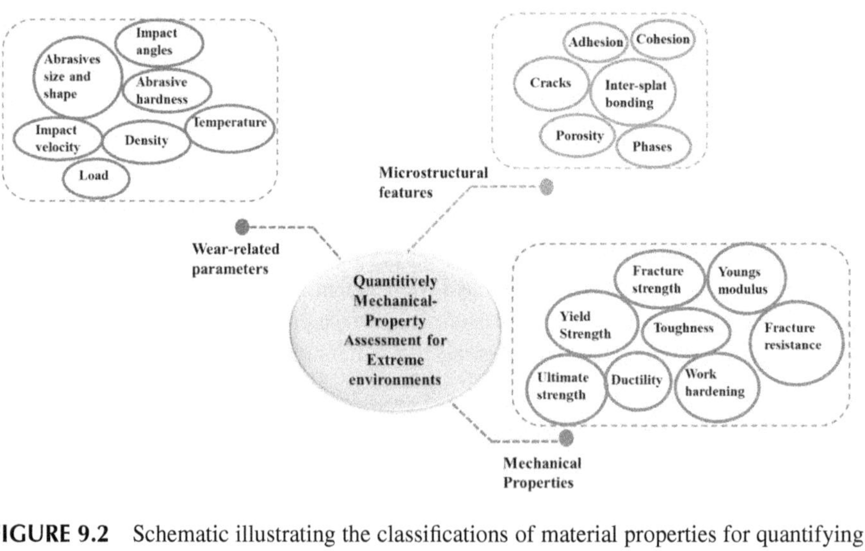

FIGURE 9.2 Schematic illustrating the classifications of material properties for quantifying the mechanical performance for extreme environmental applications [3].

Among the typical HEA systems for coatings, AlCoCrFeNi has been intensively investigated because it can accommodate FCC, BCC, or ordered B2 phases depending on processing conditions. Dense coatings with enhanced wear resistance compared to standard alloys are achieved through atmospheric plasma spraying and high-velocity oxy-fuel (HVOF) deposition of AlCoCrFeNi but porosity and oxide inclusions are still issues [12]. Experiments on AlCoCrFeNi-based coatings have demonstrated that Al, because it has a larger atomic size, enhances lattice strain and hardness, and consequently decreases wear rates. Lattice distortion also increases strength by generating atomic-scale stress fields that hinder slip. The sluggish diffusion effect, a second characteristic of HEAs, decelerates grain growth, phase transformations, and coarsening processes and leads to thermal stability and high-temperature wear resistance over the long term. Heat treatment work on FeCoCrMoNi coatings illustrated that phase transformations including precipitation of σ and μ phases happen gradually with temperature, and wear resistance is maximized at intermediate annealing conditions as a result of a balance between phase stability and strengthening [13]. HEAs provide a multipurpose palette of alloy systems for coatings. Refractory HEAs offer superior hardness and thermal stability in high-temperature conditions, lightweight HEAs decrease density without losing strength, and corrosion- or wear-resistant HEAs provide durability in harsh chemical or tribological environments [14].

9.3 HVOF SPRAYING OF HEAS

High velocity oxy-fuel spraying has shown to be a flexible thermal spray approach for putting down high entropy alloy (HEA) coatings. This is because it can make dense, well-adhering layers with little oxidation compared to typical plasma spraying. HVOF works by burning fuel like kerosene, propane, or hydrogen with oxygen to make a jet of hot gases that moves faster than sound. The alloy particles are melted and propelled to velocities exceeding 600–1000 m/s by this flame. Particles that are still partially or fully melted will smash into splats and solidify upon impact with a substrate. As time passes, this microstructure grows into a lamellar coating. The absence of phase separation and oxidation is prevented by HVOF's cooler flame compared to plasma spray. Because of this, HVOF is an excellent material for HEAs that require corrosion-resistant, robust metastable solid-solution phases [15]. The quality of HEA coatings produced by HVOF is greatly affected by factors such as fuel type, spray distance, powder feeding speed, and the thermal and kinetic characteristics of the particles in flight. A variety of fuel types and oxygen-to-fuel ratios affect flame speed and temperature. Because of this, the melting and oxidation properties of particles are defined. Coating density and corrosion resistance in FeCrMnCoNi HEAs can be improved by using fuel-rich flames in conjunction with suitable feed rates to reduce porosity and oxide inclusions [16]. Particles' temperatures and motion histories are significantly impacted by the standoff distance, the distance between the target and the spray. Although particles travel faster and reach higher temperatures at shorter distances, the substrate may be scorched and oxidized to an excessive degree. Longer distances, on the other hand, don't melt enough and don't bond splats as well since the particles cool down while they are flying. The powder feed rate needs to be just right. If it's too low, the particles may get too hot and splash around, and if it's too high, the particles may not melt and the material may be more porous. In-flight particle velocity and temperature are still the most critical parameters, where higher velocities enhance dense splat deformation and improved adhesion and higher velocities and temperature enhance stronger adhesion, whereas the best thermal input ensures adequate melting without unnecessary oxidation [17]. Figure 9.3 shows the schematic of an HVOF spray gun.

Microstructural control of HEA coatings deposited by HVOF is an active field of research because the technique can regulate porosity, splat morphology, grain refinement, and oxide content. Well-optimized parameters provide coatings with porosities usually below 1%, significantly lower than atmospheric plasma sprayed coatings [19]. From the viewpoint of grain structure, the high cooling rates involved in HVOF solidification retain metastable single-phase solid solutions like face-centred cubic (FCC) or body-centred cubic (BCC) structures, inhibiting undesirable precipitation. For example, AlCrFeCoNi coatings deposited by HVOF retained metastable BCC phases with minimal segregation, whereas post-deposition annealing could provoke dual-phase FCC + BCC structures, impacting hardness and corrosion behaviour. Oxide formation continues to be an ongoing problem, since HEAs tend to have constituents with varying oxidation affinities. Nevertheless, fuel-rich spray conditions, shorter spray distance, and particle temperature control have been shown to inhibit oxide inclusions, resulting in enhanced electrochemical passivation and corrosion resistance. Still, certain oxide stringers are usually unavoidable in HVOF-sprayed HEAs, and their management is dependent on careful optimization of both feedstock properties and flame chemistry.

A critical comparison between HVOF and other thermal spray methods brings out its strengths and weaknesses in HEA applications. In comparison to plasma spraying, HVOF has lower particle temperatures but significantly higher velocities, resulting in more dense coatings with smaller microstructures and less oxidation. Plasma-sprayed coatings are generally thicker but have greater porosity and oxide content, as shown in comparison testing of AlCoCrFeNi coatings where APS coatings showed more

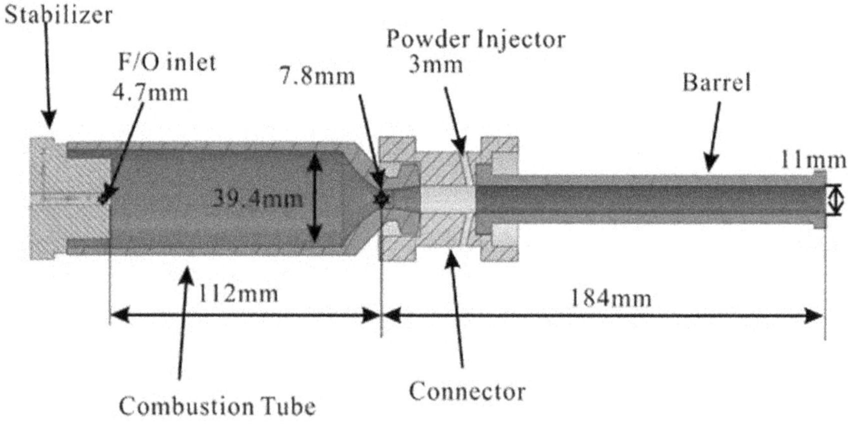

FIGURE 9.3 Schematic of HVOF spray gun [18].

oxide inclusions and lower hardness than their HVOF counterparts [20]. Cold spray, however, solely depends on the application of high kinetic energy to deposit solid-state particles without melting. This suppresses oxidation and phase transformation but dense bonding is harder to achieve in high-melting-point HEAs. Though cold spray yields oxide-free coatings, HVOF is more flexible as regards coating hardness and adhesion because of partial feedstock melting [21]. HVOF spraying in general offers a compromise between the high coating density and deposition efficiency of plasma spray and the low phase retention and oxidation of cold spray. In an HEA, the capacity to deposit a dense, adherent film without losing metastable solid solutions makes it an excellent choice for wear- and corrosion-resistant applications [22].

9.4 MICROSTRUCTURE ANALYSIS OF HVOF-SPRAYED HEA COATINGS

The phase distribution and grain structure of coatings are controlled both by the feedstock chemistry and the spraying conditions. Because of high kinetic energies and comparatively moderate temperatures of particles, HVOF coatings tend to have finer grains with minimal elemental segregation relative to plasma-sprayed equivalents. XRD scans commonly identify the existence of single-phase solid solutions, either face-centred cubic (FCC) or body-centred cubic (BCC), depending on the composition of alloys, whereas in certain systems, mixed FCC + BCC microstructures are stabilized by rapid solidification. For instance, AlCoCrFeNi and FeCrMnCoNi coatings preserve FCC/BCC solid-solution phases after spraying, although limited phase decomposition into ordered arrays or intermetallic has been noted for particles with excessive heating [23]. SEM/FESEM observations indicate lamellar microstructures with well-bonded splats, with higher magnification indicating nanoscale cellular or dendritic microstructure in the splats due to rapid solidification. Transmission electron microscope (TEM) work further supports fine sub-grain microstructures, networks of dislocations, and sporadic nano-precipitates in coatings, which are the cause of their increased hardness and resistance to wear [24]. Figure 9.4 shows SEM secondary electron images of CoCrFeNiMn high-entropy alloy powders: deposited CS powders (a,c), HVOF powders (b,d). Figure 9.5 shows the XRD patterns of CoCrFeNiMn high-entropy alloy powders: CS powders (a) and HVOF powders (b).

Elemental distribution in HVOF-sprayed HEAs has been extensively analyzed with energy-dispersive spectroscopy (EDS) and electron backscatter diffraction (EBSD). Near-homogeneous distribution of major elements between splats is usually exhibited by these analyses, with mere marginal segregation of Al or Cr in Al-containing HEAs. EBSD mapping reveals information on crystallographic texture and grain boundary features, typically in the form of fine equiaxed grains with high-angle grain boundaries that are favourable for strengthening. In addition, EBSD phase maps assist in verifying the coexistence of FCC and BCC domains in certain multi-principal element systems. Phase stability after spraying is not necessarily absolute even under rapid solidification; metastable FCC or BCC phases can change under subsequent thermal exposure. For instance, Al-rich coatings of AlxCoCrFeNi exhibit a tendency to undergo FCC → BCC or FCC → ordered B2 transformations upon spraying or heat treatment, whereas some Co- or Mn-enriched HEAs maintain their

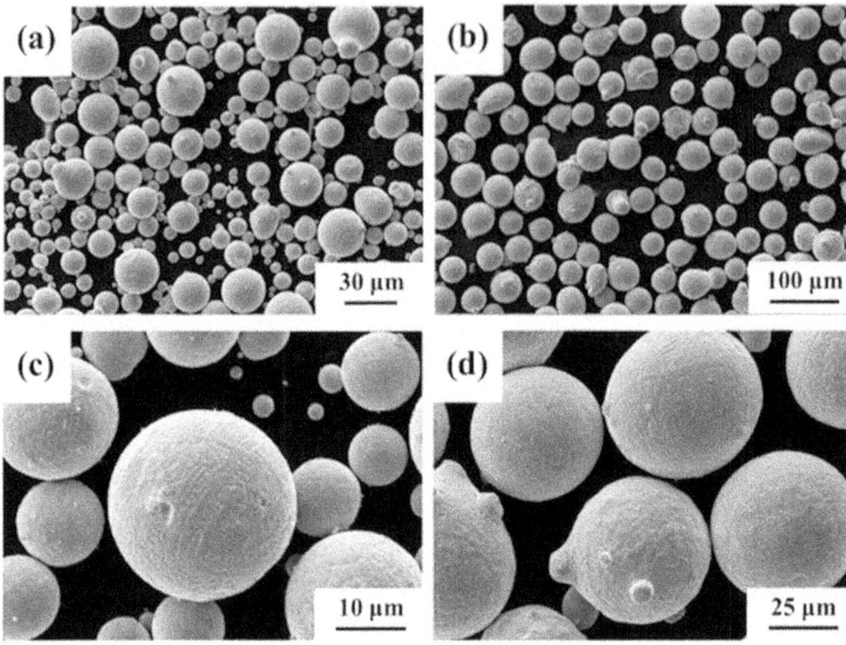

FIGURE 9.4 SEM secondary electron images of CoCrFeNiMn high-entropy alloy powders: (**a**,**c**) CS powders; (**b**,**d**) HVOF powders [25].

FCC structures more strongly. These transformations are important because they have a direct impact on hardness, ductility, and corrosion resistance of the coatings [26].

Oxides, carbides, and sometimes borides are also significant constituents developed during HVOF spraying as a result of the reactive condition and alloying elements in HEAs. The presence of oxygen from the flame or ambient atmosphere can result in fine oxide stringers or nanometric oxide particles distributed in the matrix. Although excessive oxide development is harmful, controlled oxide dispersion can increase hardness by dispersion strengthening. Cr- and Al-rich HEAs are particularly susceptible to the formation of protective oxides like Cr_2O_3 or Al_2O_3 upon spraying, which are responsible for passivation and better corrosion performance. Analogously, carbides like (Cr,Fe)7C3 or (Nb,Ti)C have been observed in certain carbide-forming HEAs, especially when carbon is incorporated through the feedstock or substrate reaction [27]. The stabilization of FCC, BCC, or mixed phases is contingent upon alloy chemistry and spray conditions, with post-deposition transition being common in Al-rich systems. The controlled growth of oxides, carbides, and borides during spraying can either help or hurt performance, depending on how much and how evenly they are spread out [28].

9.5 WEAR BEHAVIOUR OF HEA COATINGS

High-entropy alloy coatings formed by thermal spray methods, especially HVOF spraying, have attracted more and more interest because of their excellent wear

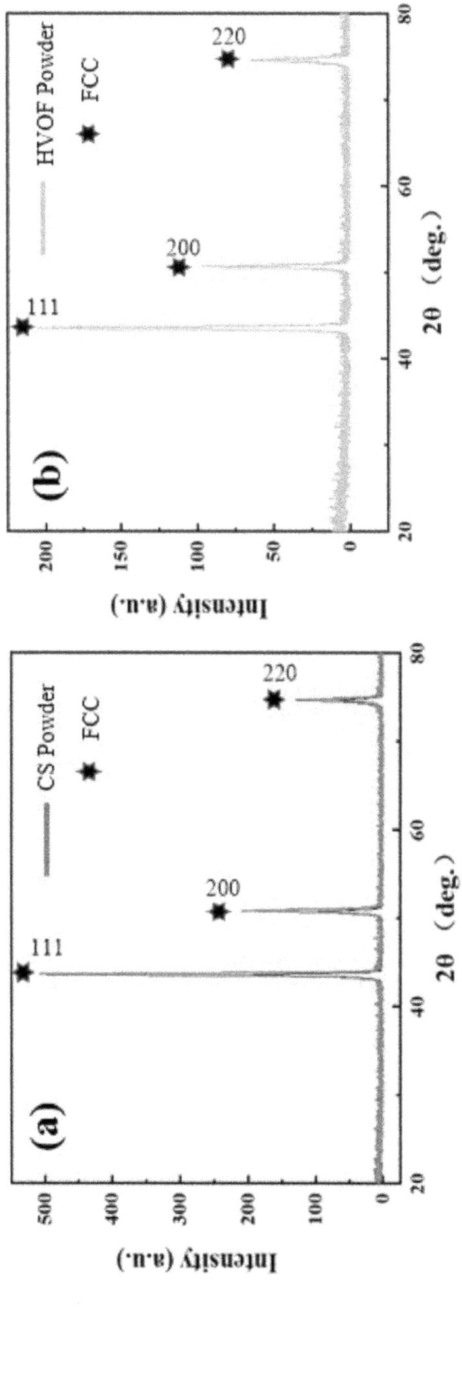

FIGURE 9.5 XRD patterns of CoCrFeNiMn high-entropy alloy powders: (**a**) CS powders; (**b**) HVOF powders [25].

resistance over traditional coatings. Their tribological performance has been comprehensively studied under a variety of operating conditions, such as dry sliding, high-temperature wear, corrosive medium, and slurry erosion conditions. Under conditions of dry sliding, HEA coatings like AlCoCrFeNi, CoCrFeMnNi, and FeCoCrNiMo generally have lower friction coefficients and wear rates compared to common coatings because of their fine-grained microstructures, high hardness, and solid-solution strengthening. Most HEAs at high temperatures exhibit superior wear stability due to the sluggish diffusion effect, which inhibits coarsening as well as softening of microstructures. In addition, Al- and Cr-doped HEAs also develop protective films of oxides at elevated temperatures that limit direct metal–metal contact and exhibit self-lubrication. In corrosive media like saline solutions or acidic solutions, HEA coatings exhibit superior wear–corrosion resistance than that of Ni-based alloys due to stable passive films of mixed oxides such as Cr_2O_3, Al_2O_3, and spinel-type oxides. Slurry erosion tests indicate that HEA coatings exhibit a stronger resistance to material loss than WC-Co or Cr_3C_2-NiCr coatings, mainly as a result of their high toughness and crack-arresting lamellar structures generated during spraying [29]. Figure 9.6 shows SEM images (BSD) of the CrFeCoNi surface in untreated (left) and borided (right) states following a reciprocating wear test.

Wear mechanisms in HEA coatings are complex and are a function of the environment. Under dry sliding conditions, abrasive and adhesive wear prevail with the

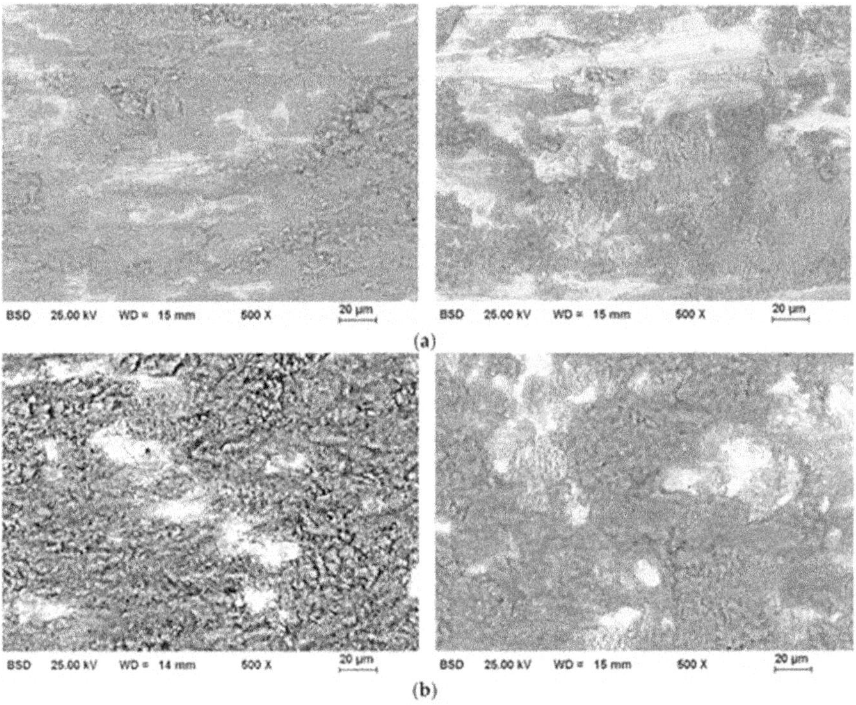

FIGURE 9.6 SEM images (BSD) of CrFeCoNi surface in untreated (left) and borided (right) state after reciprocating wear test: (**a**) casting and (**b**) HVOF coating [30].

occurrence of grooves and micro-ploughing on SEM analysis. Severe abrasive wear is inhibited by the HEAs' high hardness, whereas their multi-component matrix behaves more effectively against adhesive transfer than softer Ni-based coatings. Oxidative wear gains prominence at higher temperatures, where metal-to-metal contact is inhibited by the development of continuous oxide films but can also cause spallation if the film is brittle. Fatigue wear is important in cyclic or impact-loaded situations, such as slurry erosion, where continuous particle impacts induce the nucleation and growth of subsurface cracks [31].

Recent literature has put a lot of stress on the link between the microstructure of HEA coatings and how well they wear. HVOF-sprayed coatings feature high porosity and thick lamellar microstructures that make them better at bonding between splats. This makes them resistant to delamination when sliding or hitting something. Grain refinement, which happens when the material cools quickly while spraying, makes the material harder and helps it withstand wear better by making the plastic deformation under the wear track thinner. The structure of the phase is also very important: FCC-majority HEAs are usually more ductile and robust, and they are also more resistant to fatigue and adhesive wear. On the other hand, BCC- or duplex FCC + BCC structures are usually harder, which makes them more resistant to abrasive wear. For instance, Al-based HEAs with both FCC and BCC phases are harder and more resistant to wear than alloys with only FCC phases. Also, the formation of oxide films during wear, notably Cr- and Al-rich oxides, creates a second layer of protection by reducing direct contact and keeping tribological performance up under oxidative or corrosive wear conditions [32].

Because of their unique microstructural features and phase stability, HEA coatings show great wear behaviour in a wide range of circumstances. Their tribological behaviour is regulated by a combination of abrasive, adhesive, oxidative, and fatigue wear processes, which are mitigated by high hardness, grain refinement, and stable oxide formation. The microstructure-performance connection shows how important phase composition, porosity control, and secondary phase production are for getting better wear resistance. Compared to older coatings like WC-Co, Cr_3C_2-NiCr, and Ni-based alloys, HEAs are the greatest in resisting wear and corrosion. This makes them the next generation of protective coatings for tough industrial uses [33].

9.6 SYNERGISTIC PROPERTIES

High-entropy alloy coatings, especially those made by HVOF spraying, have a unique mix of hardness, toughness, and resistance to breaking because of their multicomponent solid-solution microstructures and fine-grained structures. HEAs are different from ordinary alloy systems because they find a compromise between hardness and toughness by using solid-solution strengthening, grain refinement, and severe lattice distortion to work together. For example, AlCoCrFeNi and CoCrFeMnNi coatings sprayed with HVOF usually have hardness levels above 700–900 HV because they have fine FCC + BCC multiphase microstructures and are robust enough to resist catastrophic brittle failure under load. Dispersing heterogeneous microstructures with nanoscale sub-grains and high-angle grain boundaries makes fracture resistance even better by redirecting cracks and spreading out stress. This makes the damage-tolerance profile much better than that of monolithic carbides or ceramic-rich coatings [34]. In corrosive sliding or slurry conditions, corrosion is accelerated by wear through the degradation of passive films, while corrosion undermines the microstructure, thus facilitating wear a process referred to as corrosion–wear synergy. HEA coatings counteract this action through their multiprincipal element chemistry, which favours the development of stable, adherent Cr-, Al-, or Ti-rich oxide films. This is different from typical alloys, which expand or coarsen quickly. For instance, Al- and Cr-based HEA coatings have oxide scales that are continuous and stick to the surface. These scales are made up of Al_2O_3 and Cr_2O_3 and can protect against wear and oxidation at high temperatures. Studies on AlCoCrFeNi and FeCoCrNi coatings have shown that these protective oxides not only keep oxygen from getting in at temperatures below 800–1000°C, but they also have a self-lubricating effect that lowers the coefficient of friction. At the same time, the solid-solution phases underneath stay hard and don't get too soft, which means they can still resist wear at high temperatures [35].

9.7 CHALLENGES AND FUTURE SCOPE

In the past few years, optimized HEAs with hard phases have shown great improvements in wear, hardness, and resistance to oxidation. For instance, TiC reinforcements in FeCoCrNiCu HEA matrices have been incorporated either by the direct addition of TiC powder or by in situ synthesis of TiC from precursors such as Ti powder or B_4C, Cr_3C_2, etc. These not only act as second phases to make things harder, but they also help make the grains finer by speeding up the nucleation rate of the solid solution matrix when it cools. In the same way, oxide dispersoids (such as spinels and mixed oxides) or oxide reinforcements have been used to make things more resistant to corrosion and wear or more stable at high temperatures. Nanostructured HEA coatings are another step forward. They use nanoscale metal or ceramic reinforcements, or feedstocks that make very fine grain size (in some cases down to tens of nm), to make coatings that are much harder, less porous, more resistant to erosion, and tougher than coarse coatings [40]. Functionally graded HEA coatings are being investigated to customize properties through thickness. For instance, by grading composition or reinforcement material, one can have a hard, ductile bonding layer close to substrate grading into a hard, wear-resistant layer at the surface. Although still in its infancy in HEA thermal-sprayed coatings (fewer publications than in other coating systems), the principle is being taken up in other neighbouring disciplines and should also work well for HEAs. Hybrid methods are among the more useful recent techniques: such as combining HVOF spraying with laser remelting, or post-spraying heat treatment, or nitriding/plasma treatments. Laser remelting of HVOF coatings (e.g. WC-NiCr coatings; though not necessarily HEA) has shown significant reductions in porosity (down to ~70-plus %), surface smoothing, microhardness improvement, and enhanced wear performance [36].

This study investigates the microstructure and electrochemical characteristics of WC-Co-Cr coatings deposited on 316L boiler steel using HVOF spraying. The coatings were less likely to wear down and corrode because they had a dense microstructure and low porosity. Electrochemical experiments showed that the materials were less likely to rust at high temperatures because they had stable carbide phases with cobalt and chromium in them. After coating, heat treatment greatly improved the microstructure, which made it operate better in harsh settings. These results indicate that WC-Co-Cr coatings applied using HVOF are suitable for corrosive, high-temperature settings [37]. Research has assessed the slurry erosion resistance of thermally sprayed CrO_3 and AlO_3 coatings on turbine steels. The research indicated that CrO_3 coatings exhibited superior erosion resistance compared to AlO_3 coatings, attributable to their increased hardness and enhanced microstructural durability. The coatings were tested on turbines where erosion is a huge problem to see if they would function. They did this by hitting them at varied angles and speeds. The results indicate how crucial it is to choose the right coating materials to make turbine parts survive longer and operate better [38].

A study investigated the utilization of HVOF-sprayed WC-Co-Cr coatings to improve the corrosion resistance of SS 304 stainless steel. The study discovered that the coatings make the material far less likely to corrode from chloride, which makes it safe to use in chemical processing and on ships. Electrochemical tests demonstrated that the coatings make a protective barrier that slows down corrosion and makes steel last longer in harsh conditions. The findings indicate that WC-Co-Cr coatings are highly effective in preventing corrosion in environments with elevated chloride levels [39]. Research examined the bioactivity and corrosion resistance of hydroxyapatite (HA)-based coatings applied to surgical-grade stainless steel utilizing Low-Velocity Oxy-Fuel (LVOF) spraying technology. The study contrasted pure HA coatings with HA composites incorporating micrometric and nanometric Al_2O_3. The findings demonstrate that HA-AlO_3 composites have improved mechanical properties and corrosion resistance, making them more appropriate for biomedical applications [40].

In addition to that, the HEA design workflows based on ML are also being utilized to predict phases and mechanical behaviour from large datasets. In the future advancements in the integration of designed reinforcements, nano-structuring, graded architectures, and hybrid post-treatments will yield HEA coatings with performance exceeding most of the conventional coatings under extreme wear, corrosion, and high-temperature conditions. But to make this promise a reality, efforts need to be continued in acquiring reproducible, high-quality powders; comprehending the full details of the in-flight and deposit microstructural development (phase transformations, oxidation, residual stresses); determining optimal post-processing conditions; and scaling up processes [41].

The biomedical and high-temperature corrosion-resistant systems slurry processing sectors have commenced the use of thermal spray coatings and innovative surface engineering techniques. This is particularly true when combined with emerging technologies such as additive manufacturing, magnetron sputtering thin-film deposition, and microwave-assisted fabrication [42–55]. Additionally, the coating structures exhibit high strength (tensile, compressive, and flexural) and functional performance in solar energy conversion coating systems, as well as substantial potential in surface topography. Finally, the developed surfaces serve as a foundation for advanced computational fluid dynamics of pipes and conduits with bends, as well as ergonomic seating systems that incorporate anthropometric, mechanical, and structural connections [56–68].

9.8 CONCLUSION

The link between microstructure and wear behaviour in HVOF-sprayed HEA coatings is what makes them work so well. When deposition happens quickly, it forms fine-grained, dense lamellar microstructures with metastable solid-solution phases. This makes the material harder, tougher, and more resistant to wear from abrasives, adhesives, and oxidation. The production of stable oxide scales makes corrosion–wear synergy even better, making sure that the material lasts in harsh conditions. HEAs are different from regular coatings because they combine high hardness with ductility and corrosion resistance. This makes them useful for protecting materials under demanding operating circumstances. While promising, there are still some big problems to solve when it comes to making compositionally uniform powders, keeping phase stability during spraying, lowering porosity and oxides, and making processes bigger so they can be used in industry. Nanostructuring, graded coating topologies, reinforcement methods, and the use of computational tools like CALPHAD and machine learning can help fill in these gaps and speed up progress. So, HVOF-sprayed HEA coatings are ready to change the face of surface engineering by providing long-lasting, high-performance options for tough jobs in the aerospace, marine, and energy industries.

REFERENCES

1. Zhou YK, Kang JJ, Zhang J, Zhu S, Fu ZQ, Zhu LN, She DS. Effect of nitriding on microstructure and wear behavior of HVOF sprayed AlxCoCrFeNi (x= 0.4, 0.7, 1.0) high-entropy alloy coatings. *Intermetallics.* 2022 Dec 1;151:107709.
2. Löbel M, Lindner T, Mehner T, Lampke T. Microstructure and wear resistance of AlCoCrFeNiTi high-entropy alloy coatings produced by HVOF. *Coatings.* 2017 Sep 12;7(9):144.
3. Bhaskaran Nair R, Supekar R, Morteza Javid S, Wang W, Zou Y, McDonald A, Mostaghimi J, Stoyanov P. High-entropy alloy coatings deposited by thermal spraying: A review of strengthening mechanisms, performance assessments and perspectives on future applications. *Metals.* 2023 Mar 13;13(3):579.
4. Meghwal A, Singh S, Anupam A, King HJ, Schulz C, Hall C, Munroe P, Berndt CC, Ang AS. Nano-and micro-mechanical properties and corrosion performance of a HVOF sprayed AlCoCrFeNi high-entropy alloy coating. *Journal of Alloys and Compounds.* 2022 Aug 15;912:165000.
5. Chen L, Bobzin K, Zhou Z, Zhao L, Oete M, Königstein T, Tan Z, He D. Wear behavior of HVOF-sprayed Al0. 6TiCrFeCoNi high entropy alloy coatings at different temperatures. *Surface and Coatings Technology.* 2019 Jan 25;358:215–22.
6. Silvello A, Cavaliere P, Yin S, Lupoi R, García Cano I, Dosta S. Microstructural, mechanical and wear behavior of HVOF and cold-sprayed high-entropy alloys (HEAs) coatings. *Journal of Thermal Spray Technology.* 2022 Apr;31(4):1184–206.
7. Zhou YK, Kang JJ, Zhang J, Fu ZQ, Zhu LN, She DS, Yue W. Microstructure and sliding wear behavior of HVOF sprayed Al (1− x) CoCrFeNiTix high-entropy alloy coatings. *Materials Letters.* 2022 May 1;314:131929.
8. Da Q, Kang JJ, Ma GZ, Zhou YK, Fu ZQ, Zhu LN, She DS, Wang HD. Investigation on the microstructure, hardness, wear, and corrosion properties of HVOF-sprayed AlCoCrFeNiB high-entropy alloy coating. *Journal of Thermal Spray Technology.* 2025 Aug 8;34(7):1–7.
9. Patel P, Alidokht SA, Sharifi N, Roy A, Harrington K, Stoyanov P, Chromik RR, Moreau C. Microstructural and tribological behavior of thermal spray CrMnFeCoNi high entropy alloy coatings. *Journal of Thermal Spray Technology.* 2022 Apr;31(4):1285–301.

10. Chen L, He D, Han B, Guo Z, Zhang L, Lu L, Wang X, Tan Z, Zhou Z. Effect of laser remelting on wear behavior of HVOF-sprayed FeCrCoNiTiAl0. 6 high entropy alloy coating. *Applied Sciences*. 2020 Oct 16;10(20):7211.
11. Zhang D, Li Q, Chang C, Zhang G, Ma X, Sun R. Effects of Mn addition on wear and corrosion resistances of AlCoCrFeNi high-entropy alloy coating sprayed by HVOF. *Journal of Materials Research and Technology*. 2025 Jan 1;34:627–38.
12. Xiao JK, Tan H, Wu YQ, Chen J, Zhang C. Microstructure and wear behavior of FeCoNiCrMn high entropy alloy coating deposited by plasma spraying. *Surface and Coatings Technology*. 2020 Mar 15;385:125430.
13. Wei Z, Wu Y, Hong S, Cheng J, Qiao L, Cheng J, Zhu S. Ultrasonic cavitation erosion behaviors of high-velocity oxygen-fuel (HVOF) sprayed AlCoCrFeNi high-entropy alloy coating in different solutions. *Surface and Coatings Technology*. 2021 Mar 15;409:126899.
14. Löbel M, Lindner T, Mehner T, Rymer LM, Björklund S, Joshi S, Lampke T. Microstructure and corrosion properties of AlCrFeCoNi high-entropy alloy coatings prepared by HVAF and HVOF. *Journal of Thermal Spray Technology*. 2022 Jan;31(1):247–55.
15. Patel P, Munagala VN, Roy A, Sharifi N, Alidokht SA, Makowiec M, Chromik RR, Moreau C, Stoyanov P. Dry sliding friction and wear behavior of HVOF sprayed FeCrMnCoNi high entropy coatings (HECs) at room and elevated temperatures. *Surface and Coatings Technology*. 2025 Jul 1;507:132138.
16. Rymer LM, Lindner T, Lampke T. Nb and Mo influencing the high-temperature wear behavior of HVOF-sprayed high-entropy alloy coatings. *Coatings*. 2022 Dec 21;13(1):9.
17. Da Q, Kang JJ, Ma GZ, Zhou YK, Fu ZQ, Zhu LN, She DS, Liang J, Wang HD. Microstructure, mechanical property, reciprocating sliding wear and electrochemical corrosion of AlCoCrFeNi high entropy alloy coatings with Si addition prepared by HVOF spraying. *Journal of Iron and Steel Research International*. 2025 Mar 13:1–8.
18. Kannan S, Ba T, Wan S, Kang CW, Pan J, Zhang Z. Numerical and experimental investigations of CoNiCrAlY particle suspension dynamics in kerosene-oxygen high velocity oxygen fuel spraying. *Coatings*. 2023 Mar 23;13(4):668.
19. Silvello A, Torres Diaz E, Rúa Ramirez E, García Cano I. Microstructural, mechanical and wear properties of atmospheric plasma-sprayed and high-velocity oxy-fuel AlCoCrFeNi equiatomic high-entropy alloys (HEAs) coatings. *Journal of Thermal Spray Technology*. 2023 Mar;32(2):425–42.
20. Abhijith NV, Kumar D, Kalyansundaram D. Development of single-stage TiNbMoMnFe high-entropy alloy coating on 304L stainless steel using HVOF thermal spray. *Journal of Thermal Spray Technology*. 2022 Apr;31(4):1032–44.
21. Lei Y, Jiang Z, Yang K, Tong B, Liu X, Zhang S, Yang Y. Microstructure evolution and high temperature performance of AlCrCu1-xMoxNi high-entropy alloy coatings prepared by HVOF. *Materials Characterization*. 2025 May 1;223:114953.
22. Patel P, Nair RB, Supekar R, McDonald A, Chromik RR, Moreau C, Stoyanov P. Enhanced wear resistance of AlCoCrFeMo high entropy coatings (HECs) through various thermal spray techniques. *Surface and Coatings Technology*. 2024 Feb 15;477:130311.
23. Yang ML, Xu JL, Huang J, Zhang LW, Luo JM. Wear resistance of N-doped CoCrFeNiMn high entropy alloy coating on the Ti-6Al-4V alloy. *Journal of Thermal Spray Technology*. 2024 Oct;33(7):2408–18.
24. Samoilova O, Shaburova N, Samodurova M, Pashkeev K, Ostovari Moghaddam A, Trofimov E. Microstructure and wear behavior of Al0. 25CoCrFeNiSi0. 6 high-entropy

alloy coating deposited on stainless steel by detonation spraying. *Journal of Thermal Spray Technology*. 2023 Jun;32(5):1220–9.

25. Wang H, Kang J, Yue W, Jin G, Li R, Zhou Y, Liang J, Yang Y. Microstructure and corrosive wear properties of CoCrFeNiMn high-entropy alloy coatings. *Materials*. 2022 Dec 21;16(1):55.
26. Majji BG, Vasudev H, Bansal A. A review on the oxidation and wear behavior of the thermally sprayed high-entropy alloys. *Materials Today: Proceedings*. 2022 Jan 1;50:1447–51.
27. Xiao JK, Wu YQ, Chen J, Zhang C. Microstructure and tribological properties of plasma sprayed FeCoNiCrSiAlx high entropy alloy coatings. *Wear*. 2020 May 15;448:203209.
28. Kumar H, Bhaduri GA, Manikandan SG, Kamaraj M, Shiva S. Microstructural characterization and tribological properties of atmospheric plasma sprayed high entropy alloy coatings. *Journal of Thermal Spray Technology*. 2022 Aug;31(6):1956–74.
29. Chen L, Wang X, Ji Y, Yao H, He D, Tan Z, Zhou Z, Bi G. Elevated-temperature wear behavior and mechanisms of APS-sprayed Al0. 6TiCrFeCoNi high entropy alloy coating. *Surface and Coatings Technology*. 2024 Feb 15;477:130294.
30. Löbel M, Lindner T, Hunger R, Berger R, Lampke T. Precipitation hardening of the HVOF sprayed single-phase high-entropy alloy CrFeCoNi. *Coatings*. 2020 Jul 20;10(7):701.
31. Meghwal A, Anupam A, Murty BS, Berndt CC, Kottada RS, Ang AS. Thermal spray high-entropy alloy coatings: A review. *Journal of Thermal Spray Technology*. 2020 Jun;29(5):857–93.
32. Liu S, Peng Y, Zhang Y, Wang Y, Fan W, Wang A, Zhang W, Tan Y, Ma Q, Lan Y. Effect of nanostructure on wear and corrosion behavior of HVAF-sprayed eutectic high-entropy alloy coatings. *Journal of Thermal Spray Technology*. 2022 Apr;31(4):1252–62.
33. Li T, Liu Y, Liu B, Guo W, Xu L. Microstructure and wear behavior of FeCoCrNiMo0. 2 high entropy coatings prepared by air plasma spray and the high velocity oxy-fuel spray processes. *Coatings*. 2017 Sep 20;7(9):151.
34. Kiape S, Glava M, Georgatis E, Kamnis S, Matikas TE, Karantzalis AE. CoCrFeMnNi0. 8V/Cr3C2-Ni20Cr high-entropy alloy composite thermal spray coating: Comparison with monolithic CoCrFeMnNi0. 8V and Cr3C2-Ni20Cr coatings. *Coatings* 2024, 14, 402.
35. Zhang LW, Xu JL, Chen J, Huang J, Luo JM. Microstructure and corrosion resistance of AlCoCrFeNi high-entropy alloy coating on sintered NdFeB magnets prepared by HVOF spraying. *Journal of Magnetism and Magnetic Materials*. 2022 Jun 1;551:169136.
36. Abhijith NV, Kumar D. Influence of postprocessing on microstructural and tribological behavior of HVOF-sprayed HEA coating. In Sujeet Kumar Sinha, Deepak Kumar, Nitya Nand Gosvami and Prathima Nalam (Eds.) *International Conference on Industrial Tribology* 2022 Dec 12 (pp. 183–95). Singapore: Springer Nature Singapore.
37. Singh SP, Verma PK, Singh S, Kapoor M, Balu R. Microstructural characterization and electrochemical behaviour of HVOF sprayed WC-Co-Cr coatings on 316L boiler steel stainless steel. *Results in Surfaces and Interfaces*. 2024 Aug 1;16:100248.
38. Verma PK, Minhas AS, Singh P, Kumar S. Slurry erosion behaviour of thermal sprayed Al2O3 and Cr2O3 coatings for turbine steels. In: *AIP Conference Proceedings 2024 Feb* 20 (Vol. 2986, No. 1, p. 020012). AIP Publishing LLC.
39. Verma PK, Vasudev H, Singh G. Improving the Corrosion Performance of SS 304 Stainless Steel Using HVOF-Sprayed WC-Co-Cr Coatings in Chloride Environments. *Journal of Bio-and Tribo-Corrosion*. 2025 Dec;11(4):1–3.

40. Verma PK, Vasudev H, Kumar V. Bioactivity and corrosion performance of low-velocity oxy-fuel-sprayed hydroxyapatite-based coatings on a surgical-grade stainless steel. *Transactions of the Indian Institute of Metals*. 2025 Aug;78(8):185.
41. Du P, Liu C, Hu H, Zhang C, Fan M, Gao M, Chen T. Study of the tribological properties of HVOF-Sprayed Ni-based coatings on Ti6Al4V titanium alloys. *Coatings*. 2022 Dec 16;12(12):1977.
42. Verma PK, Vasudev H, Singh, G. Improving the corrosion performance of SS 304 stainless steel using HVOF-sprayed WC-Co-Cr coatings in chloride environments. *Journal of Bio-and Tribo-Corrosion*. 2025;11(4):1–13.
43. Dutta V, Thakur L, Singh B, Vasudev H. A study of erosion–corrosion behaviour of friction stir-processed chromium-reinforced NiAl bronze composite. *Materials*. 2022;15(15):5401.
44. Singh M, Vasudev H, Kumar R. Corrosion and tribological behaviour of BN thin films deposited using magnetron sputtering. *International Journal of Surface Engineering and Interdisciplinary Materials Science (IJSEIMS)*. 2021;9(2):24–39.
45. Sharanabasava H, Raviprakash M, Prasad CD, Ramesh MR, Phanibhushana MV, Vasudev H, Kumar S. Microstructure, mechanical and wear properties of SiC and Mo reinforced NiCr microwave cladding. *Advanced Materials and Processing Technologies*. 2023;10(4):3620–33.
46. Verma PK, Minhas AS, Singh P, Kumar S. Slurry erosion behaviour of thermal sprayed Al2O3 and Cr2O3 coatings for turbine steels. *AIP Conference Proceedings*. 2024;2986(1):020012.
47. Singh M, Vasudev H, Singh M. Surface protection of SS-316L with boron nitride based thin films using radio frequency magnetron sputtering technique. *Journal of Electrochemical Science and Engineering*. 2022;12(5):851–63.
48. Prashar G, Vasudev H, Thakur L, Bansal A. Performance of thermally sprayed hydroxyapatite coatings for biomedical implants: A comprehensive review. *Surface Review and Letters*. 2023;30(1):2241001.
49. Vasudev H. Wear characteristics of Ni-WC powder deposited by using a microwave route on mild steel: Microwave cladding of Ni-WC. *International Journal of Surface Engineering and Interdisciplinary Materials Science (IJSEIMS)*. 2020;8(1):44–54.
50. Parkash J, Saggu HS, Vasudev H. A short review on the performance of high velocity oxy-fuel coatings in boiler steel applications. *Materials Today: Proceedings*. 2022;50:1442–6.
51. Verma PK, Vasudev H, Kumar V. Biocompatibility and corrosion performance of HA-based LVOF sprayed and post-annealed coatings on SS 254 surgical-grade bio-implant material. *Ceramics International*. 2025;51(10):13207–25.
52. Singh P, Vasudev H, Bansal A. Effect of post-heat treatment on the microstructural, mechanical, and bioactivity behavior of the microwave-assisted alumina-reinforced hydroxyapatite cladding. *Proceedings of the Institution of Mechanical Engineers, Part E: Journal of Process Mechanical Engineering*. 2024;238(3):1060–71.
53. Verma PK, Vasudev H, Singh G, Singh S. A review on the microstructural and biomedical properties of implants manufactured using additive manufacturing. *Journal of The Institution of Engineers (India): Series D,* 2024;106:1554–601.
54. Prashar, G., Vasudev, H. and Thakur, L., 2021. Influence of heat treatment on surface properties of HVOF deposited WC and Ni-based powder coatings: a review. *Surface Topography: Metrology and Properties*, 9(4):043002.
55. Verma PK, Kumar V, Vasudev H. Bioactivity and corrosion analysis of thermally sprayed hydroxyapatite based coatings. *Journal of Electrochemical Science and Engineering*. 2024;14(4):419–39.

56. Kulkarni SD, Manjunatha, Chandrasekhar U, Manjunath KV, Prasad CD, Vasudev H. Design and optimization of polyvinyl-nitride rubber for tensile strength analysis. *International Journal on Interactive Design and Manufacturing (IJIDeM)*. 2024;18(8):5893–908.
57. Verma PK, Vasudev H. Importance of hydroxyapatite coatings in improving new bone osteogenesis: A review. In Hitesh Vasudev, Chander Prakash, Mohammad Khalid, Kavindra Kesari (Eds.) *Advanced Manufacturing Processes*, 2025. pp.284–99, Boca Raton, CRC Press.
58. Kattimani MA, Venkatesh PR, Masum H, Math MM, Bahadurdesai VN, Mustafkhadri S, Prasad CD, Vasudev H. Design and numerical analysis of tensile deformation and fracture properties of induction hardened inconel 718 superalloy for gas turbine applications. *International Journal on Interactive Design and Manufacturing (IJIDeM)*. 2024;18(8):5939–49.
59. Poojari M, Hanumanthappa H, Prasad CD, Jathanna HM, Ksheerasagar AR, Shetty P, Shanmugam BK, Vasudev H. Computational modelling for the manufacturing of solar-powered multifunctional agricultural robot. *International Journal on Interactive Design and Manufacturing (IJIDeM)*. 2024;18(8):5725–36.
60. Prakash C, Gupta LR, Mehta A, Vasudev H, Tominov R, Korman E, Fedotov A, Smirnov V, Kesari KK. Computing of neuromorphic materials: An emerging approach for bioengineering solutions. *Materials Advances*. 2023;4:5882–919.
61. Singh J, Gill HS, Vasudev H. Computational fluid dynamics analysis on role of particulate shape and size in erosion of pipe bends. *International Journal on Interactive Design and Manufacturing (IJIDeM)*. 2023;17(5):2631–46.
62. Rao GS, Mukkamala U, Hanumanthappa H, Prasad CD, Vasudev H, Shanmugam B, KishoreKumar KC. Evaluating and optimizing surface roughness using genetic algorithm and artificial neural networks during turning of AISI 52100 steel. *International Journal on Interactive Design and Manufacturing (IJIDeM)*. 2024;18(8):6151–60.
63. Singh J, Singh R, Singh S, Vasudev H, Kumar S. Reducing scrap due to missed operations and machining defects in 90PS pistons. *International Journal on Interactive Design and Manufacturing (IJIDeM)*. 2023;17(5):2527–39.
64. Nagabhushana P, Ramprasad S, Prasad CD, Vasudev H, Prakash C. Numerical investigation on heat transfer of a nano-fluid saturated vertical composite porous channel packed between two fluid layers. *International Journal on Interactive Design and Manufacturing (IJIDeM)*. 2024;18(5):2927–44.
65. Mahantesh MM, Rao KR, Chandra AP, Vijayakumar MN, Nandini B, Prasad CD, Vasudev H. 2024. Design and modeling using finite element analysis for the sitting posture of computer users based on ergonomic perspective. *International Journal on Interactive Design and Manufacturing (IJIDeM)*. 18(8):5875–91.
66. Verma PK, Vasudev H, Kumar V. Dental implant materials and designs and failure during implant placement and loading: A review. In Hitesh Vasudev, Chander Prakash, Mohammad Khalid, Kavindra Kesari (Eds.) *Advanced Manufacturing Processes*, 2025. pp.138–50, Boca Raton, Imprint: CRC Press.
67. Singh, P., Vasudev, H. and Bansal, A., 2024. Effect of post-heat treatment on the microstructural, mechanical, and bioactivity behavior of the microwave-assisted alumina-reinforced hydroxyapatite cladding. *Proceedings of the Institution of Mechanical Engineers, Part E: Journal of Process Mechanical Engineering*, 238(3):1060–71.
68. Verma PK, Vasudev H, Kumar V. Effect of electro-corrosion behavior of different grades of stainless steel used in bio-implants: A review. In Hitesh Vasudev, Chander Prakash, Mohammad Khalid, Kavindra Kesari (Eds.) *Advanced Manufacturing Processes,* 2025. pp. 269–83, Boca Raton, CRC Press.

10 Oxidation Behaviour of Thermally Sprayed Composite Coatings for Boiler Steel Protection

Praveen Kumar Verma, Hitesh Vasudev, Durga Prasad, Sule Emmanuel Oshioriameh, and Harminder Singh Saggu

10.1 INTRODUCTION

Boiler systems are of primary importance in the functioning of thermal and coal-fired power generation. Internal boiler components, including superheater tubes, reheaters, and economizers, function in oxidizing atmospheres and in high temperatures of more than 800°C. Such high temperatures and oxidizing atmospheres can cause serious and aggressive corrosion and oxidation of the boiler steels, thus leading to wastage, inefficiencies, and unscheduled downtime. Brittle oxide scales formed during thermal cycling are particularly problematic because the scales oxidize and spall off, exposing fresh metal to oxidation. This process creates successive cycles of oxidation and thermal cycling, thus leading to diminished operating life and accelerated degradation. At temperatures above and around 600°C, the conventional methods of protection such as corrosion-resistant alloying with chromium or nickel provide basal protective oxidation. High Cr ferritic steels (T22, T91), poor oxidation ferritic-austenitic steels, and austenitic steels (beyond 600°C) suffer from scale instability and internal oxidation. Such protective corrosion problems can be dealt with by the use of advanced surface designing methods, specifically thermal spray coatings. The use of thermal spray coatings leads to the positive and unique capability of depositing thick and metallurgically adherent protective oxidation coatings on steel without adversely affecting the bulk microstructure of the steel [1].

Thermally sprayed composite coatings offer both the benefits of metallic matrices and the appropriately enhanced reinforced ceramics. The metallic constituents—Ni, NiCr, and FeCr—provide ductility, while diffusion barriers are provided by metallic constituents; and hardness, thermal stability, and oxidation resistance are provided by the ceramics: Al_2O_3, Cr_2O_3, and YSZ. The result of this dual-phase structure is exceptional performance under cyclic oxidation, thermal shock, and erosion, all in the extreme conditions encountered in boilers. The oxidation behaviour of such coatings

DOI: 10.1201/9781003730637-10

is determined by the combined influence of composition, microstructure, porosity, and residual stresses. In addition, the type of sprayed process—flame spray, plasma spray, high-velocity oxy-fuel (HVOF), and cold spray—determines coating density, oxide dispersion, and ultimately, the oxidation kinetics. Interfacial diffusion has recently been minimized through varied feedstock formulation and the addition of nano-reinforced composites, along with graded coating architectures designed to improve performance and promote the formation of stable, protective oxides [2]. Figure 10.1 provides a mechanistic outline of thermal spray and how various adjustable settings of air flow and temperature can be tailored to obtain coatings of varying requirements. The outline also provides a description of powder injection, heating of particles, gas stream acceleration, and the deposition of particles to produce a dense coating layer on the substrate.

This chapter investigates the mechanisms of high-temperature oxide layer formation on boiler steels which have been covered using thermally sprayed composite coatings from the perspectives of thermodynamics, kinetics, and the evolution of the microstructure. The chapter explains the fundamental design principles concerning the resistance of materials to high-temperature oxidation, and oxidation on composite coatings, described in terms of the results derived from experimental studies and theoretical studies. The section on the principles of high-temperature oxidation resistance focuses on the contribution of particular methods of characterization, including thermogravimetric analysis (TGA), X-ray diffraction (XRD), and scanning electron microscopy (SEM/EDS), to the understanding of the dynamics of the layers of oxides that grow during high temperature exposure. The aim of this discussion, in the final analysis, is to demonstrate the principles of design in composite coatings whose configuration and composite structure are designed to optimize the longevity of boiler parts in service during high-temperature, severe oxidizing conditions [4].

10.2 FUNDAMENTALS OF OXIDATION IN BOILER ENVIRONMENTS

The interaction between metals and oxygen gases favours the formation of oxides. It is not just steam that contributes to the complicating factors during oxidation processes within the boiler environment. Oxidation behaviour is further complicated by the

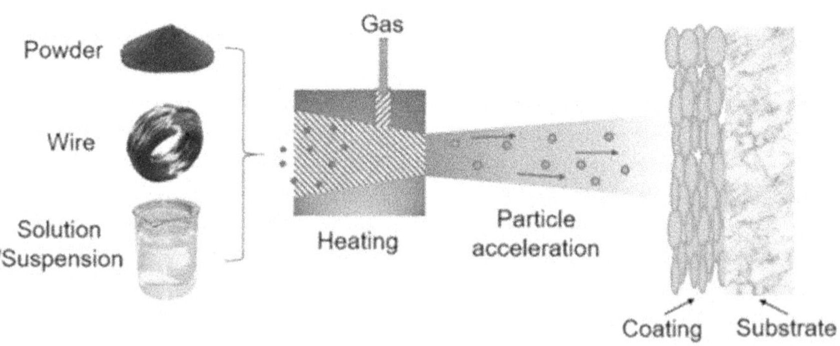

FIGURE 10.1 Schematic of a thermal spray process. Controllable air flow and temperature meet different requirements [3].

presence of sodium sulfate, vanadium pentoxide, and other combustion products as they form mixed oxides and sulfates. Oxidation can only be understood fully by looking at the thermodynamic viability of oxides and kinetic control. The phenomenon of metal oxidation can be forecast through the use of an Ellingham diagram which analyzes the temperature dependence of ΔG°, the Gibbs free energy of formation for oxides. For metals such as aluminium and chromium, the oxides they form, Al_2O_3 and Cr_2O_3 respectively, are associated with very low and negative values of ΔG°, and as such, are very stable in comparison to the oxides of iron and nickel, which are much less stable, Fe_2O_3 and NiO. In Cr- and Al-containing composite coatings, layers of adherent, continuous, diffusion-oxygen diffusion barrier, Cr_2O_3 or Al_2O_3 scales are formed through selective oxidation. The temperature and the partial pressure of oxygen of the surrounding atmosphere can control the oxidation state and thus the interleaved structure of the oxide layers of the Fe-oxides (Fe_3O_4, Fe_2O_3, and FeO), which ultimately plays a role in the multi-layered structure of oxide scales found on boiler steels. The growth of oxide scales generally follows parabolic kinetics, described by the equation [5]:

$$x^2 = k_p t$$

where x is the oxide thickness, k_p is the parabolic rate constant, and t is the time. It means that the oxide film can grow in thickness over time. That is, the oxide film grows under diffusion control, and there are conditions under which the film grows linearly. This linear growth can occur at high temperature and under cyclic exposure conditions, due to scale cracking and spallation, After the scales are spalled, they can also grow back and become spalled again. In the case of the boiler, water vapour is present which means that there is also bath evaporation. Under such conditions, boiling is also present which means that there is a high partial pressure of water vapour. These conditions facilitate the production of volatile hydroxides and therefore enhance oxidation. Instead of being protective, the oxide can become volatile and undermine the protective characteristics of the oxide. Scales containing chromium are continuously volatile under such conditions. Therefore, coating design must be balanced around the content of chromium to obtain closure scales and ensure that the scales are not volatilizing under such conditions. Figure 10.2 depicts a boiler water wall tube, its surroundings, and how the high-temperature flue gases come into contact with the tube and inner circulating pressurized water/steam in the tube and how thermal stress, oxidation, and corrosion interact.

Fly ash deposits and impurities such as Na_2SO_4, K_2SO_4, and V_2O_5 accelerate oxidation by forming molten salts that dissolve protective oxides and increase ionic transport. This process, known as hot corrosion, is often synergistic with oxidation and cannot be ignored in boiler operation. Composite coatings containing oxides like Al_2O_3 or YSZ provide chemical stability against such molten salts by preventing infiltration and maintaining a refractory surface.

10.3 THERMAL SPRAY TECHNIQUES FOR OXIDATION-RESISTANT COATINGS

Thermal spray technology consists of several processes that create coatings by projecting molten or semi-molten feedstock materials onto a substrate that has been prepared, where they rapidly cool, forming a dense layer that adheres to the

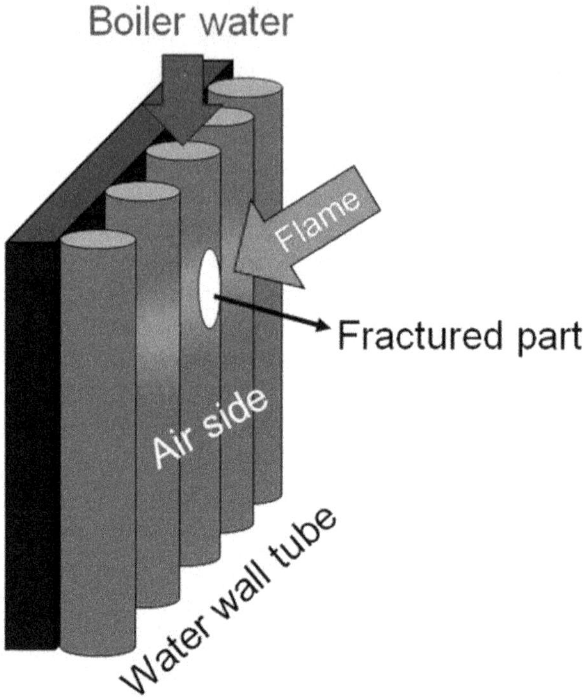

FIGURE 10.2 Schematic of a boiler water wall tube and the operating environment [6].

substrate surface. For high-temperature applications such as the protection of boiler steels from oxidation and corrosion, these techniques stand out because advanced materials (metals, alloys, ceramics, and composites) can be used without changing the bulk properties of the substrate. The greatest strength of thermal spraying is its versatility—almost any material that can be melted can be sprayed—and the control over coating thickness, microstructure, and composition that can be achieved to meet particular functional requirements [7]. Figure 10.3 illustrates the standard configuration of thermal barrier coatings (TBCs), depicting the bond coat, thermally grown oxide (TGO) layer, and ceramic topcoat.

Thermal spray processes can be generally grouped into four main categories: (1) conventional flame spray, (2) plasma spray, (3) high-velocity oxy-fuel, and (4) cold spray. For each varied methodology, the spray parameters of each of these techniques, which can be tuned to produce various particle speeds and temperatures, as well as coating characteristics, profoundly affect adhesion and oxidation resistance. In boiler applications, the low oxidation environment makes process choice very critical as coating and substrate oxidation content and density are the most important factors that determines long-term protection [9].

Flame spraying is one of the oldest thermal spray techniques with the use of oxygen and a fuel gas (acetylene, propane, or hydrogen) to produce a flame which will melt either feedstock powder or wire. The melting particles get accelerated by the hot

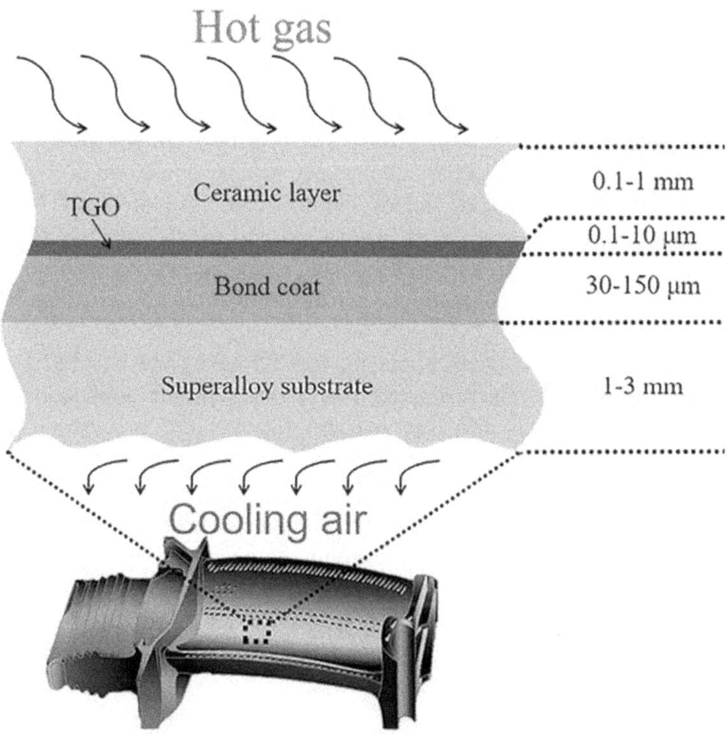

FIGURE 10.3 Schematic of thermal barrier coatings (TBCs) structure [8].

expanding gas and impinge on the surface. Even though it is simple and inexpensive to perform, the flame sprayed coatings will tend to have higher porosity (10–20%) and higher oxide inclusions than the more sophisticated techniques. The presence of this porous microstructure will allow the passage of oxygen which will promote substrate oxidation and reduce oxidation resistance. Even so, flame-sprayed coatings are acceptable for use in moderate temp applications, or as bond coats under dense top layers. Post-spray sealing or remelting treatments, which reduce porosity, increase oxidation resistance, and are of great value when used with alloy systems such as Ni–Cr or Fe–Cr–Al to provide useful protection [10].

Among the various techniques for coating ceramics and composites, Atmospheric Plasma Spraying (APS) is the most utilized. In this technique, an electric arc ionizes a mixture of gases (typically argon, hydrogen, or nitrogen) to form a plasma jet which can reach temperatures of 10,000 K. While achieving this, the technique also utilizes high temperature jets which can melt the feedstock powders. The molten powders are then sprayed onto the surface at speeds of 150 to 300 m/s. The coating is formed by the solidification of successive splats. These splats form a lamellar structure. When discussing oxidation protective measures plasma spraying shows versatility and flexibility in the deposition of both metal and ceramic protective materials, such as NiCrAlY, Al_2O_3, Cr_2O_3, and YSZ-based composites. Among the various thermal spraying techniques, the unique feature of plasma spraying is the ability to create protective coatings with specific microstructural and oxide dispersion and oxidation control. Due to the very high coating plasma spraying parameters, some oxidation may occur in-flight and lead to coating oxide stringers or inclusions. Depending on the size and shape, inclusions may promote or impede oxidation of the final protective layer. Use of VPS and LPPS techniques to achieve consistent in-flight coating oxidations and create dense, clean, and less oxidized coatings is preferred. Minimizing oxidations in-flight during spraying of plasma coatings of NiCrAlY and NiCr–Al_2O_3 composites protect and oxidize perfectly during extended exposure periods of 900–1000°C making these types of coatings very useful as protective coatings for boiler components [11].

Innovative hybrid laser systems pioneered by CST have contributed to developing proprietary laser systems incorporated into several patent applications. Dense microstructures behave as diffusion barriers for oxygen, preventing oxidation and improving oxidation resistance. Composite systems like Cr_3C_2–NiCr, NiCr–Al_2O_3, and NiCrAlY–YSZ—ceramic reinforcements stabilized with ductile metal matrices—are particularly successful with HVOF spraying. HVOF deposition's rapid cooling of splats refines the grain size, resulting in fine and uniform microstructures, thus improving diffusion resistance and mechanical stability during thermal cycling. Researchers have reported that HVOF-sprayed NiCrAlY coatings on T22 and P91 steels, when exposed to 900°C for 100 hours, had oxidation rates an order of magnitude lower than uncoated substrates. In addition, the coatings, exhibiting minimal spallation, maintained their structure due to strong adhesive and cohesive bonds [12].

10.3.1 Cold Spraying

Among newer techniques for solid-state coating application, the cold spray (CS) method consists of propelling metallic and composite powders at supersonic velocities (300–1200 m/s) and impacting the powders on the substrate at temperatures below the melting point of the added powders, using high-pressure gas (usually nitrogen and helium) as a propellant. Although solid coatings are applied by gas dynamic spraying, the bonding mechanism consists of deformation plus mechanical interlocking and hence, the coatings are not fused. The coatings are high-purity and oxidant free, which makes them suitable for applications where oxide inclusions are not desirable [13]. Figure 10.4 shows the diagram showing the thermal spray process, with the feedstock material (in powder, wire, or rod form) being heated and accelerated by a hot gas stream before being deposited onto the substrate surface to create a dense coating layer.

Dense, adherent coatings with very low porosity ($<0.5\%$) can be made using NiCrAl, FeCrAl, and Cu–Al_2O_3 composites. As no thermal degradation or phase transformation occurs during deposition, the intended composite microstructure remains intact. Nevertheless, cold spray deposition of pure ceramic coatings remains problematic, as ceramics are brittle and difficult to deform plastically. To solve this issue, hybrid powders or mechanically alloyed metal–ceramic feedstocks are incorporated, which result in composite coatings with superb bonding. Cold-sprayed NiCrAl coatings achieve remarkable cyclic oxidation resistance up to 950°C,

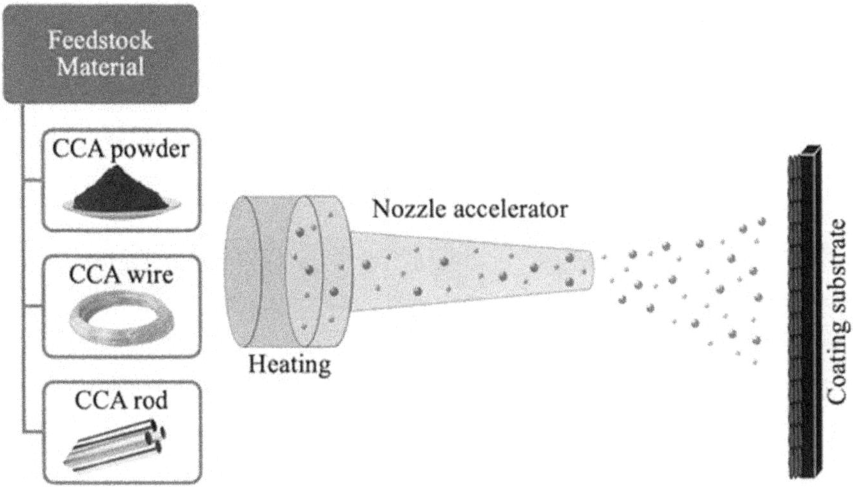

FIGURE 10.4 Schematic showing the thermal spray process [14].

comparable to that of HVOF-sprayed coatings, due to the development of a continuous, slow-growing Al_2O_3 scale [15].

10.3.2 Process–Microstructure–Property Relationship

The oxidation performance of thermally sprayed coatings is fundamentally governed by their microstructural attributes—porosity, oxide inclusion content, splat morphology, and interface quality—all of which depend on process parameters such as particle velocity, temperature, and standoff distance [16]. In plasma spray, the high-temperature environment promotes partial oxidation and lamellar structures with inter-splat voids, while in HVOF and cold spray, the high kinetic energy and relatively low temperature produce dense, diffusion-resistant coatings. The following were the thermal spray coatings microstructure and deposition characteristics on the occasion of the testing optical micrographs (see Figure 10.5): (a) NiCr – uniform lamellar structure, (b) IN625, a dense HVOF-sprayed coating with fine splat boundaries; (c) NiCrTi, an arc-sprayed coating exhibiting larger splat formation; (d) NiCrW, an HVOF-sprayed coating with refined lamellae and minimal porosity; and (e) FeCr, a relatively coarse microstructure with visible inter-splat gaps.

Dense microstructures reduce oxygen ingress, while finely dispersed oxide particles, if stable, can act as diffusion barriers and reinforce the matrix. Moreover, coating–substrate adhesion plays a critical role during cyclic oxidation. Strong interfacial bonding, typically achieved in HVOF and cold-sprayed coatings, suppresses spallation by minimizing thermal expansion mismatch stresses. Post-deposition heat treatments may further enhance bonding by promoting interdiffusion and healing microvoids. In boiler environments, where repeated heating and cooling cycles are inevitable, the coating must sustain mechanical integrity and chemical stability. Hence, selecting the appropriate thermal spray technique involves balancing deposition cost,

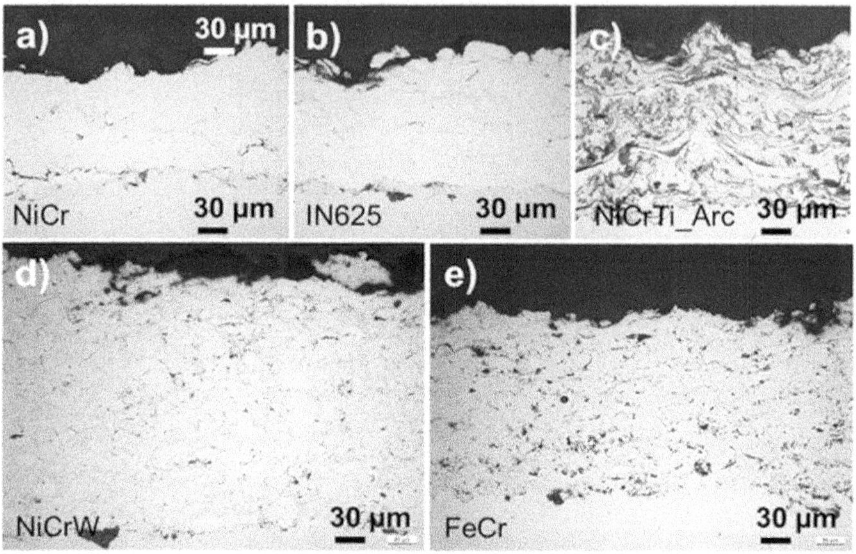

FIGURE 10.5 Optical micrographs of the thermal spray coatings before testing: (**a**) HVOF sprayed NiCr and (**b**) IN625; (**c**) arc sprayed NiCrTi; (**d**) HVOF sprayed NiCrW; and (**e**) FeCr [17].

coating density, and operational temperature. Plasma and HVOF spraying remain the preferred options for high-temperature oxidation resistance, while cold spray provides an emerging pathway for near-oxide-free coatings in advanced applications [18].

10.4 COMPOSITE COATING SYSTEMS FOR OXIDATION RESISTANCE

Composite coatings have emerged as an advanced class of protective layers for high-temperature boiler steels because they combine the benefits of metallic and ceramic constituents within a single architecture. The metallic matrix provides ductility, toughness, and good adhesion to the substrate, while the ceramic or intermetallic reinforcement enhances hardness, chemical stability, and oxidation resistance. Such coatings exploit the synergistic effects of both phases, yielding superior thermal and environmental stability compared to monolithic metallic or ceramic coatings. In oxidation-prone environments, particularly those involving steam and combustion gases, these composite systems act as efficient barriers to oxygen and aggressive species by forming adherent, slow-growing oxide scales that resist cracking and spallation during cyclic heating. The choice of coating composition plays a critical role in determining oxidation performance. Among various systems, NiCr-based composites, NiCrAlY-based composites, Cr_3C_2–NiCr, and FeCrAl-based composites are the most extensively employed for high-temperature oxidation protection in power plant applications. Each system offers distinct advantages in terms of phase stability, oxide scale characteristics, and mechanical compatibility with boiler steels [19].

10.4.1 NiCr–Al_2O_3 Composite Coatings

Nickel–chromium alloys are well-known for their oxidation resistance due to their ability to form stable Cr_2O_3 scales. When reinforced with aluminium oxide (Al_2O_3), the oxidation resistance is further enhanced by introducing an inert ceramic phase that inhibits diffusion and reinforces the microstructure. The Al_2O_3 particles act as physical barriers to the outward diffusion of metal ions and inward diffusion of oxygen, thereby slowing oxide scale growth. Additionally, the high melting point and chemical inertness of Al_2O_3 provide thermal stability and mechanical rigidity to the coating at temperatures beyond 900°C. During oxidation, the NiCr matrix selectively oxidizes to form a continuous Cr_2O_3 layer, which can be further stabilized by the presence of dispersed Al_2O_3 particles. The interface between the metallic and ceramic phases promotes the formation of fine-grained oxides that adhere well to the substrate. HVOF- or plasma-sprayed NiCr–Al_2O_3 coatings typically exhibit parabolic oxidation kinetics with rate constants an order of magnitude lower than uncoated steel substrates. Microstructural studies reveal that the addition of Al_2O_3 refines the splat boundaries and reduces interconnected porosity, effectively limiting oxygen penetration. The improved adherence of the oxide scale also reduces spallation during cyclic exposure. A further advantage of the NiCr–Al_2O_3 system is its compatibility with various substrate steels, including ferritic and martensitic grades. The coefficient of thermal expansion (CTE) of NiCr alloys closely matches that of boiler steels, thereby minimizing interfacial stresses during thermal cycling. Moreover, post-deposition annealing of such coatings can lead to partial diffusion bonding and grain boundary healing, further enhancing oxidation resistance [20]. Figure 10.6 shows SEM images illustrating the top-surface morphologies of Al_2O_3 coatings deposited using various thermal spray techniques.

10.4.2 NiCrAlY–YSZ Composite Coatings

The NiCrAlY–YSZ (yttria-stabilized zirconia) composite system represents one of the most advanced designs for high-temperature oxidation and thermal barrier protection. NiCrAlY alloys serve as bond coats in turbine and boiler applications due to their excellent ability to form slow-growing Al_2O_3 scales. The addition of YSZ introduces a thermally stable ceramic reinforcement with low thermal conductivity and excellent oxidation resistance. This composite configuration provides a dual function: the NiCrAlY matrix ensures adherence and diffusion barrier characteristics, while YSZ particles enhance the thermal insulation and stabilize the oxide scale against spallation. The oxidation mechanism in NiCrAlY–YSZ coatings primarily involves selective oxidation of aluminium, leading to the formation of a thin, continuous α-Al_2O_3 layer at the coating surface or within inter-splat regions. This alumina scale acts as a diffusion barrier to both oxygen and metal ions, significantly reducing oxidation rates. The presence of yttrium (Y) in the alloy composition plays a vital role by improving the adhesion of the oxide layer through the reactive element effect (REE). Y segregates at oxide grain boundaries, reducing cation diffusion and suppressing void formation at the oxide–metal interface [22].

YSZ reinforcement provides an additional barrier effect and improves mechanical durability during cyclic oxidation by accommodating strain through its partially

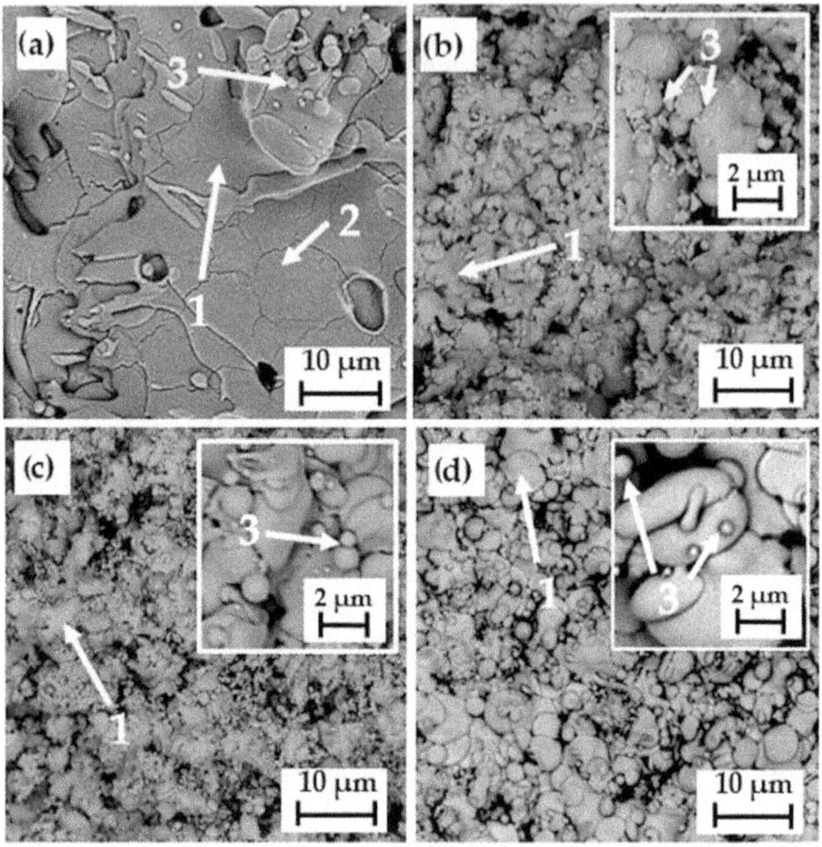

FIGURE 10.6 SEM top-surface morphologies of Al_2O_3 coatings sprayed by: (**a**) APS, (**b**) SPS, (**c**) A-SPS; and (**d**) S-HVOF. 1—splat; 2—cracks; 3—not fully melted or re-solidified particles [21].

stabilized tetragonal structure. Furthermore, YSZ inhibits crack propagation through crack deflection and energy dissipation mechanisms. Experimental investigations have shown that plasma-sprayed NiCrAlY–YSZ coatings on boiler steels exhibit oxidation weight gains of less than 2 mg/cm² after 100 hours of exposure at 1000°C, compared to 20–30 mg/cm² for uncoated steels. These coatings also retain integrity during repeated thermal cycles, demonstrating excellent spallation resistance [23].

10.4.3 Cr_3C_2–NiCr Composite Coatings

Among cermet systems, chromium carbide–nickel–chromium (Cr_3C_2–NiCr) composites are widely used for oxidation and wear protection under high-temperature conditions. The Cr_3C_2 phase provides hardness, wear resistance, and chemical stability, while the NiCr binder ensures ductility and adhesion. In oxidation environments, the NiCr matrix forms protective Cr_2O_3 scales, while the carbides undergo limited oxidation to form chromium oxide, which integrates into the overall oxide scale structure. This dual-phase oxidation mechanism results in dense, adherent surface layers that resist further oxidation. HVOF-sprayed Cr_3C_2–NiCr coatings have low porosity (<1%) and increased inter-splat cohesion because of the particles deposition with high kinetic energy. The Cr_3C_2 only partially oxidizing in-flight preserved the unwanted oxides from forming and kept the coating chemically untainted. When exposed to a temperature in the range of 800–1000°C a composite oxide Cr_2O_3 and (Ni,Cr) spinels were formed. The oxides adhered strongly to the underlying layer and had a slow growing rate [24].

Cr_3C_2–NiCr coatings are some of the finest for not only oxidation and hot corrosion, but for general corrosion as well. Dual carbides in the composite also increase the level of chemical inertness, decrease the likelihood of oxide scale cracking, and ultimate strain energy absorption. The metallic matrix and oxide bonded strongly which accounts for little to no delamination in oxidation. Research states that T91 Cr_3C_2–NiCr steels are 80% less oxidative when uncoated steels are held for 100 hours at 900°C. The slight addition of reactive elements, Y or Ce, increases adhesion, suppresses spallation, and enhances oxidation [25].

10.4.4 FeCrAl-based Composite Coatings

Another collection of oxidation-resistant substances, FeCrAl alloys, produces continuous, adherent alumina (Al_2O_3) scales oxidation. When utilized as thermally sprayed coatings, FeCrAl-based composites perform excellently in oxidizing environments in steam and air up to 1100°C. Fe and Cr together give structural integrity and resistance to oxidation, and Al leads to protective α-Al_2O_3 formation. Because alumina is thermodynamically stable compared to Cr_2O_3 as well as Fe oxides, protective alumina is assured over time. As for composites, oxide ceramics like Al_2O_3 or YSZ can be used to reinforce FeCrAl coatings to achieve better mechanical and thermal shock resistance. Al's selective oxidation yields an alumina layer that is thin and uniform. Cr stabilizes this oxide and prevents the in-diffusion of Fe and O. The thermal cycling has no effect on the coating microstructure because of the ductile Fe-based matrix that holds the thermal stresses. According to thermogravimetric analyses, FeCrAl coatings exhibit low oxidation rate constants (10^{-14}–10^{-13} $g^2\ cm^{-4}\ s^{-1}$), which is one order of magnitude lower than those of NiCr or Cr_3C_2–NiCr coatings etched in the same conditions. The alumina scale formed upon FeCrAl also does not volatilize in water vapour, which is a major plus for boiler use. Cold-sprayed FeCrAl coatings have outstanding cyclic oxidation resistance due to the dense microstructure and the limited oxide inclusions produced during deposition [26].

10.4.5 Microstructural and Interfacial Considerations

The factors influencing the oxidation resistance of composite coatings are not limited to their composition but also include the microstructural characteristics of porosity, splat boundaries, phase distribution, and interfacial bonding. In optimally designed coatings, reinforcement particles uniformly dispersed within the metallic matrix serve as micro-barriers that inhibit the diffusion of oxygen. Consequently, fine-grained and continuous oxide layers are formed. The decrease of open porosity and the presence of stable interfaces between splats are crucial to halting oxygen ingress. Thermally sprayed coatings often exhibit lamellar structures with inter-splat voids, which can become initiation sites for oxidation. Post-spray treatments such as vacuum annealing, laser remelting, or sol–gel sealing can significantly improve the microstructural integrity by reducing porosity and healing microcracks. Additionally, controlled oxidation during the initial stages of exposure can create a pre-formed oxide layer that enhances long-term stability [27].

The interface between the coating and substrate also plays a decisive role in determining performance under thermal cycling. Differential thermal expansion between the coating and steel substrate can lead to stress accumulation and eventual delamination. Therefore, coatings with intermediate thermal expansion coefficients (e.g. NiCr-based systems) are generally preferred for ferritic and martensitic steels to ensure stress compatibility. A comparative evaluation of different composite coatings reveals that NiCrAlY–YSZ and FeCrAl-based systems exhibit the lowest oxidation rates due to the formation of stable Al_2O_3 scales. NiCr–Al_2O_3 coatings provide a balance between oxidation resistance and mechanical toughness, while Cr_3C_2–NiCr coatings are favoured when simultaneous wear and oxidation resistance are required. The choice of coating thus depends on the specific service temperature, environment, and mechanical demands of the boiler component. Overall, composite coatings outperform monolithic metallic coatings by providing multi-mechanistic protection: diffusion barrier effect, stress relaxation, and reinforcement-induced microstructural refinement. The controlled combination of metallic and ceramic phases ensures that oxide scales remain adherent and protective even after prolonged cyclic exposure at high temperatures. These characteristics make composite coatings indispensable for extending the lifespan of boiler steels operating in oxidizing atmospheres [28].

10.5 OXIDATION MECHANISMS AND KINETICS OF THERMALLY SPRAYED COMPOSITE COATINGS

For long-term predictions regarding the behaviour of thermally sprayed composite coatings for high-temperature boiler applications, one needs to understand the oxidation mechanisms and kinetics. The coating's composition and microstructure, the processes used to make the coating (which lead to porosity and oxide inclusions), and the exposed environment dictate the rate of oxidation. The formation and growth of protective oxide layers occur through the action of diffusion processes, phase transformations, and interfacial reactions. The protective quality of an oxide coating is determined by adhesion-related processes, morphology, and spallation resistance at cyclic thermal stresses, in addition to the thermodynamic stability of the oxide [29].

10.5.1 Thermodynamics of Oxide Formation

At higher temperatures, oxidation occurs as metallic constituents react with oxygen, forming surface or internal oxides within the coating. The thermodynamic prediction of oxide formation can be determined through the Gibbs free energy of formation

(ΔG°). For Al, Cr, and Si, the oxides Al_2O_3, Cr_2O_3, and SiO_2 are very stable due to the highly negative values of ΔG°. As a result, coatings with these elements acquire adherent protective oxide films that obstruct further oxidation. For example, Cr_2O_3 in NiCr-based coatings is the main protective oxide up to about 1000°C. Above that temperature or in oxygen-rich atmospheres, however, Cr_2O_3 can lose its protective qualities as CrO_3 volatilizes. In contrast, Al_2O_3 created in NiCrAlY or FeCrAl coatings remains stable above 1100°C due to its high thermodynamic stability and low oxygen permeability. The presence of reactive alloys Y, Zr, and Hf promotes the adherence and continuity of the alumina layer through interfacial defect structure modification. Thermal spray coatings, particularly composites, benefit from the presence of both Cr and Al in the matrix, as they form a duplex protective scale of Cr_2O_3 and Al_2O_3, providing multi-stage oxidation resistance. Initially, Cr_2O_3 forms rapidly, offering immediate protection, while Al_2O_3 grows slowly and eventually dominates as the outer scale stabilizes over time. This dual-layer structure significantly improves oxidation kinetics compared to single-phase systems [30].

10.5.2 Oxidation Kinetics and Parabolic Rate Law

In practice, the oxidation rate of thermally sprayed coatings is typically several orders of magnitude lower than that of bare steels due to the protective nature of the coatings. Composite coatings such as NiCr–Al_2O_3 and Cr_3C_2–NiCr often exhibit even lower values because the dispersed ceramic phases further restrict ionic diffusion. The oxide growth mechanism is typically governed by the diffusion of oxygen anions through the oxide scale or by the diffusion of metal cations outward. In Al_2O_3-forming systems, the rate-limiting step is the inward diffusion of oxygen ions through the dense alumina layer, whereas in Cr_2O_3-forming systems, both metal and oxygen diffusion can occur through grain boundaries and defects. Therefore, coatings that promote the formation of α-Al_2O_3 (the most stable phase) tend to show superior oxidation resistance [31].

10.5.3 Role of Microstructure and Defects

The microstructural characteristics of thermally sprayed coatings—porosity, oxide inclusions, splat boundaries, and interfacial cracks—strongly influence oxidation kinetics. Porosity acts as a channel for oxygen diffusion, accelerating localized oxidation and scale spallation. Therefore, coatings with low porosity (<2%), as typically achieved by HVOF and cold spray processes, demonstrate superior oxidation performance compared to conventional plasma-sprayed coatings with higher porosity levels. Inter-splat boundaries are diffusion limits. The boundaries may get sealed during high-temperature sintering due to oxide formation and subsequently reducing oxygen permeability. The presence of ceramic reinforcements in composite coatings is advantageous since they act as diffusive barriers and grain boundary pins which stabilize the grain boundaries. Coating inclusions may benefit or detract from the properties. Fine-spaced oxides are effective at increasing diffusion resistance in the material but large oxide stringers or clusters facilitate crack propagation and reduce cohesion. Inter-lamellar voids are closed and metallurgical bonding is promoted via post-deposition heat treatment, thus enhancing resistance to oxidation. Functional boiler components are exposed to imbricated heating and cooling cycles [32].

Cyclic oxidation testing, therefore, provides a realistic evaluation of coating durability. The performance of thermally sprayed coatings under cyclic conditions depends on oxide scale adhesion, thermal expansion compatibility, and stress relaxation mechanisms. HVOF-sprayed NiCr–Al_2O_3 coatings have demonstrated excellent cyclic oxidation resistance, maintaining oxide adherence after more than 200 cycles between 200 and 900°C. The combination of Cr_2O_3 and Al_2O_3 scales forms a graded protective layer that can accommodate thermal stresses. Similarly, Cr_3C_2–NiCr coatings exhibit self-healing behaviour: microcracks formed during thermal cycling are filled by the oxidation of Cr_3C_2 to Cr_2O_3, sealing the surface and preventing further degradation. Cold-sprayed NiCrAl coatings, due to their low porosity and high ductility, exhibit superior cyclic stability, retaining their oxidation resistance even after 500 cycles between room temperature and 1000°C. The ductile metallic matrix effectively absorbs strain, while the adherent alumina scale resists spallation. In contrast, plasma-sprayed coatings with higher porosity or oxide stringers are more prone to spallation due to stress concentration at inter-splat boundaries [33].

10.5.4 Diffusion Mechanisms in Oxide Scale Growth

The growth of protective oxide scales is controlled by the diffusion of ions or vacancies through the oxide film. In Al_2O_3-forming systems, oxygen anions diffuse inward through the oxide lattice or along grain boundaries, while aluminium cations migrate outward. However, the diffusivity of oxygen in α-Al_2O_3 is extremely low (~10^{-18} cm^2/s at 1000°C), making the layer an excellent diffusion barrier. For Cr_2O_3-forming systems, the diffusion rates are higher (~10^{-14}–10^{-16} cm^2/s), explaining why Cr_2O_3 scales grow faster and are less protective over long durations. The incorporation of Al_2O_3 or YSZ particles into the matrix effectively suppresses Cr_2O_3 grain boundary diffusion by forming complex oxide networks. Reactive elements like Y and Ce further inhibit cation diffusion by blocking vacancy movement, leading to slower oxide growth and improved scale adherence. The microstructural integrity of the coating–substrate interface also influences diffusion behaviour. A metallurgically bonded interface, as obtained in HVOF or post-annealed coatings, minimizes interfacial voids and prevents oxidation along the coating–substrate boundary—a common failure mechanism in plasma-sprayed coatings [34]. Figure 10.7 is a schematic representation of the high-temperature oxidation process. Figure 10.7(a) shows the mechanism of oxide film formation on the metal surface, illustrating oxygen and metal ion diffusion through the oxide scale, and Figure 10.7(b) shows three typical stages of oxidation reaction with increasing exposure time—initial fast oxidation, parabolic growth by diffusion control, and steady-state or saturation stage.

10.6 EXPERIMENTAL CASE STUDIES AND CHARACTERIZATION TECHNIQUES

The oxidation behaviour of thermally sprayed composite coatings on boiler steels is assessed through some experimental research along with some sophisticated

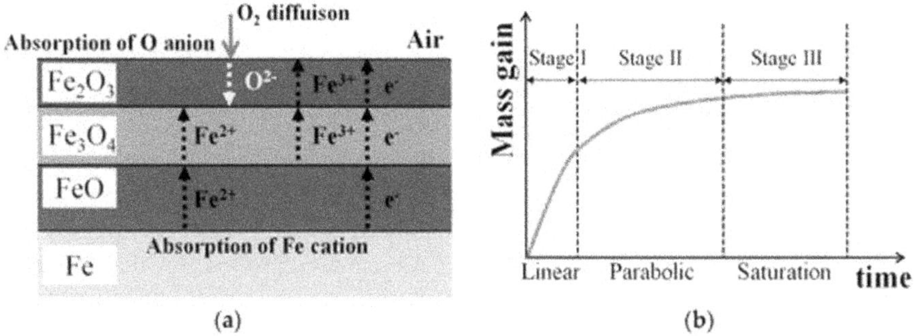

FIGURE 10.7 Illustration of high-temperature oxidation: (**a**) mechanism of oxide formation; and (**b**) three stages of oxidation reaction according to exposure time [35].

characterization techniques. The goal of these investigations is to understand how the microstructure, the quality of coating–substrate interface, and the oxidation rate interact when the conditions are either simulated or when they are at actual boiler operating conditions. Knowing how long the coatings would behave satisfactorily at high temperatures is necessary to refine the process parameters and predict service life with confidence. For the coating, the experimental investigations are usually based on cyclic oxidation, isothermal exposure, and thermogravimetric analysis, followed by some analytical techniques to understand the microstructure for the defined exposures [36].

10.6.1 Experimental Methodology and Testing Environments

Testing the oxidation of coated and uncoated steels used in boilers is done in environments that mirror real flue gases and steam conditions with the focus on the main two procedures which are isothermal and cyclic oxidation. For isothermal oxidation, the steel samples are maintained at a set constant temperature which is between 600 and 1000°C for hours on end, which could be between 50 and 500 hours. This aids in sharpening the understanding of the kinetics of oxidation in addition to the mechanisms of growth scale. In cyclic oxidation, the thermal transients which occur during the start and shut down of the boilers are taken into account. In these tests, the samples are elevated to the set temperature, maintained for a certain period of time, which is 1–2 hours, and then cooled to room temperature in air. Multiple cycles are done to assess the extent of spallation and cracking of the scale, as well as the overall integrity of the coating [37].

Test environments are carefully designed to reproduce the oxygen and steam partial pressures found in superheater or reheater zones. In many studies, laboratory oxidation is performed in air or synthetic flue gas mixtures containing O_2, CO_2, H_2O, and SO_2 to mimic industrial exposure. For example, a typical simulated flue gas composition might include 10 vol.% O_2, 15 vol.% CO_2, 5 vol.% H_2O, and 100 ppm SO_2, balanced with N_2. The presence of alkali sulfates or chlorides can also be introduced to replicate hot-corrosion conditions in coal-fired plants. Such carefully controlled atmospheres ensure that the obtained data are representative of real-world degradation mechanisms [38].

10.6.2 Weight Change and Kinetic Analysis

The most common metric for quantifying oxidation resistance is the specific weight change (ΔW/A) as a function of exposure time. This parameter reflects the net mass gain or loss per unit surface area due to oxide formation and spallation. Thermogravimetric analysis (TGA) or periodic weighing during cyclic exposure provides quantitative data that can be fitted to kinetic models. For protective coatings, the oxidation behaviour typically follows a parabolic rate law, where the square of the weight gain is proportional to time, indicating diffusion-controlled growth of the oxide layer. Conversely, a linear or accelerated mass change trend signifies non-protective oxide formation or scale spallation. In composite coatings, a lower parabolic rate constant (k_p) compared to the bare steel indicates improved oxidation resistance. For instance, NiCr–Al_2O_3 and Cr_3C_2–NiCr coatings have been shown to exhibit a significant reduction in oxidation rate due to the formation of continuous, slow-growing Cr_2O_3 or Al_2O_3 scales. The kinetic data are often analyzed using Arrhenius-type plots to determine the activation energy (Q) for oxidation, which reflects the diffusion barrier effectiveness of the coating. Coatings with higher Q values typically offer greater resistance to oxygen transport, confirming their thermochemical stability under severe service conditions [39].

10.6.3 Microstructural and Morphological Characterization

Detailed microstructural examination provides essential insights into the oxidation mechanisms at play. Scanning electron microscopy (SEM) coupled with energy-dispersive X-ray spectroscopy (EDS) is the most frequently used tool for analyzing surface and cross-sectional morphologies after exposure. SEM micrographs reveal oxide scale thickness, uniformity, crack density, and porosity evolution. Cross-sectional imaging distinguishes between the top oxide scale, the intermediate thermally grown oxide (TGO), and the underlying metallic coating or substrate. In well-designed coatings, the oxide layer is dense, adherent, and free of discontinuities. Conversely, porous or delaminated scales indicate ineffective diffusion barriers or thermal expansion mismatch. EDS line scanning and mapping help identify compositional gradients across the oxide–coating–substrate interface. For instance, an outward diffusion of Fe from the substrate into the oxide layer, coupled with inward diffusion of O and Cr, is typical in uncoated steels. In coated systems, however, the migration of Fe is significantly suppressed, demonstrating the effectiveness of barrier coatings. The enrichment of Cr, Al, or Si near the surface indicates selective oxidation and the establishment of protective scales such as Cr_2O_3 or Al_2O_3 [40].

X-ray diffraction (XRD) is employed to identify crystalline oxide phases formed after oxidation. In thermally sprayed composite coatings, phase analysis before and after exposure reveals transformations such as spinel formation or metastable oxide stabilization. For example, the oxidation of NiCr–Al_2O_3 coatings often leads to the development of spinel phases like $NiAl_2O_4$ or $NiCr_2O_4$, which contribute to chemical stability and adherence of the oxide scale. The detection of Fe_2O_3 or Fe_3O_4 peaks in the oxide layer of uncoated steels suggests poor protection, whereas their absence in coated samples indicates effective diffusion blocking. Raman spectroscopy further supplements XRD by detecting amorphous or nanocrystalline oxide species that are difficult to identify through diffraction alone. Raman signatures of chromia (Cr_2O_3), alumina (Al_2O_3), and magnetite (Fe_3O_4) provide qualitative evidence of selective oxidation. The combination of XRD and Raman data allows a deeper understanding of phase evolution during cyclic exposure and the stability of oxide scales formed at various temperatures. X-ray photoelectron spectroscopy (XPS) and auger electron spectroscopy (AES) allow quantitative analysis of surface chemistry and oxidation states. XPS also determines the oxidation states of some metals, such as Cr^{3+} in Cr_2O_3, or Al^{3+} in Al_2O_3, thus confirming the presence of protective oxides. With ion sputtering, one may perform depth profiling, which facilitates the determination of oxide thickness and analysis of interdiffusion patterns. In the case of NiCrAlY coatings, the sequence of Al and Cr concentrations suggests their preferential oxidation near the surface, thus forming the stable alumina-chromia bilayer, which tremendously facilitates the bilayer dual barrier as it suppresses oxidation and maintains the adhesion of the coating. AES depth profiling is also advantageous because of its increased depth resolution and therefore better distinction of compositional differences at the coating–substrate interface. The importance of these findings lies in the diffusion of Fe, Ni, and Cr which determines the long-term stability of the system, particularly in high-temperature conditions [41].

10.6.4 Cross-Sectional Analysis and Interfacial Stability

Examinations of interfaces at the atomic layer through FIB (focused ion beam) sectioning and sequence of images via TEM (Transmission Electron Microscopy) elucidates the structure and the chemical composition of the oxide scales at the atomic level. Imaging via TEM depicts the presence of nanocrystalline oxide grains and dislocation networks along with phase boundaries that are of central importance to the diffusion kinetics. For advanced coatings like NiCrAlY–YSZ composites, observations via TEM frequently reveal the presence of a continuous alumina layer to be situated underneath a YSZ topcoat, which effectively acts to prevent the topcoat from further oxidation. SAED (selected area electron diffraction) patterns illustrate the crystallographic orientation and the definitions of the epitaxial relationship which exist between the oxide phase and the metal phase as well as scales, explaining the adherence of scales. At a larger level, oxidation being the first in a series of tests, scratch testing or gold standard pull-off tests are done to assess the adhesion of the coatings. A minimal reduction in adhesion strength after prolonged oxidation indicates excellent coating–substrate compatibility and minimal thermal mismatch. Coatings that maintain adhesion values above 25 MPa even after 500 h of oxidation at 900°C are considered robust for boiler applications [42].

Laboratory experiments are complemented by field exposure tests in actual boiler environments to validate coating performance under real operational stresses. Such studies have demonstrated that HVOF-sprayed NiCr–Al_2O_3 and plasma-sprayed NiCrAlY coatings can extend the oxidation life of ferritic steels such as T22 and martensitic steels like T91 by 3–5 times compared to uncoated substrates. Post-service inspection typically reveals intact, adherent oxide scales with negligible signs of spallation or cracking. Furthermore, the coating's ability to resist erosion-corrosion from fly ash and molten salts enhances its multifunctional performance. In some industrial case studies, composite coatings have also been tested on superheater tubes and water walls of pulverized coal-fired boilers operating at 650–750°C. The coated samples exhibited significantly lower weight gain and minimal microstructural degradation even after 1000 hours of exposure [43].

10.7 COMPARATIVE ASSESSMENT WITH UNCOATED STEELS

To develop a clear picture of how protective surface treatments work under high-temperature oxidation environments, one has to evaluate the thermal spray composite coatings and compare them to uncoated boiler steels. SA213-T22, T91, T92, and SS310 boiler steels are considered economically viable options for "superheaters" and "reheaters" because of their particular mechanical strength and creep resistance. However, without protective coatings, it still will be exposed to accelerated oxidation, scaling, and metal wastage, particularly in steams and flue gases containing oxygen, sulfur oxide, and water vapour. Next to thermally sprayed composite coatings, most studies focusing on improving the performance of boiler steels examined the thermally sprayed NiCr, NiCrAlY, and Cr_3C_2–NiCr composite coatings. All of them were demonstrated to be substantially more effective in preventing oxide growth and other destructive processes as compared to bare steels [44].

One of the distinguishing characteristics between coated and uncoated steels is the oxidation kinetics, which is usually explained through the oxidation rate, in terms of weight change per unit area and time. For uncoated steels, e.g. T22 or T91, with enough time, the oxidation pattern is parabolic then linear, and may even accelerate. This change suggests the formation of porous oxide layers which crack and spall, thus allowing the ingress of oxygen and subsequent oxidation of the underlying metal. In contrast, thermally sprayed composite coatings tend to maintain a steady parabolic growth rate, indicating the formation of a stable and adherent oxide scale that restricts diffusion of oxygen and metal ions. The parabolic rate constants (k_p) for coated specimens are generally an order of magnitude lower than those for uncoated substrates, emphasizing the diffusion-limiting effect of dense composite microstructures. For example, in HVOF sprayed NiCr–Al_2O_3 coatings tested at 900°C for 1000 h, the specific weight gain was found to be approximately 2.5–3.0 mg/cm^2, compared to 18–20 mg/cm^2 for bare T22 steel under identical conditions. Similarly, Cr_3C_2–NiCr coatings exhibited a weight gain of 1.8–2.2 mg/cm^2 at 800°C, almost 90% lower than the corresponding uncoated steel. These reductions can be attributed to the formation of dense and continuous Cr_2O_3 and spinel-type oxides ($NiCr_2O_4$, $NiAl_2O_4$) that act as highly stable diffusion barriers. The presence of ceramic reinforcements such as Al_2O_3 or YSZ further refines the microstructure, closing interconnected pores and suppressing fast diffusion paths. In comparison, the oxide scales on uncoated steels typically contain hematite (Fe_2O_3) and magnetite (Fe_3O_4) phases that are non-protective and exhibit poor adherence due to thermal expansion mismatch and high defect density [45].

Microscopic observations provide visual evidence of the superior oxidation resistance of thermally sprayed coatings. Uncoated steels show considerable, thick, multi-layered and cracked oxide scales with voids and scales detaching from the substrate. Internal oxidation and the uneven diffusion of iron in the substrate also contribute to an uneven oxide–metal interior interface. These features facilitate spallation to occur during thermal cycling, exposing fresh metal to the oxidizing atmosphere. In contrast, coated specimens show scales that are thin, uniform, and adherent. Even after prolonged exposure, the scales are only 2–10 μm thick. There are virtually no signs of delamination, detachment porosity, or delamination. Projections of the protective action of coatings are revealed by cross-sectional analyses. For instance, NiCr–Al_2O_3 coatings are structured as duplex oxides with a spinel outer layer and an inner chromia or alumina layer. Such a bilayer formation efficiently halts outward metal cation diffusion while simultaneously restricting inward oxygen flow. In uncoated steels, the oxide growth front viciously advances into the substrate, depleting and weakening the surface region. Such deep oxide layers lead to a manifest weakening of the structure, a gross loss of material, and a reduction of service life. Moreover, the superior adhesion of thermally grown oxides (TGOs) in coated systems minimizes spallation even under cyclic heating, whereas uncoated steels experience recurrent oxide shedding with every thermal cycle [46].

The chemical and phase compositions of the oxides formed on coated and uncoated steels differ substantially. X-ray diffraction and Raman spectroscopy analyses indicate that uncoated steels predominantly form Fe_2O_3 and Fe_3O_4, which are thermodynamically less stable and prone to rapid volatilization or cracking. In contrast, thermally sprayed composite coatings, depending on their composition, develop protective oxides such as Cr_2O_3, Al_2O_3, or spinel-type oxides that remain stable even at high temperatures. The formation of these oxides is thermodynamically favoured due to their low oxygen ion diffusivity and high melting points. For example, the presence of Al in NiCrAlY coatings promotes the selective oxidation of Al to form α-Al_2O_3, which is one of the most adherent and slow-growing oxides known for high-temperature protection. Similarly, the Cr component ensures the formation of a continuous Cr_2O_3 subscale, which further enhances the oxidation resistance. EDS and XPS analyses corroborate these findings by showing enrichment of Al^{3+} and Cr^{3+} at the outer surface of coated samples, confirming selective oxidation. In uncoated steels, Fe^{3+} is the dominant species, and the rapid diffusion of Fe cations through the oxide layer contributes to continuous scale growth. This difference in oxidation mechanism explains the superior long-term stability of coated systems. Additionally, composite coatings exhibit better chemical compatibility between metallic and ceramic phases, reducing differential oxidation rates and internal stresses. The incorporation of refractory oxide reinforcements such as TiO_2 or YSZ further enhances stability by acting as diffusion barriers and anchoring sites for the thermally grown oxide layer [47].

Mechanical integrity after oxidation is a key factor in determining the suitability of coatings for boiler applications. Coatings must not only resist oxidation but also maintain adhesion and mechanical strength after extended exposure to high temperatures. Pull-off adhesion and scratch testing performed on oxidized samples reveal that thermally sprayed coatings retain high adhesion strength even after prolonged thermal cycling. For instance, NiCr–Al_2O_3 and NiCrAlY coatings maintain adhesion values above 25–30 MPa after 500 hours of oxidation at 900°C, while uncoated steels often exhibit surface spallation and significant mass loss. The high ductility of the metallic matrix in composite coatings accommodates thermal expansion mismatch between the oxide scale and substrate, preventing crack initiation. Cyclic oxidation testing further emphasizes this advantage. During repeated heating and cooling, oxide scales on uncoated steels spall due to poor adherence and high compressive stresses, leading to catastrophic weight fluctuations [48].

10.8 CHALLENGES AND FUTURE RESEARCH DIRECTIONS

Despite significant advancements in thermally sprayed composite coatings for high-temperature boiler applications, several challenges remain that limit their full potential. Current research and industrial practice have demonstrated that coatings such as NiCr–Al_2O_3, NiCrAlY, and Cr_3C_2–NiCr can effectively reduce oxidation and corrosion rates in ferritic and martensitic steels. However, evolving boiler operating conditions, particularly in ultra-supercritical and advanced ultra-supercritical (A-USC) plants, impose more severe thermal, chemical, and mechanical stresses, necessitating further innovations in coating design, process optimization, and predictive modelling. One of the primary challenges is the mismatch in the coefficient of thermal expansion (CTE) between the coating and substrate. During cyclic thermal loading, differential expansion generates tensile and compressive stresses at the coating–substrate interface, leading to microcrack formation and oxide spallation. Even coatings with dense, adherent oxides may suffer partial delamination after prolonged cyclic exposure, reducing their protective efficiency. Advanced multilayer or functionally graded coatings (FGCs) have been proposed to mitigate this issue, where compositional gradients gradually transition from the substrate to the outer oxide-forming layer, thereby reducing thermal stress accumulation.

Porosity, inter-splat boundaries, and oxide inclusions are inherent to thermal spray processes. While HVOF and cold spray techniques produce relatively dense coatings, residual microstructural defects can serve as pathways for oxygen ingress, accelerating localized oxidation. Large oxide clusters or agglomerates in composite coatings may also initiate crack propagation under thermal cycling. Minimizing such defects through optimized spray parameters, post-deposition heat treatment, or novel feedstock preparation remains an ongoing challenge [49].

In steam-rich and flue gas environments containing alkali salts or SO_2, protective oxides such as Cr_2O_3 can volatilize as CrO_3 or react with molten salts to form sulfates, leading to accelerated hot corrosion. While Al_2O_3-based scales are more resistant, their slow initial formation can leave the substrate temporarily unprotected during start-up. Coatings must therefore be engineered to provide rapid formation of protective scales or incorporate additives (e.g. reactive elements such as Y, Hf, or Ce) that enhance oxide adherence and retard volatilization. Nanostructured coatings, produced using nano-sized feedstocks or suspensions, provide refined grain boundaries and high-density microstructures that impede diffusion. The incorporation of ceramic nanoparticles (e.g. Al_2O_3, TiO_2, YSZ) into metallic matrices enhances oxidation resistance by acting as physical diffusion barriers, pinning the oxide layer, and limiting crack propagation. Nano-composite coatings also exhibit superior cyclic oxidation behaviour due to reduced defect sizes and improved scale adherence.

The addition of reactive elements such as yttrium, hafnium, or cerium to the coating matrix has been shown to significantly improve oxidation performance. These elements segregate at the oxide–metal interface, reduce vacancy concentrations, enhance adhesion, and stabilize α-Al_2O_3 formation [50].

The coating structures demonstrate elevated strength (tensile, compressive, and flexural) and effective performance in solar energy conversion systems, along with significant potential in surface topography. The constructed surfaces provide a basis for sophisticated computational fluid dynamics of pipes and conduits with bends, as well as ergonomic seating solutions that integrate anthropometric, mechanical, and structural linkages [51–63]. The biomedical and high-temperature corrosion-resistant systems in slurry processing have begun utilizing thermal spray coatings and advanced surface engineering methods. This is especially accurate when integrated with upcoming technologies like additive manufacturing, magnetron sputtering thin-film deposition, and microwave-assisted fabrication [64–77].

10.9 CONCLUSIONS

Research on thermally sprayed composite coatings on boiler steels shows promising results for limiting the impacts of oxidation, corrosion, and wear on significantly high-temperature material degradation. Coated steels (uncoated steels for comparative purposes) and microstructural studies (plastics for comparative purposes) demonstrate the effectiveness of the coatings regulating oxidation when dense and adherent scales of Cr_2O_3, Al_2O_3, and Cr spinel oxide were formed. These scales effectively interrupted the diffusion of oxygen and the metallic ions causing a reduction in the gain of mass of up to 90% when compared with scales formed on uncoated substrates. SEM, TEM, and in general microscopic studies show the plated steels retained microstructural recovery during the maintenance of a regime of continuous duplex oxide layers which lent themselves to mechanical and chemical stability for wear. On the other hand, uncoated steels exhibited no recovery of microstructure control during maintenance of continuous duplex oxide layers which lent themselves to mechanical and chemical stability for wear. The performance of coatings is strongly influenced by deposition methods, feedstock composition, and post-deposition treatments. Techniques such as HVOF and plasma spraying yield dense coatings with minimal porosity, while the inclusion of ceramic reinforcements and reactive elements enhances scale stability, adhesion, and resistance to cyclic thermal stresses. Functionally graded or multilayer designs offer additional advantages by reducing thermal expansion mismatch between coating and substrate, thereby mitigating stress accumulation and oxide spallation. However, current systems face challenges, including residual porosity, microstructural defects, limited hot corrosion resistance, and incomplete predictive capability for service life, particularly in ultra-supercritical boilers where temperatures exceed 650–700°C and aggressive flue gas chemistries prevail.

Based on these observations, several practical recommendations can be made for optimizing coating performance. The selection of coating materials should prioritize oxidation-resistant metallic alloys combined with ceramic reinforcements to ensure chemical stability and mechanical compliance. Functionally graded or multi-layer coatings are recommended for components exposed to cyclic thermal loading to minimize interfacial stresses and enhance adherence. Deposition parameters must be carefully optimized to reduce porosity, oxide inclusions, and inter-splat boundaries, and post-deposition heat treatment can further improve metallurgical bonding and the formation of protective thermally grown oxides. Incorporation of reactive elements such as yttrium, hafnium, or cerium is beneficial in promoting scale adherence, reducing vacancy concentrations, and stabilizing α-Al_2O_3 formation; proper distribution and content must be ensured to avoid negative effects on mechanical properties.

REFERENCES

1. Praveen AS, Singh H, Prakash S. High-temperature oxidation and erosion of HVOF sprayed NiCrSiB reinforced with Al_2O_3 and WC single bond Co on SS304. *Surface and Coatings Technology*. 2022;417:127246.
2. Bai M, Zhang J, Zhang L, et al. Preparation of MCrAlY–Al_2O_3 composite coatings with high-temperature oxidation resistance. *Journal of Thermal Spray Technology*. 2019;28(5):1011–21.
3. Wu D, Yuan Z, Liu S, Zheng J, Wei X, Zhang C. Recent development of corrosion factors and coating applications in biomass firing plants. *Coatings*. 2020 Oct 19;10(10):1001.
4. Jafari R, Ghorbani M, Shamanian M, et al. KCl-induced high-temperature corrosion behavior of NiCrAlY coatings. *Journal of Thermal Spray Technology*. 2018;27(4):732–41.
5. Singh G, Singh R, Singh H, et al. Microstructural analysis and hot corrosion behavior of HVOF-sprayed NiCrAlY coatings on T-22 boiler steel. *Journal of Thermal Spray Technology*. 2020;29(2):334–45.
6. Kim YS, Kim WC, Jain J, Huang EW, Lee SY. Hydrogen embrittlement of a boiler water wall tube in a district heating system. *Metals*. 2022 Jul 29;12(8):1276.
7. Sadeghi E, Ghorbani M, Shamanian M, et al. Advances in corrosion-resistant thermal spray coatings for boiler applications. *Journal of Thermal Spray Technology*. 2019;28(3):506–18.
8. Zhang B, Zheng S, Dong J, Yin W, Wu H, Tian L, Liu G. Effect of pre-heat-treatment on the oxidation resistance of MCrAlY coatings: A review. *Coatings*. 2023 Jul 8;13(7):1222.
9. Branagan DJ, Breitsameter M, Meacham BE, Belashchenko V. High-performance nanoscale composite coatings for boiler applications. *Journal of Thermal Spray Technology*. 2005 Jun;14(2):196–204.
10. Singh S, Goyal K, Bhatia R. A review on protection of boiler tube steels with thermal spray coatings from hot corrosion. *Materials Today: Proceedings*. 2022 Jan 1;56:379–83.
11. Szymański K, Hernas A, Moskal G, Myalska H. Thermally sprayed coatings resistant to erosion and corrosion for power plant boilers: A review. *Surface and Coatings Technology*. 2015 Apr 25;268:153–64.
12. Dhand D, Kumar P, Grewal JS. A review of thermal spray coatings for protection of steels from degradation in coal fired power plants. *Corrosion Reviews*. 2021 Jun 25;39(3):243–68.
13. Singh G, Bala N, Chawla V. High temperature oxidation behaviour of HVOF thermally sprayed NiCrAlY coating on T-91 boiler tube steel. *Materials Today: Proceedings*. 2017 Jan 1;4(4):5259–65.

14. Klenam D, Asumadu T, Bodunrin M, Vandadi M, Bond T, van der Merwe J, Rahbar N, Soboyejo W. Cold spray coatings of complex concentrated alloys: Critical assessment of milestones, challenges, and opportunities. *Coatings*. 2023 Mar 1;13(3):538.
15. Rezakhani D. Corrosion behaviours of several thermal spray coatings used on boiler tubes at elevated temperatures. *Anti-Corrosion Methods and Materials*. 2007 Jul 3;54(4):237–43.
16. Galedari SA, Mahdavi A, Azarmi F, Huang Y, McDonald A. A comprehensive review of corrosion resistance of thermally-sprayed and thermally-diffused protective coatings on steel structures. *Journal of Thermal Spray Technology*. 2019 Apr 17;28(4):645–77.
17. Oksa M, Metsäjoki J, Kärki J. Corrosion testing of thermal spray coatings in a biomass co-firing power plant. *Coatings*. 2016 Nov 24;6(4):65.
18. Kumar M, Singh H, Singh N. Fire side erosion–corrosion protection of boiler tubes by nanostructured coatings. *Materials and Corrosion*. 2015 Jul;66(7):695–709.
19. Singh J, Vasudev H, Singh S. Performance of different coating materials against high temperature oxidation in boiler tubes: A review. *Materials Today: Proceedings*. 2020 Jan 1;26:972–8.
20. Kumar R, Kumar R, Kumar S. Erosion corrosion study of HVOF sprayed thermal sprayed coating on boiler tubes: A review. *IJSMS*. 2018;1(3):1–6.
21. Michalak M, Sokołowski P, Szala M, Walczak M, Łatka L, Toma FL, Björklund S. Wear behavior analysis of Al2O3 coatings manufactured by APS and HVOF spraying processes using powder and suspension feedstocks. *Coatings*. 2021 Jul 23;11(8):879.
22. Goyal K, Singh H, Bhatia R. Current status of thermal spray coatings for high temperature corrosion resistance of boiler steel. *Journal of Material & Metallurgical Engineering*. 2016;6(1):29–35.
23. Heath GR, Heimgartner P, Irons G, Miller RD, Gustafsson S. An assessment of thermal spray coating technologies for high temperature corrosion protection. *Materials Science Forum*. 1997 Oct 27;251:809–16.
24. Kumar K, Kumar S, Gill HS. Role of carbide-based thermal-sprayed coatings to prevent failure for boiler steels: A review. *Journal of Failure Analysis and Prevention*. 2024 Aug;24(4):1628–63.
25. Oksa M. Nickel-and iron-based HVOF thermal spray coatings for high temperature corrosion protection in biomass-fired power plant boilers. *Fuel Processing Technology*. 2015;125:236–45.
26. Gupta V, Srivastava A, Asthana N, Rathinam V, Rajendran S, Khan AA, Fatima S. High temperature oxidation mitigation efficacy of Al2O3 and Ni-20Cr ceramic coating on boiler steel application. *Journal of Molecular Structure*. 2025 Feb 5;1321:139831.
27. Fauchais P, Vardelle A. Thermal sprayed coatings used against corrosion and corrosive wear. *Advanced Plasma Spray Applications*. 2012 Mar 21;10:34448.
28. Lakkannavar V, Yogesha KB, Prasad CD, Mruthunjaya M, Suresh R. A review on tribological and corrosion behaviour of thermal spray coatings. *Journal of the Institution of Engineers (India): Series D*. 2025 Apr;106(1):753–69.
29. Pokhmurskii V, Student M, Gvozdeckii V, Stypnutskyy T, Student O, Wielage B, Pokhmurska H. Arc-sprayed iron-based coatings for erosion-corrosion protection of boiler tubes at elevated temperatures. *Journal of Thermal Spray Technology*. 2013 Jun;22(5):808–19.
30. Sahil, Singh S, Goyal K, Bhandari D, Krishan B. Hot corrosion behaviour of thermal-sprayed TiO2-reinforced Cr2O3 composite coatings on T-22 boiler steel at elevated temperature. *Journal of Bio-and Tribo-Corrosion*. 2021 Sep;7(3):100.
31. Goyal R, Goyal P. High temperature boiler steel corrosion protection with thermal spray coating powders and processes. *AIP Conference Proceedings* 2025 Mar 12;3227(1):020004.

32. Singh G, Bala N, Chawla V, Singla YK. Hot corrosion behavior of HVOF-sprayed carbide based composite coatings for boiler steel in Na2SO4–60% V2O5 environment at 900° C under cyclic conditions. *Corrosion Science*. 2021 Sep 1;190:109666.
33. Lakkannavar V, Yogesha KB, Prasad CD, Suresh R, Lakshmikanthan A, Hanumanthlal S. Investigation of cyclic oxidation and corrosion behavior of plasma-sprayed NiCrAlY/Cr3C2/h-BN coatings on T22 boiler steel alloy. *Surface and Coatings Technology*. 2024 Sep 30;492:131205.
34. Senthilkumar V, Thiyagarajan B, Duraiselvam M, Karthick K. Effect of thermal cycle on Ni–Cr based nanostructured thermal spray coating in boiler tubes. *Transactions of Nonferrous Metals Society of China*. 2015 May 1;25(5):1533–42.
35. Kwon GH, Park H, Choi B, Lee YK, Moon K. Influence of Cr content on the high-temperature oxidation behavior and mechanism of low-alloy steels. *Materials*. 2023 Jul 12;16(14):4964.
36. Ansari MS, Chawla V, Bansal A, Aggarwal V. Comparative study of high-temperature oxidation behavior of bare and plasma sprayed Al2O3–40% TiO2 coated T-91, A-1 boiler steel and Superfer800H superalloy in air. *Journal of Materials Engineering and Performance*. 2022 Jan;31(1):753–68.
37. Mahdavi A, Medvedovski E, Leal Mendoza G, McDonald A. Corrosion resistance of boronized, aluminized, and chromized thermal diffusion-coated steels in simulated high-temperature recovery boiler conditions. *Coatings*. 2018 Jul 24;8(8):257.
38. Lakkannavar V, Yogesha KB, Prasad CD, Prasad C. Investigation of cyclic oxidation and hot corrosion behaviour of plasma-sprayed NiCrAlY/Cr3C2/h-BN/Mo coatings on T22 boiler steel alloy. *Journal of Bio-and Tribo-Corrosion*. 2024 Dec;10(4):85.
39. Fauchais PL, Heberlein JV, Boulos MI. Industrial applications of thermal spraying technology. In: PL Fauchais, JVR Heberlein, MI Boulos (Eds.), *Thermal spray fundamentals: From powder to part* 2013 (pp. 1401–566). Boston, MA: Springer US.
40. Hidalgo VH, Varela FB, Menéndez AC, Martınez SP. A comparative study of high-temperature erosion wear of plasma-sprayed NiCrBSiFe and WC–NiCrBSiFe coatings under simulated coal-fired boiler conditions. *Tribology international*. 2001 Mar 1;34(3):161–9.
41. Singh A, Sharma V, Mittal S, Pandey G, Mudgal D, Gupta P. An overview of problems and solutions for components subjected to fireside of boilers. *International Journal of Industrial Chemistry*. 2018 Mar;9(1):1–5.
42. Singh J, Vasudev H, Singh S. A study on the prevention of hot corrosion of boiler steel with a high-velocity oxy-fuel spray-coating process. In: A Sachdeva, C Bhargava, P Kumar (Eds.), *Smart nanotechnology with applications* 2020 Dec 15 (pp. 105–13). Boca Raton: CRC Press.
43. Agnihotri A, Kalsi SS, Kansal H. Effect of burnishing on HVOF coated boiler steel against hot corrosion in actual boiler environment. *Surface Engineering*. 2023 May;39(5):559–71.'
44. Mehta J, Mittal VK, Gupta P. Role of thermal spray coatings on wear, erosion and corrosion behavior: A review. *Journal of Applied Science and Engineering*. 2017 Dec;20(4):445–52.
45. Szczucka-Lasota B, Szymański K. Oxidation resistance of surface-modified coatings for energy boilers. *Archives of Materials Science*. 2012 Aug;76:76.
46. Verstak A, Wang B, Baranovski V, Beliaev A. Composite coatings for elevated temperature erosion-corrosion protection in fossil-fueled boilers. In NACE Technical Committees, *Corrosion* 1998 1998 Mar 22 (pp. 1–15). Houston, TX: Association for Materials Protection and Performance.

47. Milan Shahana S, Bakshi SR, Kamaraj M. High-temperature oxidation and hot corrosion of thermal spray coatings. In: Khanna PK (Ed.), *A treatise on corrosion science, engineering and technology* 2022 May 5 (pp. 407–20). Singapore: Springer Nature Singapore.
48. Boulos MI, Fauchais PL, Heberlein JV. Industrial applications of thermal spray technology. In: Boulos MI, Fauchais PL, Heberlein JV (Eds.), *Thermal spray fundamentals: from powder to part* 2021 Oct 20 (pp. 997–1096). Cham: Springer International Publishing.
49. Sadeghi E, Markocsan N, Joshi S. Advances in corrosion-resistant thermal spray coatings for renewable energy power plants. Part I: Effect of composition and microstructure. *Journal of Thermal Spray Technology*. 2019 Dec;28(8):1749–88.
50. Czupryński A, Adamiec J, Adamiak M, Żuk M, Kříž A, Mele C, Kciuk M. High-temperature corrosion of flame-sprayed power boiler components with nickel alloy powders. *Materials*. 2023 Feb 16;16(4):1658.
51. Kulkarni SD, Manjunatha, Chandrasekhar U, Manjunath KV, Prasad CD, Vasudev H. Design and optimization of polyvinyl-nitride rubber for tensile strength analysis. *International Journal on Interactive Design and Manufacturing (IJIDeM)*. 2024;18(8):5893–908.
52. Kattimani MA, Venkatesh PR, Masum H, Math MM, Bahadurdesai VN, Mustafkhadri S, Prasad CD, Vasudev H. Design and numerical analysis of tensile deformation and fracture properties of induction hardened inconel 718 superalloy for gas turbine applications. *International Journal on Interactive Design and Manufacturing (IJIDeM)*. 2024;18(8):5939–49.
53. Poojari M, Hanumanthappa H, Prasad CD, Jathanna HM, Ksheerasagar AR, Shetty P, Shanmugam BK, Vasudev H. Computational modelling for the manufacturing of solar-powered multifunctional agricultural robot. *International Journal on Interactive Design and Manufacturing (IJIDeM)*. 2024;18(8):5725–736.
54. Prakash C, Gupta LR, Mehta A, Vasudev H, Tominov R, Korman E, Fedotov A, Smirnov V, Kesari KK. Computing of neuromorphic materials: An emerging approach for bioengineering solutions. *Materials Advances*. 2023;4:5882–919.
55. Singh J, Gill HS, Vasudev H. Computational fluid dynamics analysis on role of particulate shape and size in erosion of pipe bends. *International Journal on Interactive Design and Manufacturing (IJIDeM)*. 2023;17(5):2631–46.
56. Rao GS, Mukkamala U, Hanumanthappa H, Prasad CD, Vasudev H, Shanmugam B, KishoreKumar KC. Evaluating and optimizing surface roughness using genetic algorithm and artificial neural networks during turning of AISI 52100 steel. *International Journal on Interactive Design and Manufacturing (IJIDeM)*. 2024;18(8):6151–60.
57. Singh J, Singh R, Singh S, Vasudev H, Kumar S. Reducing scrap due to missed operations and machining defects in 90PS pistons. *International Journal on Interactive Design and Manufacturing (IJIDeM)*. 2023;17(5):2527–39.
58. Nagabhushana P, Ramprasad S, Prasad CD, Vasudev H, Prakash C. Numerical investigation on heat transfer of a nano-fluid saturated vertical composite porous channel packed between two fluid layers. *International Journal on Interactive Design and Manufacturing (IJIDeM)*. 2024;18(5):2927–44.
59. Mahantesh MM, Rao KR, Chandra AP, Vijayakumar MN, Nandini B, Prasad CD, Vasudev H. Design and modeling using finite element analysis for the sitting posture of computer users based on ergonomic perspective. *International Journal on Interactive Design and Manufacturing (IJIDeM)*. 2024;18(8):5875–91.
60. Verma PK, Vasudev H, Kumar V. Dental implant materials and designs and failure during implant placement and loading: A review. In: H Vasudev, C Prakash, M

Khalid, K Kesari (Eds.), *Advanced manufacturing processes* (pp. 138–50). Boca Raton: CRC Press.

61. Verma PK, Vasudev H, Kumar V. Bioactivity and corrosion performance of low-velocity oxy-fuel-sprayed hydroxyapatite-based coatings on a surgical-grade stainless steel. *Transactions of the Indian Institute of Metals*. 2025;78(8):185.
62. Verma PK, Vasudev H, Kumar V. Effect of electro-corrosion behavior of different grades of stainless steel used in bio-implants: A review. In: H Vasudev, C Prakash, M Khalid, K Kesari (Eds.), *Advanced manufacturing processes* 2025 (pp. 269–83). Boca Raton: CRC Press.
63. Verma PK, Vasudev H. Importance of hydroxyapatite coatings in improving new bone osteogenesis: A review. In: H Vasudev, C Prakash, M Khalid, K Kesari (Eds.), *Advanced manufacturing processes* 2025 (pp. 284–99). Boca Raton: CRC Press.
64. Dutta V, Thakur L, Singh B, Vasudev H. A study of erosion–corrosion behaviour of friction stir-processed chromium-reinforced NiAl bronze composite. *Materials*. 2022;15(15):5401.
65. Singh M, Vasudev H, Kumar R. Corrosion and tribological behaviour of BN thin films deposited using magnetron sputtering. *International Journal of Surface Engineering and Interdisciplinary Materials Science (IJSEIMS)*. 2021;9(2):24–39.
66. Sharanabasava H, Raviprakash M, Prasad CD, Ramesh MR, Phanibhushana MV, Vasudev H, Kumar S. Microstructure, mechanical and wear properties of SiC and Mo reinforced NiCr microwave cladding. *Advanced Materials and Processing Technologies*. 2023; 10(4):3620–33.
67. Singh M, Vasudev H, Singh M. Surface protection of SS-316L with boron nitride based thin films using radio frequency magnetron sputtering technique. *Journal of Electrochemical Science and Engineering*. 2022;12(5):851–63.
68. Prashar G, Vasudev H, Thakur L, Bansal A. Performance of thermally sprayed hydroxyapatite coatings for biomedical implants: A comprehensive review. *Surface Review and Letters*. 2023;30(1):2241001.
69. Vasudev H. Wear characteristics of Ni-WC powder deposited by using a microwave route on mild steel: Microwave cladding of Ni-WC. *International Journal of Surface Engineering and Interdisciplinary Materials Science (IJSEIMS)*. 2020;8(1):44–54.
70. Parkash J, Saggu HS, Vasudev H. A short review on the performance of high velocity oxy-fuel coatings in boiler steel applications. *Materials Today: Proceedings*. 2022;50:1442–6.
71. Singh P, Vasudev H, Bansal A. Effect of post-heat treatment on the microstructural, mechanical, and bioactivity behavior of the microwave-assisted alumina-reinforced hydroxyapatite cladding. *Proceedings of the Institution of Mechanical Engineers, Part E: Journal of Process Mechanical Engineering*. 2024;238(3):1060–71.
72. Verma PK, Vasudev H, Singh G, Singh S. A review on the microstructural and biomedical properties of implants manufactured using additive manufacturing. *Journal of The Institution of Engineers (India): Series D*. 2024;106:1587–601.
73. Singh SP, Verma PK, Singh S, Kapoor M, Balu R. Microstructural characterization and electrochemical behaviour of HVOF sprayed WC-Co-Cr coatings on 316L boiler steel stainless steel. *Results in Surfaces and Interfaces*. 2024;16:100248.
74. Verma PK, Minhas AS, Singh P, Kumar S. Slurry erosion behaviour of thermal sprayed Al2O3 and Cr2O3 coatings for turbine steels. *AIP Conference Proceedings*. 2024;2986(1):020012.
75. Verma PK, Kumar V, Vasudev H. Bioactivity and corrosion analysis of thermally sprayed hydroxyapatite based coatings. *Journal of Electrochemical Science and Engineering*. 2024;14(4):419–39.

76. Verma PK, Vasudev H, Kumar V. Biocompatibility and corrosion performance of HA-based LVOF sprayed and post-annealed coatings on SS 254 surgical-grade bio-implant material. *Ceramics International*. 2025;51(10):13207–25.
77. Verma PK, Vasudev H, Singh G. Improving the corrosion performance of SS 304 stainless steel using HVOF-sprayed WC-Co-Cr coatings in chloride environments. *Journal of Bio-and Tribo-Corrosion*. 2025;11(4):1–13.

11 Case Study on High Temperature Oxidation Behaviour of Superalloy-Based Composite Coatings

Gaurav Prashar and Hitesh Vasudev

11.1 INTRODUCTION

Superalloys are widely used in industries such as power generation, furnaces, nuclear, and petrochemical manufacturing where extreme temperatures, thermal stresses, and corrosive environments are present. They were created using the idea of selective oxidation to promote oxidation resistance [1]. Cr, Al, Ni, and Si are used to increase oxidation resistance because they oxidize selectively and generate protective oxide [2]. Though Cr_2O_3 is primarily responsible for oxidation resistance, it frequently affects mechanical characteristics negatively and is expensive in high-temperature environments [3]. In order to solve this issue, alloying elements, such as Mo, Ta, and Nb through the solid solution hardening process, are typically added to improve mechanical properties. But making the whole of the boiler tubes from superalloy is an expensive process. Therefore, the easiest way is to deposit coatings by using thermal spray techniques. Due to its inexpensive cost and superior effectiveness over alternative thermal spraying systems like detonation guns, vacuum plasma sprays, high velocity oxygen fuels, or cold spraying, plasma spraying has become widely used. Plasma spray uses the principle of thermal and kinetic energy where particles strike the surface of the substrate with impact, flatten, and are then deposited in the form of splats on the substrate. By using the plasma spray technology, a porous structure with discontinuity holes, cracks, and lamellar inter-pore formations is produced [4, 5].

The integration of a ductile phase with a hard phase has been recognized as an effective approach to mitigate the harmful effects of oxidation in metallic coatings. Scholars have proposed the implementation of multi-phase coatings in academic discourse to mitigate material degradation [5–7]. Thermal spray coatings based on Ni-Cr and NiCr-Cr_3C_2 are frequently recommended for components functioning under elevated temperature conditions.

Researchers have also investigated composite Ni-Al_2O_3 coating systems [8–13]. In our previous research, we explored the impact of incorporating Al_2O_3 reinforcement particles into the Inconel-718 matrix. A coating composition of 30 wt% demonstrated

DOI: 10.1201/9781003730637-11

the highest resistance to oxidation and erosion. This phenomenon can be attributed to the increased hardness conferred by a higher Al_2O_3 content, which restricts plastic deformation and subsequently improves the coating's overall performance [14–16]. According to Kim et al. [17], there may be between 20 and 40 wt% of hard phase reinforcement in ductile nickel matrix.

The goal of the current research work is to create composite coatings made of Inconel625 and 30%Al_2O_3 (In625+Al). In order to offer the requisite strength, hardness, and endurance at high temperatures, Inconel625 contain a significant quantity of alloying elements, mainly (Ni-Cr-Mo). The superalloy-based composite coating In625+Al underwent cyclic oxidation, which involved heating for an hour in a silicon carbide tube furnace and cooling for 20 minutes in room air. After each cycle, a digital electronic weighing balance was used to measure weight change. X-ray diffraction (XRD), energy-dispersive X-ray spectroscopy (EDS), and field emission scanning electron microscopy (FESEM) were used to characterize the exposed specimens. The findings may enhance the resistance of boiler bed coils to oxidation under elevated temperature conditions.

11.2 MATERIALS AND METHODS

11.2.1 Coatings Development

In625+Al (size: 40–50 μm) composite mixture was sprayed using the plasma spray process available at Metalizing equipment Co-pvt. Ltd. Jodhpur. The substrate material was ASTM SA210 GrA1. For oxidation testing the required specimen's size of 20×15×5 mm^3 was prepared using EDM. Figure 11.1 shows an image of the cross-section of the deposited In625+Al coating. A good contact of In625+Al coating with the substrate indicates a better bonding of In625+Al coating with a coating thickness 250 μm.

11.2.2 Methods

Examining the oxidation characteristics of the manufactured composite coating is the main goal of this work. Using a specially built silicon carbide tube furnace, the

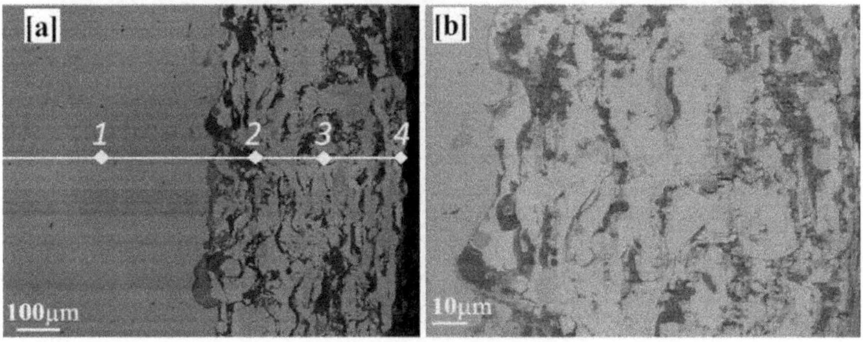

FIGURE 11.1 Cross-sectioned image of the deposited In625+Al coating.

specimens (coated and uncoated) were oxidized, and the oxidized surfaces of the specimens were inspected using XRD, EDS, and FESEM. The superalloy-based In625+Al composite coating underwent cyclic oxidation, which involved heating for an hour in a tube furnace and cooling for 20 minutes in room air. The samples were then weighed after each cycle using an electronic scale. Finally, the parabolic rate constant of specimens was determined using Eqn. 11.1.

$$x^2 = (Kp)(t) \tag{11.1}$$

where,

x= weight gain/surface area in mg/cm^2
Kap= parabolic rate constant
t= time in sec

11.3 RESULTS AND DISCUSSIONS

11.3.1 Visual Observations

The uncoated substrate's colour changed immediately after the first cycle, beginning with gray-bluish, and continuing until the fifth cycle finished, when the substrate's colour changed to grayish-black. The colour of the substrate fully changed to black

FIGURE 11.2 a) Oxidized sample, and b) oxidized In625+Al coating after 50 cycles.

TABLE 11.1
Kp values of specimens

Specimen	Kp (10^{-4} mg^2 cm^{-4} s^{-1})
Substrate	619.35
In625+Al	6.36

after eight cycles, and a white layer of oxidized scale was also seen on the surface. After 50 cycles, the bare substrate's substance spalled, peeled off with a popping noise, producing tiny chips that resulted in a significant mass gain. The In625+Al coating specimen generally supports the parabolic oxidation rate rule. After the 15th cycle, cracks started to appear along the coating thickness. As the number of cycles increased, the crack width also widened and a network of cracks appeared along the coating thickness (Figure11.2b). But the deposited coating didn't peel off and reduce the weight gain of the substrate by 90% up to 50 cycles. The results thus demonstrate the In625+Al composite structure's value in preventing oxidation of the substrate. The values of Kp for uncoated and coated specimen are shown in Table 11.1.

11.4 ANALYSIS OF SCALE (FESEM/EDS)

The SEM analysis of the boiler steel without coating shows the formation of a granulated scale [18, 19], and the EDS analysis further confirms that this granular scale is composed primarily of Fe and O. This significantly backs up the theory that a weak Fe_2O_3 phase could form, which has also been demonstrated by XRD analysis. The In625+Al scale displays a distinct pyramidal and crystalline structure [20]. The In625+Al coating's major elements, as determined by the EDS study, are Ni and Cr, with a significant amount of O also present. Thus, even after oxidation tests, the feed-stock powder's composition does not change.

11.5 ANALYSIS OF SCALE (XRD)

To determine the phases included in the scale generated after 50 cycles, an XRD graph is plotted relating intensity and diffraction angle 2θ. This can be accomplished by comparing intensity peak d values to the d values provided in the JCPDS files for the relevant phase. Figure 11.3 shows X-ray diffractograms of the scale developed on the oxidized specimens (both coated and uncoated). Fe_2O_3 is the primary phase identified in the uncoated substrate, whereas for the oxidized In625+Al coating the prime phase formed is Ni-Cr, NiO, Cr_2O_3, Al_2O_3, and spinel $NiCr_2O_4$, which provides resistance against oxidation [21–23].

Slurry processing systems that are biomedical and resistant to corrosion at high temperatures have started using thermal spray coatings and other modern surface engineering techniques. When combined with emerging technologies such as microwave-assisted fabrication, additive manufacturing, and magnetron sputtering thin-film deposition, this becomes even more precise [24–37]. The coating structures

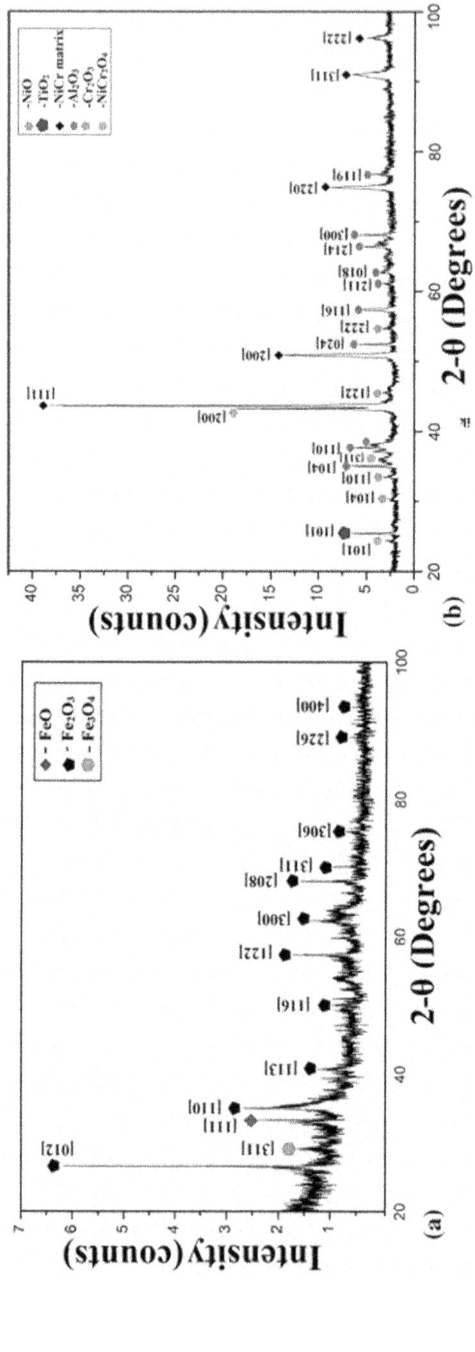

FIGURE 11.3 XRD patterns: a) oxidized sample, and b) oxidized In625+Al coating after 50 cycles.

show great promise in surface topography, as well as increased strength (tensile, compressive, and flexural) and efficient operation in solar energy conversion systems. Anthropometric, mechanical, and structurally linked seating solutions, as well as complex computational fluid dynamics of curved pipes and conduits, can be built on top of the designed surfaces [38–49].

11.6 DISCUSSION

It is obvious that the initial cycles have greater weight gain. But with passage of time due to the development of the protective and coherent oxides on the surface of In625+Al coating the weight gain doesn't occur at rapid rate. The observations show that the oxide coating appeared to be compact, protective, and nearly consistent in thickness over the specimen's surface. Similar results are also found in the literature [21, 22]. In625+Al has a strong oxidation resistance related to the protective oxide layers.

11.7 CONCLUSION

- The In625+Al coating was developed using the plasma spray technology, and the oxidation resistance under high-temperature conditions was analyzed. Important findings are summarized as below:
- The developed coatings' cross-section revealed a coating thickness of 200–250 μm and good bonding at the interface without any evidence of element diffusion.
- In625+Al has strong oxidation resistance related to the protective oxide layers.
- The prime phase formed in the In625+Al coating is Ni-Cr, NiO, Cr_2O_3, Al_2O_3, and spinel $NiCr_2O_4$.
- The coating reduces the weight gain of the substrate by 90% up to 50 cycles.

REFERENCES

1. Pettit, F.S., Meier, G.H., Gell, M., Kartovich, C.S., Bricknel, R.H., Kent, W.B. and Radovich, J.F., 1984. Oxidation and hot corrosion of superalloys. *Superalloys* 85(1), pp.651–687.
2. Meier, G.H., 1989. A review of advances in high-temperature corrosion. *Materials Science and Engineering: A*, 120, pp.1–11.
3. Chen, H.C., Liu, Z.Y. and Chuang, Y.C., 1993. Degradation of plasma-sprayed alumina and zirconia coatings on stainless steel during thermal cycling and hot corrosion. *Thin Solid Films*, 223(1), pp.56–64.
4. Prashar, G., Vasudev, H., 2022. A review on the influence of process parameters and heat treatment on the corrosion performance of Ni-based thermal spray coatings. *Surface Review and Letters*, 29(1), p.2230001.
5. Prashar, G., Vasudev, H., Thakur, L., 2022. High-temperature oxidation and erosion resistance of Ni-based thermally-sprayed coatings used in power generation machinery: A review. *Surface Review and Letters,* 29(3), p.2230003.
6. Tobar, M.J., Álvarez, C., Amado, J.M., Rodríguez, G. and Yáñez, A., 2006. Morphology and characterization of laser clad composite NiCrBSi–WC coatings on stainless steel. *Surface and Coatings Technology*, 200(22–23), pp.6313–6317.

7. Vasudev, H., Thakur, L., Singh, H. and Bansal, A., 2020. A study on processing and hot corrosion behaviour of HVOF sprayed Inconel718-nano Al2O3 coatings. *Materials Today Communications*, 25, p.101626.
8. Prashar, G. and Vasudev, H., 2022. Structure–property correlation of plasma-sprayed Inconel625-Al2O3 bimodal composite coatings for high-temperature oxidation protection. *Journal of Thermal Spray Technology*, 31, pp.1–24.
9. Sundararajan, T., Kuroda, S. and Abe, F., 2005. Steam oxidation of 80Ni-20Cr high-velocity oxyfuel coatings on 9Cr-1Mo steel: Diffusion-induced phase transformations in the substrate adjacent to the coating. *Metallurgical and Materials Transactions A*, 36(8), 2165–2174.
10. Yin, Z., Tao, S., Zhou, X. and Ding, C., 2007. Tribological properties of plasma sprayed Al/Al2O3 composite coatings. *Wear*, 263(7–12), 1430–1437.
11. Ramanujam, N. and Nakamura, T., 2009. Erosion mechanisms of thermally sprayed coatings with multiple phases. *Surface and Coatings Technology*, 204(1–2), pp.42–53.
12. Praveen, A.S., Sarangan, J., Suresh, S. and Subramanian, J.S., 2015. Erosion wear behaviour of plasma sprayed NiCrSiB/Al2O3 composite coating. *International Journal of Refractory Metals and Hard Materials*, 52, pp.209–218.
13. Grewal, H.S., Singh, H. and Agrawal, A., 2013. Microstructural and mechanical characterization of thermal sprayed nickel–alumina composite coatings. *Surface and coatings Technology*, 216, pp.78–92.
14. Vasudev, H., Thakur, L., Singh, H. and Bansal, A., 2022. Effect of addition of Al2O3 on the high-temperature solid particle erosion behaviour of HVOF sprayed Inconel-718 coatings. *Materials Today Communications*, 30, p.103017.
15. Vasudev, H., Thakur, L., Singh, H. and Bansal, A., 2021. Erosion behaviour of HVOF sprayed alloy718-nano Al2O3 composite coatings on grey cast iron at elevated temperature conditions. *Surface Topography: Metrology and Properties*, 9(3), p.035022.
16. Vasudev, H., Prashar, G., Thakur, L. and Bansal, A., 2021. Microstructural characterization and electrochemical corrosion behaviour of HVOF sprayed Alloy718-nanoAl2O3 composite coatings. *Surface Topography: Metrology and Properties*, 9(3), p.035003.
17. Kim, H.J., Hwang, S.Y., Lee, C.H. and Juvanon, P., 2003. Assessment of wear performance of flame sprayed and fused Ni-based coatings. *Surface and Coatings Technology*, 172(2–3), 262–269.
18. Kumar, M., Singh, H. and Singh, N., 2014. Production of nanocrystalline Ni-20Cr coatings for high-temperature applications. *Journal of Thermal Spray Technology*, 23(4), pp.692–707.
19. Singh, H., Kumar, M. and Singh, N., 2021. High temperature oxidation behavior of a novel Ni-20Cr alloy powder coating on T22 steel. In *Proceedings of the international thermal spray conference (ITSC-2013)*, (Busan, Republic of Korea, 2013, pp.120–125).
20. Khoi, N.N., Smeltzer, W. W. and Embury, J.D., 1975. Growth and structure of nickel oxide on nickel crystal faces. *Journal of the Electrochemical Society*, 122(11), p.1495.
21. Singh, H., Puri, D., Prakash, S. and Maiti, R., 2007. Characterization of oxide scales to evaluate high temperature oxidation behavior of Ni–20Cr coated superalloys. *Materials Science and Engineering: A*, 464(1–2), pp.110–116.
22. Calvarin, G., Molins, R. and Huntz, A.M., 2000. Oxidation mechanism of Ni—20Cr foils and its relation to the oxide-scale microstructure. *Oxidation of Metals*, 53(1), pp.25–48.

23. Dutta, V., Thakur, L., Singh, B. and Vasudev, H., 2022. A study of erosion–corrosion behaviour of friction stir-processed chromium-reinforced NiAl bronze composite. *Materials*, 15(15), p.5401.
24. Singh, M., Vasudev, H. and Kumar, R., 2021. Corrosion and tribological behaviour of BN thin films deposited using magnetron sputtering. *International Journal of Surface Engineering and Interdisciplinary Materials Science (IJSEIMS)*, 9(2), pp.24–39.
25. Sharanabasava, H., Raviprakash, M., Prasad, C.D., Ramesh, M.R., Phanibhushana, M.V., Vasudev, H. and Kumar, S., 2023. Microstructure, mechanical and wear properties of SiC and Mo reinforced NiCr microwave cladding. *Advanced Materials and Processing Technologies*, 10(4), pp.3620–3633.
26. Singh, M., Vasudev, H. and Singh, M., 2022. Surface protection of SS-316L with boron nitride based thin films using radio frequency magnetron sputtering technique. *Journal of Electrochemical Science and Engineering*, 12(5), pp.851–863.
27. Prashar, G., Vasudev, H., Thakur, L. and Bansal, A., 2023. Performance of thermally sprayed hydroxyapatite coatings for biomedical implants: A comprehensive review. *Surface Review and Letters*, 30(1), p.2241001.
28. Vasudev, H., 2020. Wear characteristics of Ni-WC powder deposited by using a microwave route on mild steel: microwave cladding of Ni-WC. *International Journal of Surface Engineering and Interdisciplinary Materials Science (IJSEIMS)*, 8(1), pp.44–54.
29. Parkash, J., Saggu, H.S. and Vasudev, H., 2022. A short review on the performance of high velocity oxy-fuel coatings in boiler steel applications. *Materials Today: Proceedings*, 50, pp.1442–1446.
30. Singh, P., Vasudev, H. and Bansal, A., 2024. Effect of post-heat treatment on the microstructural, mechanical, and bioactivity behavior of the microwave-assisted alumina-reinforced hydroxyapatite cladding. *Proceedings of the Institution of Mechanical Engineers, Part E: Journal of Process Mechanical Engineering*, 238(3), pp.1060–1071.
31. Verma, P.K., Vasudev, H., Singh, G. and Singh, S., 2024. A review on the microstructural and biomedical properties of implants manufactured using additive manufacturing. *Journal of The Institution of Engineers (India): Series D*, 106, pp.1587–601.
32. Singh, S.P., Verma, P.K., Singh, S., Kapoor, M. and Balu, R., 2024. Microstructural characterization and electrochemical behaviour of HVOF sprayed WC-Co-Cr coatings on 316L boiler steel stainless steel. *Results in Surfaces and Interfaces*, 16, p.100248.
33. Verma, P.K., Minhas, A.S., Singh, P. and Kumar, S., 2024, February. Slurry erosion behaviour of thermal sprayed Al2O3 and Cr2O3 coatings for turbine steels. *AIP Conference Proceedings*, 2986(1), p.020012.
34. Verma, P.K., Kumar, V. and Vasudev, H., 2024. Bioactivity and corrosion analysis of thermally sprayed hydroxyapatite based coatings. *Journal of Electrochemical Science and Engineering*, 14(4), pp.419–439.
35. Verma, P.K., Vasudev, H. and Kumar, V., 2025. Biocompatibility and corrosion performance of HA-based LVOF sprayed and post-annealed coatings on SS 254 surgical-grade bio-implant material. *Ceramics International*, 51(10), pp.13207–13225.
36. Verma, P.K., Vasudev, H. and Singh, G., 2025. Improving the Corrosion Performance of SS 304 Stainless Steel Using HVOF-Sprayed WC-Co-Cr Coatings in Chloride Environments. *Journal of Bio-and Tribo-Corrosion*, 11(4), pp.1–13.
37. Kulkarni, S.D., Manjunatha, Chandrasekhar, U., Manjunath, K.V., Prasad, C.D. and Vasudev, H., 2024. Design and optimization of polyvinyl-nitride rubber for tensile strength analysis. *International Journal on Interactive Design and Manufacturing (IJIDeM)*, 18(8), pp.5893–5908.

38. Kattimani, M.A., Venkatesh, P.R., Masum, H., Math, M.M., Bahadurdesai, V.N., Mustafkhadri, S., Prasad, C.D. and Vasudev, H., 2024. Design and numerical analysis of tensile deformation and fracture properties of induction hardened inconel 718 superalloy for gas turbine applications. *International Journal on Interactive Design and Manufacturing (IJIDeM)*, 18(8), pp.5939–5949.
39. Poojari, M., Hanumanthappa, H., Prasad, C.D., Jathanna, H.M., Ksheerasagar, A.R., Shetty, P., Shanmugam, B.K. and Vasudev, H., 2024. Computational modelling for the manufacturing of solar-powered multifunctional agricultural robot. *International Journal on Interactive Design and Manufacturing (IJIDeM)*, 18(8), pp.5725–5736.
40. Prakash, C., Gupta, L.R., Mehta, A., Vasudev, H., Tominov, R., Korman, E., Fedotov, A., Smirnov, V. and Kesari, K.K., 2023. Computing of neuromorphic materials: An emerging approach for bioengineering solutions. *Materials Advances*. 4, pp.5882–5919.
41. Singh, J., Gill, H.S. and Vasudev, H., 2023. Computational fluid dynamics analysis on role of particulate shape and size in erosion of pipe bends. *International Journal on Interactive Design and Manufacturing (IJIDeM)*. 17(5), pp.2631–2646.
42. Rao, G.S., Mukkamala, U., Hanumanthappa, H., Prasad, C.D., Vasudev, H., Shanmugam, B. and KishoreKumar, K.C., 2024. Evaluating and optimizing surface roughness using genetic algorithm and artificial neural networks during turning of AISI 52100 steel. *International Journal on Interactive Design and Manufacturing (IJIDeM)*, 18(8), pp.6151–6160.
43. Singh, J., Singh, R., Singh, S., Vasudev, H. and Kumar, S., 2023. Reducing scrap due to missed operations and machining defects in 90PS pistons. *International Journal on Interactive Design and Manufacturing (IJIDeM)*, 17(5), pp.2527–2539.
44. Nagabhushana, P., Ramprasad, S., Prasad, C.D., Vasudev, H. and Prakash, C., 2024. Numerical investigation on heat transfer of a nano-fluid saturated vertical composite porous channel packed between two fluid layers. *International Journal on Interactive Design and Manufacturing (IJIDeM)*, 18(5), pp.2927–2944.
45. Mahantesh, M.M., Rao, K.R., Chandra, A.P., Vijayakumar, M.N., Nandini, B., Prasad, C.D. and Vasudev, H., 2024. Design and modeling using finite element analysis for the sitting posture of computer users based on ergonomic perspective. *International Journal on Interactive Design and Manufacturing (IJIDeM)*, 18(8), pp.5875–5891.
46. Verma, P.K., Vasudev, H. and Kumar, V., 2025. Dental implant materials and designs and failure during implant placement and loading: A review. In H. Vasudev, C. Prakash, M. Khalid, K. Kesari (eds.), *Advanced Manufacturing Processes* (pp.138–150). Boca Raton: CRC Press.
47. Verma, P.K., Vasudev, H. and Kumar, V., 2025. Bioactivity and corrosion performance of low-velocity oxy-fuel-sprayed hydroxyapatite-based coatings on a surgical-grade stainless steel. *Transactions of the Indian Institute of Metals*, 78(8), p.185.
48. Verma, P.K., Vasudev, H. and Kumar, V., 2025. Effect of electro-corrosion behavior of different grades of stainless steel used in bio-implants: A review. In H. Vasudev, C. Prakash, M. Khalid, K. Kesari (eds.), *Advanced Manufacturing Processes* (pp.269–283). Boca Raton: CRC Press.
49. Verma, P.K. and Vasudev, H., 2025. Importance of hydroxyapatite coatings in improving new bone osteogenesis: A review. In *Advanced manufacturing processes* (pp.284–299).

12 Future Recommendations

Gaurav Prashar and Hitesh Vasudev

12.1 THERMAL SPRAY COATINGS AND THE ASSOCIATED INDUSTRIES

Thermal spraying is an effective method for producing innovative and reliable coatings. The coatings' versatility, strength, and customization options contribute to their popularity across various applications. The method involves applying heated feedstock materials onto the prepared surface of a substrate, resulting in the formation of protective coatings (Figure 12.1). The heated particles after impact are flattened into "splats" that interlock and accumulate to form a layered coating. This method improves the surface characteristics of a component, offering resistance to corrosion, protection against wear, thermal insulation, and restoration of dimensions [1–3].

Thermal spray coatings serve as essential components in various industrial applications, enhancing the durability and performance of critical parts across sectors such as aerospace and heavy machinery. In material science, thermal spray coatings serve as effective protective solutions, offering strong defence against environmental factors. Thermal spraying is a process that entails the application of layers composed of molten metal, ceramics, or composites. Ceramic coatings demonstrate exceptional performance in extreme temperature conditions, offering superior thermal insulation and resistance to corrosion. Carbide-containing metal matrix composites provide a distinctive combination of wear resistance, corrosion resistance, and impact resistance, rendering them suitable for the demanding operating conditions typical in the oil and gas industry. Thermal spray technology is undergoing continuous development, leading to the emergence of new applications for coating materials. The five main industries associated with thermal spray are:

The aerospace sector requires components capable of enduring extreme conditions while ensuring optimal performance levels are maintained. Thermal spray coatings are widely utilized in this sector, providing aircraft parts, including turbine blades, engine parts, and landing gear, with improved wear resistance and thermal stability. Thermal barrier coatings (TBCs), composed of ceramic materials such as zirconia, are utilized on essential engine components to safeguard against high-temperature conditions and maintain optimal engine performance. Thermal spray coatings contribute to enhanced fuel efficiency and decreased maintenance costs within the aerospace sector.

DOI: 10.1201/9781003730637-12

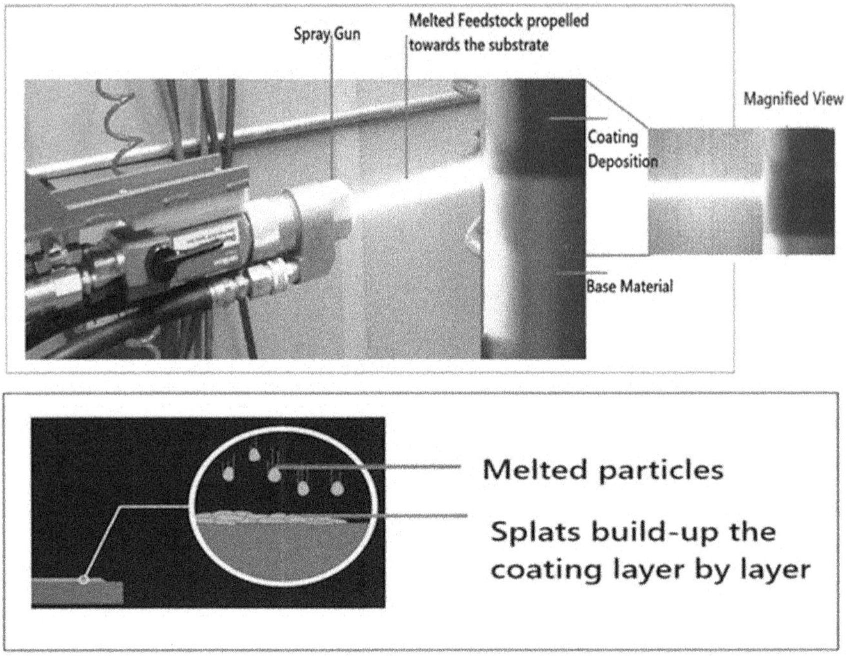

FIGURE 12.1 Generalized principle behind thermal spray coating formulation.

In the auto industry, thermal spray coatings are essential for improving engine performance, minimizing emissions, and prolonging the durability of different parts. Cylinder walls, pistons, and valves frequently undergo coating with thermal spray materials to enhance wear resistance and minimize friction, which in turn contributes to improved fuel efficiency. Thermal spray coatings offer corrosion protection for automobile components that are subject to adverse environmental conditions, including exhaust systems and suspension sections, thereby ensuring their longevity and reliability.

The energy industry, encompassing power generation as well as oil and gas companies, derives significant advantages from the utilization of thermal spray coatings. TBCs are employed on turbine blades and combustion area components to endure the elevated temperatures produced during power generation. The coatings improve turbine efficiency while providing protection against fatigue from heat and corrosion. Specially developed coatings in the oil and gas sector offer corrosion resistance for pumps, pipelines, and valves thus improving their service life and minimizing maintenance downtime.

Coatings in the healthcare sector serve specific purposes in improving the functionality and biocompatibility of medical equipment. Orthopedic implants, including replacement joints and bone screws, may be coated with bioactive materials like hydroxyapatite to enhance bone integration and minimize the likelihood of rejection. Deposited coatings contribute to the development of wear-resistant and strong surfaces for surgical equipment, thereby ensuring precision and longevity throughout medical procedures.

The electronics sector depends on coatings to deliver efficient electromagnetic interference (EMI) protection and heat control for electronic parts. High thermal conductivity coatings, including Cu and Al, are utilized on heat sinks and electronic enclosures to effectively dissipate heat, thereby avoiding overheating and maintaining the best possible performance.

Thermal spray coatings represent a versatile and efficient solution across multiple industries, offering added protection, enhanced functionality, and a longer lifespan for vital parts.

The ongoing advancements in technology and materials are anticipated to broaden the scope of thermal spray coatings, addressing the changing requirements of sustainable industrial thermal coating in various sectors. In the near future, thermal spray coatings will transition from being solely protective barriers to becoming integral components in sustainability initiatives. A future characterized by the recycling of materials, minimized energy consumption, and optimized control techniques—each aspect of the thermal spray coating technique fulfils a specific function. This concept represents a framework of innovation and creativity, wherein the boundaries of potential are consistently broadened and redefined.

12.2 THE NEED FOR SUSTAINABLE COMPOSITE COATINGS

Overcoming the serious problems caused by human impact on the environment needs the active involvement of every sphere and, certainly, the coatings industry. Although much effort has been put in to promote individual action and responsibility, the industry does not have any eco-strategy in general. The world we are living in now is one in which sustainability, specifically environmental sustainability, is part of every single agenda and everyone hopes he or she could show that what he or she does is in line with the sustainability agenda.

Thermally sprayed composite coatings have important applications in contemporary industry due to their ability to improve the surface properties of components applied in very hostile environments. They have many applications in the aerospace, biomedical, energy, and automotive areas in terms of wear resistance, anti-corrosion, thermal insulation, and biocompatibility properties. Application of these coatings is normally carried out using techniques like plasma spraying, high-velocity oxy-fuel (HVOF), high-velocity air-fuel (HVAF), solid state, or cold spraying where composites, metals, and ceramics are deposited on the substrates [1–3]. Figure 12.2 illustrates the comprehensive categorization of thermal spray processes according to their primary energy source, as well as the associated particle speeds and temperatures.

Although they have design advantages (technical), conventional thermally sprayed composite coatings have acute sustainability issues. Most of these coatings are based on feedstock materials of energy-intensive metals (such as chromium, cobalt, tungsten, ceramics) and the spraying processes are in themselves very energy-intensive (emitting large amounts of energy and particulate emissions, noise pollution, and also waste). Moreover, other thermally sprayed materials (e.g. hard chromium coatings) face

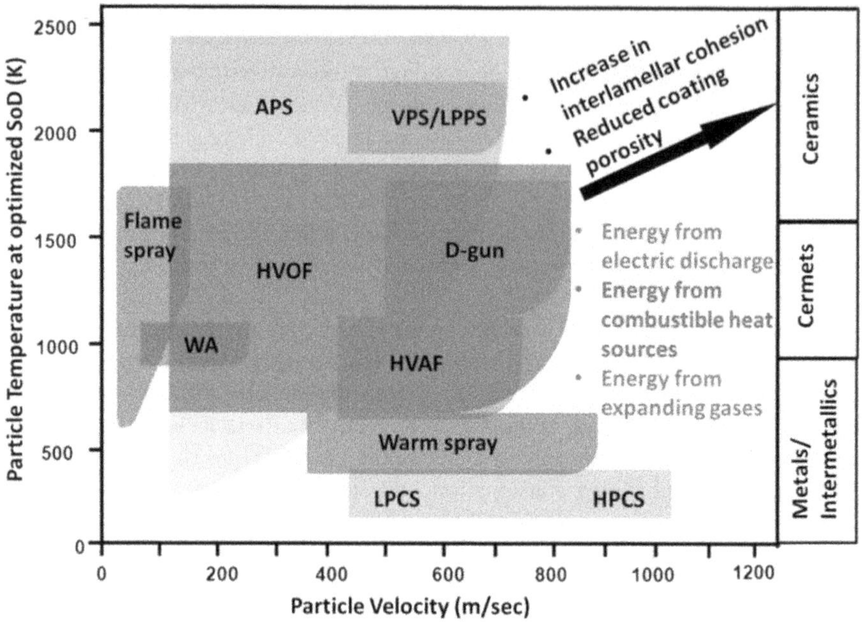

FIGURE 12.2 A breakdown of thermal spray procedures based on the sources of heat generation, particle speeds, and temperatures is presented in a simplified manner [3].

growing regulation concerns as they pollute the environment and are harmful to health. In 2003 one fatal case was reported in which the worker spent 2 days in a temporary worksite in the state of Washington, USA spraying a Ni and Cr-based thermal spray coating (Washington State Department of Labor and Industries, 2005). The employee experienced the gradual appearance of cough, dyspnea, and lethargy. Shortly after that, he died, and a postmortem found that his lungs had been damaged severely due to the inhalation of toxic metals. It was ascertained that the victim had received a Cr and Ni overdose after operating thermal metal sprays without any ventilation or a fitted respirator. Very high concentration of these metals in air, which in most cases are above the set limits in places of work, have been observed in many occupational settings. In exposed workers, some of these metals had accumulated to an abnormal level (i.e. chromium, nickel, and cobalt) when biological samples were tested [4].

Bemer et al. [5] compared the emission rates and particles size of aerosols produced by electric arc and flame spray techniques. The rates of emission were extremely high, in particular for electric arc spray coating, and they were significantly higher than in arc welding. When it comes to releasing tiny nanomaterials to the environment, it is significant to observe that particles produced in the coating process by thermal spraying are ultrafine or in the nanometre size range. Ultrafine particles with similar composition when compared on a mass basis were found to be more pneumotoxic [6, 7] and have a higher lung deposition [8, 9] based on animal toxicology research. They can bioaccumulate in organisms since they are very small, and hence potentially pose a threat to both humans and animals [10]. Furthermore, thermal spraying is likely to result in large amounts of the emissions that contain metal-bearing nanoparticles, and thus this can be dangerous to the respiratory system particularly to the workers who are exposed to the metal-bearing nanoparticles. Salmatonidis et al. [11] examined nanoparticle exposure of ceramic coating deposited using thermal spraying at the workplace settings. Observations, which did not affect the normal working conditions in the facility, were conducted in two sites in a precision engineering workshop (T.M. Comas) of an industrial precision engineering facility, situated nearby the city of Barcelona, Spain, over a period of 5 days. In both the locations, workers were being exposed. By the use of an inhalation model, it was found that particles that were released in the course of thermal spraying were mostly deposited at the alveolar level (70 %).

Therefore, with global regulations, and corporate sustainability objectives, the importance of having sustainable approaches has increased. The different components of sustainable approaches in thermal spraying are shown in Figure 12.3.

It is necessary to come up with feedstock materials that are environmentally and human friendly, that allow energy saving deposition processes, than enable the conservation of the waste and coating systems, and that can be recycled or repaired. The development of thermally sprayed coatings that achieve a balance between performance and environmental requirements is a complicated multi-parameter optimization problem that entails specification of the coating material, and optimization maintained through processing conditions and the intended functionality. AI and ML are currently becoming disruptive in this arena. Combining materials databases, process monitoring data, and performance data, AI can achieve data-driven design and predictive modelling of sustainable thermally sprayed coatings [12]. These technologies will allow faster discovery of the materials, provide a means of optimizing spray parameters to make the process energy and material efficient, as well as allow prediction of lifetime performance under service conditions to make sustainable engineering decisions more informed.

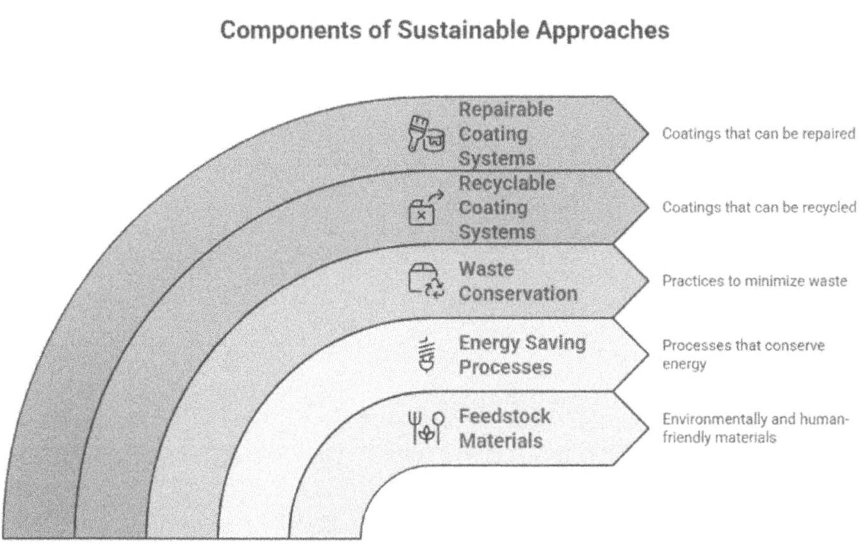

FIGURE 12.3 Various components of sustainable approaches for design of composite coatings.

12.3 SUSTAINABLE DEVELOPMENT GOALS (SDGS) AND THERMAL SPRAY SECTOR

In 2015, to ensure prosperity, peace, and collaboration among all individuals around the globe by 2030, the United Nations passed a set of 17 goals, known as sustainable development goals (SDGs), having economic, social, and ecological aspects. Considering the thermal spray value chain and market size there is a significant role for it to achieve appropriate SDGs. Relevant SDGs may include SDG7, SDG8, SDG 9, SDG 12, SDG13, and SDG17 (Figure 12.4). Similar to the other manufacturing industries, the thermal spray industry would be in a position to develop a sustainable industry structure that would hopefully serve in supporting the industry to achieve its social responsibility objectives [13]. Few data are available yet when it comes to some of the practices attached to the thermal spray value chain, but there is overexploitation of resources, among which is the generation of large amounts of waste materials. It requires giving the incentives of the SDGs to the stakeholders of the thermal spray value chain (supporting transition to circular economies—reuse, recycle, remanufacture, and recover and other framework directives).

12.4 GREENER APPROACHES FOR THERMALLY SPRAYED COATING TECHNOLOGIES

Since sustainability is actively considered one of the core areas of materials science and surface engineering, and since greener thermally sprayed coating technologies lie in the area of development, this topic is considered to be an issue of research and innovation. Such methods are focused on minimizing the environmental footprint

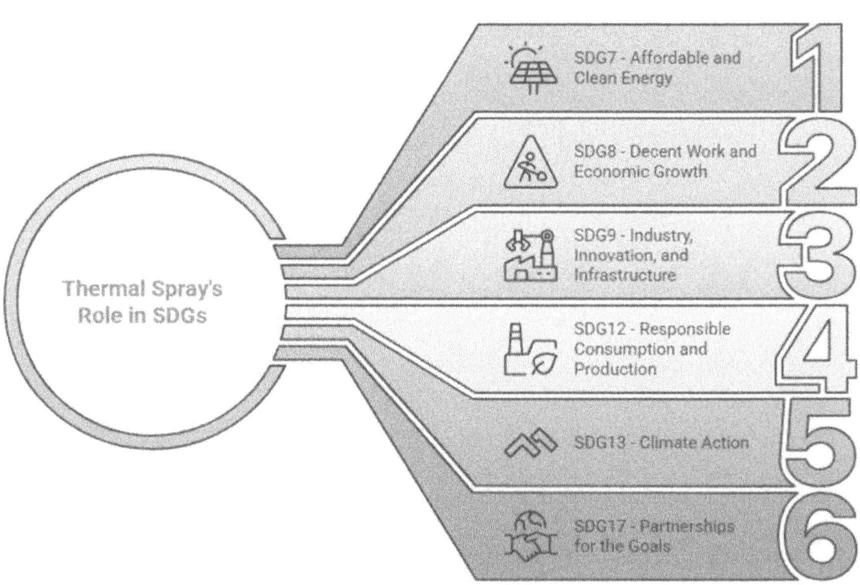

FIGURE 12.4 Thermal spray and relevant SDGs.

of the coating in lifecycles, which includes raw material extraction, feedstock preparation, in-service performance, end-of-life management, and spraying operations. Environmental regulations, climate change concerns, and increasingly independent energy- and resource-efficient manufacturing processes are the fundamental core factors affecting the green transition in thermal spraying.

12.5 USE OF NON-TOXIC AND SUSTAINABLE FEEDSTOCK MATERIALS

The thermal spray coatings are becoming of higher interest in non-toxic and renewable feedstock materials predominantly because of the growth of environmental responsibility and regulations. Such a transition entails a substitute of traditional materials, such as specific metals and ceramics, with environmentally friendly solutions, such as bio-based polymers or natural minerals, in combination with recycled solutions. The aim of these sustainable alternatives is to reduce the environmental footprint of the coating operations but to not compromise or even to enhance the performance attributes. The rare-earth zirconates-based materials are another new candidate for use as thermal barrier coatings (TBCs). These materials aim to increase the functional qualities of TBCs by producing engines that are highly efficient and environmentally friendly [14].

There should be an initiative to recycle and re-use in the thermal spray industry and to keep on showing the benefit of the fact that thermal spraying is a rather green method. The major add-ons to climate change in thermal spray can be understand with the help of the schematic in Figure 12.3. It is estimated that in the year 2015 alone, 2 million kg of YSZ [15] was utilized in coatings alone. Assuming the coating deposition efficiency of 60%, the quantity of powder wasted as far as spraying is concerned would be approximately 800 tons. Yttrium is one of the precious critical raw materials which cannot be wasted. Therefore, there is an urgent need to recycle the waste powder using chemical/mechanical approaches. The use of masks and polymeric compounds is hazardous to the environment and we should think of possible ways of reducing or recycling the masking materials. The hydrocarbons mainly arise due to the combustion processes and therefore it is relevant to replace them with clean fuels, such as hydrogen. Gases are employed to deposit coating at a commercial level through thermal spray, such as H_2, O_2, propane, acetylene, N_2, Ar, He, and propylene. Excluding hydrogen, the remaining ones are largely included in the hydrocarbon category, hence leading to greenhouse gases. In addition, liquid fuels like kerosene and Jet B fuel are used to burn and melt the coating powders. Therefore, so far as thermal spray is concerned, the inputs and outputs must be handled so as to make a positive contribution to the environment (Figure 12.5).

Nevertheless, certain elements of re-use and recycling are not practiced in thermal spray industry at present (such as the possibility of reusing the coating materials removed from the coated parts). One example is that TBC may be removed by grit blasting, chemical stripping, water jet, and laser ablation methods. All these processes are helpful to remove the coating, yet, they are not helpful in recovering the removed materials. To retrieve the useful material would prove to be a laborious and expensive operation (of low yield) because the removed mass would contain contaminants,

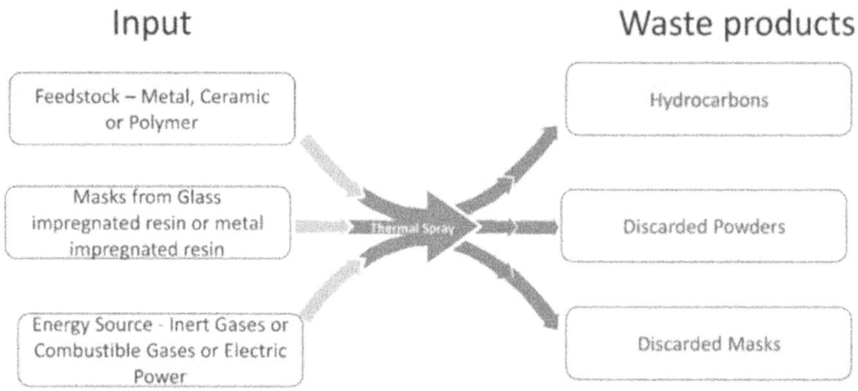

FIGURE 12.5 I/P during processing of coatings and the O/P in form of waste products [16].

rusts, stains ,and high toxicity, and this may not be the most appealing opportunity to many businesses [16].

The thermal spray industry is currently enjoying popularity with regard to hydrocarbons due to economical and geographical reasons. Although the large corporations could afford to spray using hydrogen and nitrogen, small-scale coating factories are still using fossils fuels in order to remain competitive in the market. Hydrogen generators could also be scaled up to keep the hydrogen costs to a minimum which could allow small companies to adjust at a faster rate. Governments can provide the suppliers of the clean fuels to be used in the production of the coatings with carbon credits and incentives. Although helium is the by-product of natural gas, it has been seen that the helium used in the cold spray process could become more frequent as newer inventions crop up.

Thermal spray processing normally needs fine powders as feedstock materials. The methods that are used to produce these powders include atomization, spray drying, and solid-state reduction. Each and every one of these processing techniques needs a given amount of energy to either liquefy the raw material or facilitate a chemical reaction process. There should be efforts made to ensure carbon neutrality in the manufacturing processes of feedstocks that make the overall carbon footprint low. To comprehend the circularity of the thermal spray approaches and its lifecycle carbon footprint, it should be noted that the background processes should be considered as well. The implementation of the concept of circularity could be conducted in several ways, one of which is the assessment of the possibility of the use of waste or by-product materials of other industries in thermal spray. To give an example, copper waste generation in the electronics and electrical sector may be a potential source of copper feedstock of electrically conductive coating. Other metals with low melting points like zinc and aluminium are capable of being reused via recycling with little heat input. Nevertheless, the oxidation of particles in the spray plume may be an obstacle in this case [13].

Overspray is a common feature in most thermal spray workshops. The operational philosophy of most of these facilities is basically to collect the overspray dust that can either be shipped to landfill or sold to a waste buyer who can segregate the actual powder materials and reprocess them to be resold. One way to categorize the materials and develop a framework of circularity is according to the ease of recycling. For example, the fact that it is possible to accumulate the overspray dust in separate bins based on the type of material is an advantage. It will be more convenient to recycle the waste product by the buyer later. This may make the sale value of dust high and may offer an opportunity of cost recovery to the thermal spray manufacturer. It could be effective to implement something that revolves around the product lifecycle not only of the coated product but also the feedstock, hence creating an implementation plan that will help reduce wastage [13].

The LIFE ReTSW-SINT is one of the best examples of reuse of the thermally sprayed feedstock overspray dust. The project [17], financed by the European Commission, aimed at setting the processing ways of applying the thermal spray waste to other industries. The project demonstrated successfully the recycling of nickel metal powders to form sintered products by the spark plasma sintering technique. The project also implemented the inclusion of the YSZ waste powder into ceramic frits and tiles. Another important thing that appeared to be important during the development of a recycling process was the fact that environmental safety is a crucial factor. Since most of the materials of commercial interest applicable in the thermal spray are solid wastes, it becomes important that the process of reuse of the materials should deactivate them and ensure that they have less contact with the environment. In the majority of the ceramics, there will be a leaching behaviour where the metals in them will be gradually and progressively released into the surrounding environment resulting in environmental contamination.

12.6 ROLE OF DIGITAL AND AI TOOLS

The effective application of such more sustainable strategies can frequently see the negotiation of tricky trade-offs involving the characteristics of materials, the conditions of processes, and the environmental consequences of the two. Here, ML is an enabling capability, speeding up the process of discovering new materials, the modelling of performance, the optimization of process variables, and the addition of sustainability metrics to the design process. ML and AI [18] have become prominent in other industries and thermal spray may also be one of the implementations of change to minimize emissions and wastage. The control organizations might soon demand industries provide them with the data analytics of emissions and waste. Suspension precursor processing is also being introduced with an online-diagnostics system which would be useful in monitoring the diagnostics of the particles in a liquid state [19]. This may lead to an increase in the efficiency of deposition of the solution precursor coatings. The analysis and data collection may be a smooth component of the thermal spray facility. In the near future, it would be necessary to gather and present data to the governing bodies in this form.

It would be possible to attach a hand-held X-ray fluorescence device to the thermal spray dust collection chutes to keep track of the discarded elements. Figure 12.6 presents a general arrangement of a thermal spray that has been used in industry and academia that has been endowed with multiple sensors and a waste collector. The

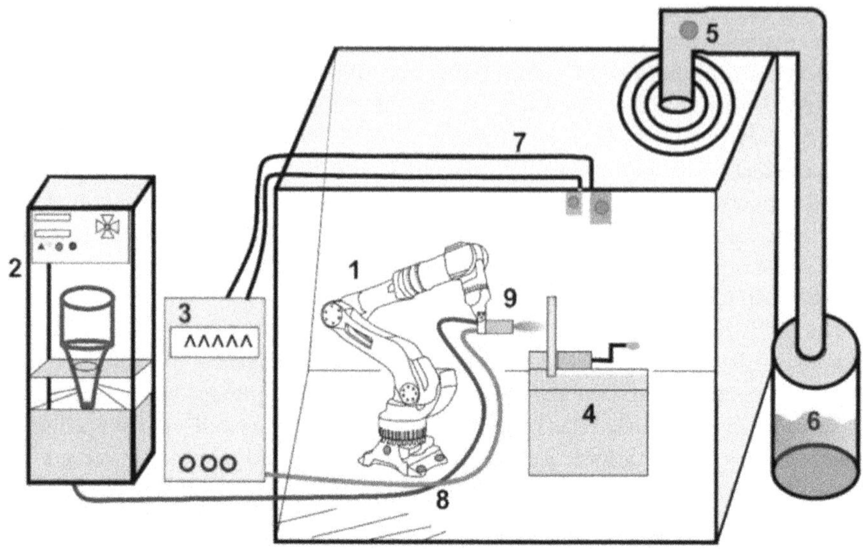

FIGURE 12.6 Future of thermal spray in terms of digitalization. Digital input data is collected in the form of red dots and has the destination of digital output to robot and guns. (1, robot; 2, powder feeder; 3, controls; 4, sample holder and controller; 5, X-ray detector; 6, collector for waste powder; 7, sensors for powder velocity and temperature; 8, feed lines; 9, plasma torch and plasma flame) [16].

work on the reduction of wastes and the transition to the energy-efficient processes begins with the need to gather data. Based on Figure 12.6, it is possible to conclude that sensors could be installed at strategic points in the booth to measure the data. There are two sets of data that can be obtained, i.e.: (i) the process parameters like particle velocity and temperature which will aid in real time optimization of gun parameters, robot speed, and process gas flows; and (ii) the resources being used and wastage which can be reported to the governing bodies. The percentage of pollutant produced in every category would assist in differentiating the industries into the categories of green, yellow, and red in terms of emissions and assist in closely tracking compliance to the goals of climate change. The data collection can be connected to servers and presented live using cloud computing. Mathematical models could also be constructed to determine the most economical fuels/oxygen ratio of any combustion spray process in order to determine the optimized mix that would melt the particles and result in the minimal consumption of fuel, thus increasing the deposition efficiency of the process [20].

12.7 ML IN MATERIALS RESEARCH

ML is becoming a transformative technology in the field of materials science research by enabling the rapid and accurate study of complex material properties, discovering new materials with specified properties, and optimizing thermal spraying processes.

Its extent also reaches the realm of material design and performance prediction [21]. ML approaches have a huge potential to transform the way we understand and discover materials science [22]. The synthesis of materials has experienced a revolution via ML. Through modelling the behavior of materials, property prediction has accelerated material selection in specific applications [23]; structure-property relationships permit controlled design [24]; high-throughput screening enhances efficiency in the review of high-capacity material libraries [25]; optimization of parameters used to develop coatings using different thermal spray techniques (maximum optimization) ensures better properties [26]; and data-driven material discovery correlates new materials with known ones to accelerate innovation by identifying future materials within data scale samples. Generative models also represent original materials in the field of material synthesis, which stimulates the imagination and experimentation as the materials have a particular property [27]. Collectively, the applications redirect the study of materials science, and acceleration and innovation are encouraged.

ML provides us with another approach/strategy to discover new materials, unlike the traditional experimentation mode whereby discovery is by way of trial and error. ML offers guidance on material design by making new possible predictions through understanding the trends in the data patterns [28]. ML has been extensively used during the last decade to produce many high-performing alloying materials including Ti, Ni, high-entropy alloys (HEAs), Al alloy, shape memory alloy, Mg alloy, Ni-based single-crystal superalloys [29, 30], and Cu alloy [31–37]. The use of ML in the design and development of new materials has expanded, presenting opportunities to improve the search for new materials with specific features. Figure 12.7 shows the applications of ML in material synthesis.

12.8 AI IN SUSTAINABLE COATING FORMULATION

ML is transforming how to formulate thermal spray coating geometries in a way that is sustainable at the level of design, control, and performance. Such data-based combinations of strategies enable coating experts to achieve high-performance surfaces using the minimum amounts of material, energy, and related environmental impacts—setting the pace towards a greener surface-engineering future (Figure 12.8).

12.9 PERFORMANCE PREDICTION AND OPTIMIZATION USING ML

ML-based inverse analysis frameworks of cold spray processes establish the most important variables in a process: critical velocity, residual stress, and powder flowability through the feature selection and regression methods. With such models, optimization of the process is made with the aim of reducing energy and reducing material utilization. Research by Hamrani et al. [38], performed an inverse analysis using a two-stage ML structure to optimize process parameters in cold spray processes, namely deposition efficiency [DE]. The framework allowed a precise forecast based on deposition occurrence and DE levels by means of jointly utilized classification and regression models as a reliable procedure in optimizing the cold spray processes. This layered modelling approach was very effective in making the model reproduce similar conditions in reality as well as

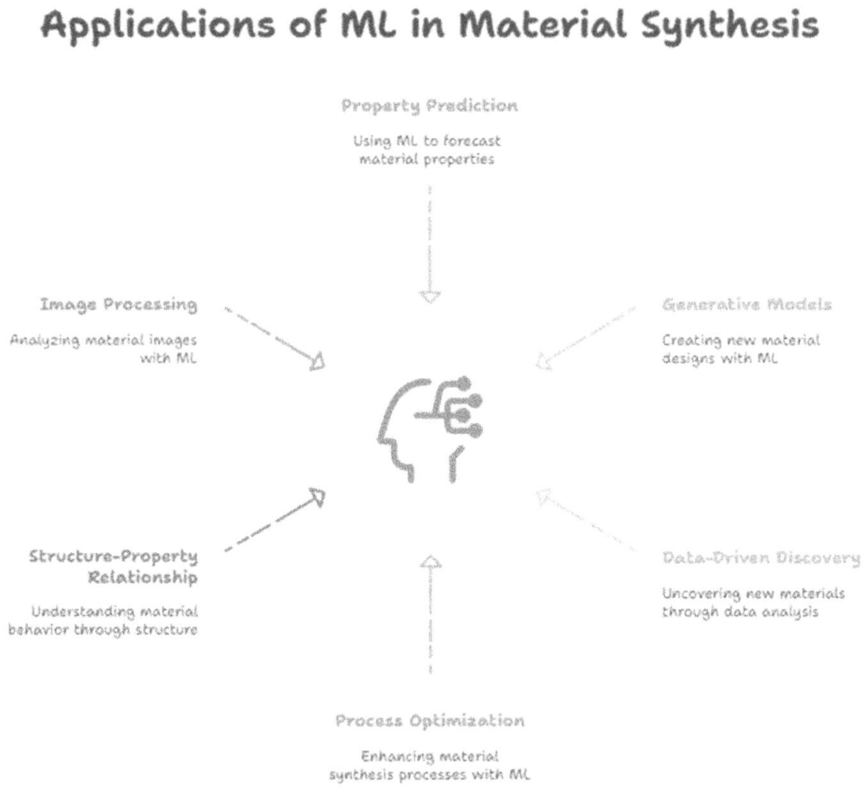

FIGURE 12.7 Role of ML in material synthesis.

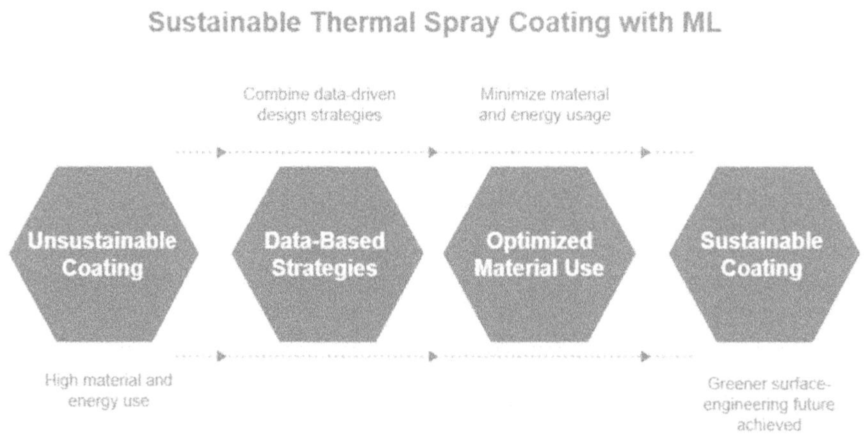

FIGURE 12.8 Role of ML in transforming unsustainable composite coating to sustainable composite coating.

fit well with reference data. ML frameworks of cold spray thermoplastics have identified powder yield strength and powder density as key drivers of a particle flattening and penetration, which brings improved deposition control based on fewer physical tests [39]. In future ML can be utilized to assess the environmental impact of various thermal spray processes, such as plasma spray, HVOF, and cold spray, in order to determine the most sustainable alternative for a specific application.

12.10 REAL-TIME MONITORING AND DIGITAL TWINS

The trends of AI and traditional manufacturing coming together are not to enhance their productivity alone but to develop compliant and flexible production processes that can be expanded in an ever-changing technical and commercial environment. End-to-end AI systems are being developed by initiatives such as CoBRAIN, which links information streams of sensors such as temperature and spray distance and gas flow streams to predictive models that can dynamically adjust spray parameters in real time to reduce defects, rework, and material wastage. Combining virtualization (digital twins) with real-time tracking helps such systems step forward into high-performance and eco-efficient autonomous thermal-spray applications. Using experimental methods and integrated computational materials engineering, CoBRAIN is set to become a game changer in the co-development of HEA-based coatings. The method does not only improve the understanding of materials behaviour, but it also speeds up the process of design and optimization of coating with better properties.

Oerlikon and MTU are developing a smart thermal spray factory on the basis of digital twins that achieve real-time monitoring, optimization, and traceability of aerospace coatings. The outcome is more intelligent, configurable control loops that maximize quality, minimize waste, and maximize productivity in the thermal spray manufacturing industry. The incorporation of particle velocity and temperature, as well as deposition rates and coating thickness, together with high-resolution physical simulation (CFD, DEM, heat flow) of the spray processes (such as HVOF and plasma), is incorporated into (DTs). These twins enable the engineer to virtually test new feedstock formulations (e.g. bimodal or hybrid powders), to optimize ratios, and spray parameter combinations in advance of physical tests [40].

ML serves as an effective tool in the design and selection process of efficient, environment-friendly thermal spray coating materials. This approach facilitates accelerated innovation, optimized resource utilization, and the development of durable, environmentally sustainable solutions, which are essential for the attainment of the SDGs. The algorithms, such as random forests and neural networks, are capable of rapidly predicting the performance of numerous material compositions. This minimizes dependence on trial-and-error experimental methodologies.

12.11 ENVIRONMENTAL AND LIFECYCLE ASSESSMENT INTEGRATION

The environmental consequences of any business operation will take precedence over its economic aspects because of the significant influence that climate change will exert on the economy. Countries will begin to consider the macroeconomic implications of climate change, moving beyond a singular focus on the effects of individual industries on the economy. Similar to other manufacturing processes, thermal spray requires raw materials and produces waste materials. The emphasis on the circular economy necessitates the quantification and tracking of resources and waste via lifecycle sustainability assessment. This approach aims to enhance environmental responsibility across the triple bottom line, encompassing environmental, economic, and social dimensions. The analysis encompasses the inventory of the process and waste, followed by an evaluation of the potential environmental impact. The process involves defining the project scope, conducting an inventory analysis, performing an environmental impact assessment, and presenting the findings in a clear format for accurate evaluation.

The application of ML-based lifecycle assessment to thermal spray processes is in its early stages. However, the potential for synergy is evident: the integration of comprehensive lifecycle inventory data with ML techniques presents a promising approach for optimizing coating processes in a predictive, efficient, and environmentally conscious manner. The domain presents significant opportunities for investigation and practical application. Integrating lifecycle assessment and lifecycle costing with ML-driven performance modelling may represent a substantial advancement. ML models can utilize existing lifecycle inventory data, such as that sourced from databases like Ecoinvent, to forecast the environmental impact of novel materials or processes by analyzing their characteristics. ML can be utilized to forecast the lifecycle impact (LCI) of a novel coating material by analyzing its chemical composition and physical properties, thereby minimizing the requirement for comprehensive laboratory testing. For instance, a study may employ ML techniques to forecast the LCI of a new thermal spray coating material by analyzing its composition, particle size, and additional pertinent parameters. The data can subsequently be utilized in a lifecycle assessment to evaluate its environmental impact. ML-driven environmental impact forecasting is shown in Figure 12.9.

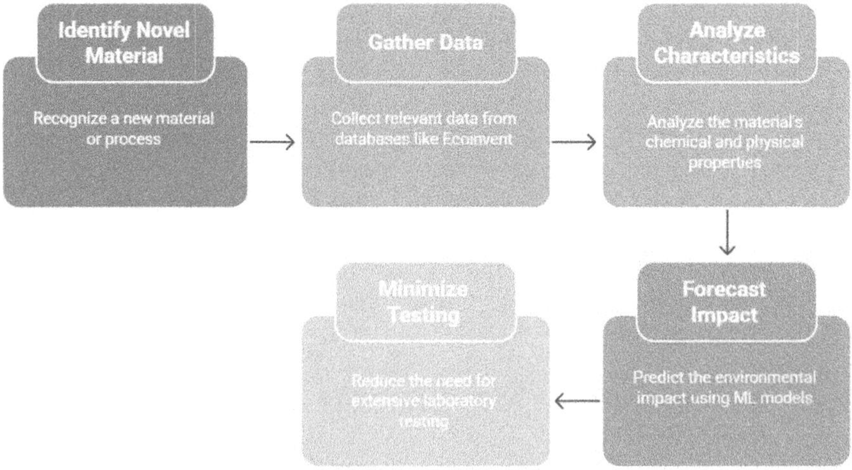

FIGURE 12.9 ML-driven environmental impact forecasting.

12.12 BARRIERS IN THE SUCCESSFUL ADOPTION OF AI IN COATING INDUSTRY

Machine learning techniques are recognized for their ability to extract features from data representations through complex neural network architectures. However, specific obstacles hinder the adoption of AI within the coating industry.

The improper use of these advanced instruments by untrained individuals presents a distinct issue. A significant number of individuals continue to lack the requisite expertise for the effective utilization of these technologies. An incorrect application typically results in inaccurate outcomes, thereby questioning the utility of such programs. Thermal spray organizations across various sectors are acknowledging the challenges associated with identifying and recruiting skilled data scientists and AI specialists. An interdisciplinary team comprising data scientists, machine learning developers, and software architects is essential for collaboration on AI projects. Numerous businesses lack access to these facilities or do not possess the financial capacity to utilize them for even a single data science project. It is essential to recognize that the practical expertise of well-trained personnel serves as a vital resource for process control and diagnostics, both in the present and moving forward. Consequently, effective methods for knowledge collection, organization, representation, and transfer will be essential.

Access to uncontaminated, relevant, and high-quality data is critical for the accomplishment of AI projects. However, achieving this on the manufacturing facility floor during coating deposition can present challenges. The collection of data typically entails a certain level of error, which may originate from various factors. One example involves sensor data collected in challenging conditions, characterized by excessive noise or vibration that could have led to inaccurate readings. One challenge in artificial intelligence is that predictions generated by AI are significantly more effective when applied to large datasets compared to smaller datasets [12]. The quantity and quality of the available data are expected to present considerable challenges. Many firms' data sets may contain between one hundred and two hundred data points for experimental formulations. The dataset in question is insufficient for effective utilization of ML and AI, which typically require the transmission of thousands or millions of data points. To optimize limited information, it is essential to develop intermediate solutions. A separate issue occurs when individuals lacking proper training operate advanced equipment incorrectly. The prevalent deficiency in the effective utilization of this technology presents a significant issue in contemporary society. Incorrect application typically results in erroneous outcomes, subsequently raising questions about the project's success.

Cultural resistance represents a significant obstacle to the adoption of AI technologies. Employees may perceive artificial intelligence as a potential threat to their positions or may exhibit resistance to the procedural changes necessary for its integration. It is essential for leaders to articulate the advantages of AI, highlighting its function in enhancing human capabilities rather than substituting them. Promoting transparency, recognizing innovation initiatives, and fostering continuous learning can progressively transform organizational culture to adopt AI.

12.13 HYBRID THERMAL SPRAY: AN APPROACH TO ACHIEVE INNOVATIVE COATING

Hybrid thermal spraying employs various feedstocks, including suspensions, powders, solution precursors wire, and suspensions to effectively integrate dissimilar materials across various length scales. This approach facilitates the development of coatings that exhibit unique properties and improved performance. This method is designed to integrate smoothly with all thermal spray approaches. The simultaneous or sequential introduction of two or more separate feedstocks, with independent control over each, enables the deposition of coatings characterized by unique architectures and innovative microstructures. The system facilitates the straightforward application of composite coatings using commercially available materials as feedstock, eliminating the requirement for intricate and expensive powder pre-alloying processes. Hybrid thermal spraying may serve as an effective screening method to facilitate the exploration of various coating compositions of interest. Hybrid thermal spray coatings are applicable in multiple domains, including wear resistance, corrosion protection, thermal barrier coatings (TBCs), and innovative applications. These coatings effectively tackle essential challenges encountered in various industries [41]. Thermal spray coatings and other contemporary surface engineering methods are now being used in biomedical slurry processing systems that are corrosion-resistant at high temperatures. This becomes even more accurate when paired with new technologies like magnetron sputtering thin-film deposition, additive manufacturing, and microwave-assisted fabrication [18, 42–54]. In addition to improved strength (tensile, compressive, and flexural) and effective performance in solar energy conversion systems, the coating structures exhibit significant promise in surface topography. On top of the planned surfaces, complicated computational fluid dynamics of curved pipes and conduits, as well as anthropometric, mechanical, and structurally related seating solutions, can be constructed [18, 42–54].

12.14 CONCLUSION

The application of AI in the green design of thermally sprayed coatings represents a significant advancement in sustainable surface engineering. Recent studies demonstrate the potential of AI, specifically machine learning methodologies such as active learning and generative models, to transform the development of thermal spray coatings. The AI-driven strategies outlined provide distinct methodologies for developing coatings that achieve superior performance while minimizing environmental impact.

Projects such as CoBRAIN demonstrate the potential of integrating AI into thermal spray workflows to facilitate real-time parameter optimization, enhance predictive process control, and provide decision support for sustainability-focused coating formulations. Through the analysis of sensor data, including temperature, pressure, and coating thickness, AI models are capable of dynamically adjusting spraying parameters. This adaptation aims to reduce waste, optimize energy consumption, and limit overspray, thereby promoting more sustainable operations.

Thermal spray technology inherently promotes sustainability by prolonging the lifespan of components, facilitating repair instead of replacement, and minimizing raw material consumption through localized application. These outcomes are consistent with lifecycle sustainability principles.

The future integration of AI in green thermal spray coatings will progress across multiple critical dimensions:

- Generative and inverse design: AI models will generate coating formulations and microstructures that are optimized for environmental metrics, including carbon footprint, material circularity, and energy reduction, while also achieving specified functional performance targets.
- Multi-objective optimization: The application of reinforcement learning and genetic algorithms facilitates the concurrent optimization of durability, process energy, emissions, and cost, resulting in eco-efficient formulation trade-offs.
- Real-time adaptive manufacturing: AI-assisted digital twins and active control can facilitate in-flight adjustments of spray parameters, minimize trial and error, and enable closed-loop sustainability feedback.
- Lifecycle-informed formulation screening: Incorporating lifecycle inventory data (e.g. for APS, HVOF, CGS using WC-Co feedstocks) into AI models will integrate sustainability quantification into coating design decisions. In summary, the integration of AI in green design for thermal spray coatings presents significant potential. Through the integration of generative design, predictive modelling, real-time process adaptation, and lifecycle assessment, the coatings industry has the potential to transition from a reactive approach to a proactive strategy in sustainability engineering. The advancement of thermal spray technology is directed towards the development of intelligent, environmentally conscious formulations that enhance performance while reducing ecological footprint—facilitating sustainable innovation on an industrial scale

The current initiatives involving hybrid spraying remain in an exploratory phase, necessitating comprehensive studies to attain a fundamental understanding.

It is essential for researchers to investigate innovative material combinations to fully realize the capabilities of hybrid thermal spray technology. The hybrid thermal spray presents considerable opportunities for future advancements. The investigation of new materials, refinement of microstructures, and utilization of functional capabilities can lead to substantial enhancements in the performance of composite coatings across various applications.

New entrants, specifically entrepreneurs, are required in the thermal spray value chain to concentrate on the large-scale implementation of next-generation technology solutions. This focus is essential for achieving short- and medium-term objectives, which include fulfilling sustainable development goals and facilitating a move to circular economies. These strategies may enhance buyer loyalty and impact brand perception.

REFERENCES

1. Prashar, G. and Vasudev, H., 2021. A comprehensive review on sustainable cold spray additive manufacturing: State of the art, challenges and future challenges. *Journal of Cleaner Production*, *310*, p.127606.
2. Prashar, G. and Vasudev, H., 2023. Deposition and fabrication of biomaterials using cold spray technique: A review on the application of biomedical-implants. *Applied Surface Science Advances*, *18*, p.100533.
3. Singh, S., Berndt, C.C., Singh Raman, R.K., Singh, H. and Ang, A.S., 2023. Applications and developments of thermal spray coatings for the iron and steel industry. *Materials*, *16*(2), p.516.
4. Antonini, J.M., McKinney, W.G., Lee, E.G. and Afshari, A.A., 2021. Review of the physicochemical properties and associated health effects of aerosols generated during thermal spray coating processes. *Toxicology and Industrial Health*, *37*(1), pp.47–58.
5. Bémer, D., Régnier, R., Subra, I., Sutter, B., Lecler, M.T. and Morele, Y., 2010. Ultrafine particles emitted by flame and electric arc guns for thermal spraying of metals. *Annals of Occupational Hygiene*, *54*(6), pp.607–614.
6. Brown, D.M., Wilson, M.R., MacNee, W., Stone, V. and Donaldson, K., 2001. Size-dependent proinflammatory effects of ultrafine polystyrene particles: A role for surface area and oxidative stress in the enhanced activity of ultrafines. *Toxicology and Applied Pharmacology*, *175*(3), pp.191–199.
7. Oberdörster, G., Ferin, J., Gelein, R., Soderholm, S.C. and Finkelstein, J., 1992. Role of the alveolar macrophage in lung injury: studies with ultrafine particles. *Environmental Health Perspectives*, *97*, pp.193–199.
8. National Institute for Occupational Safety and Health, 2007. Progress toward safe nanotechnology in the workplace: A report from the NIOSH nanotechnology research center [DHHS (NIOSH) Publication No. 2007-123], a progress report on the status of nanotechnology research conducted by the NIOSH NTRC through 2006. Available at: www.cdc.gov/niosh/docs/2007-123/pdfs/2007-123.pdf
9. Oberdörster, G., Oberdörster, E. and Oberdörster, J., 2005. Nanotoxicology: An emerging discipline evolving from studies of ultrafine particles. *Environmental Health Perspectives*, *113*(7), pp.823–839.
10. Lima, R.S. and Marple, B.R., 2007. Thermal spray coatings engineered from nanostructured ceramic agglomerated powders for structural, thermal barrier and biomedical applications: A review. *Journal of Thermal Spray Technology*, *16*(1), pp.40–63.
11. Salmatonidis, A., Ribalta, C., Sanfélix, V., Bezantakos, S., Biskos, G., Vulpoi, A., Simion, S., Monfort, E. and Viana, M., 2019. Workplace exposure to nanoparticles during thermal spraying of ceramic coatings. *Annals of Work Exposures and Health*, *63*(1), pp.91–106.
12. Thakur, L., Vasudev, H., Singh, J. and Prashar, G., 2023. *Artificial Intelligence and Machine Learning in the Thermal Spray Industry: Practices, Implementation, and Challenges*. CRC Press.
13. Faisal, N.F. and Prathuru, A., 2024. Sustainable development goals and circularity in thermal spray coating manufacturing and value chain. *Spraytoday*, *4*, pp.8–11.
14. Islam, K.S., Chowdhury, N., Islam, R., Mitra, P., Dutta, A. and Supal, A.R., 2025. Nanomaterial-based thermal spray coating technology: A review. *Heliyon*, *11*(10), p.e43389.
15. Smialek, J.L. and Miller, R.A., 2018. Revisiting the birth of 7YSZ thermal barrier coatings: Stephan Stecura. *Coatings*, *8*(7), p.255.

16. Viswanathan, V., Katiyar, N.K., Goel, G., Matthews, A. and Goel, S., 2021. Role of thermal spray in combating climate change. *Emergent Materials*, *4*(6), pp.1515–1529.
17. LIFE ReTSW-SINT: Recycling of thermal spray waste in sintered products (reference no., LIFE12 ENV/IT/000678), 2013–16. https://webgate.ec.europa.eu/life/publicWebsite/project/LIFE12-ENV-IT-000678/recycling-of-thermal-spraywaste-in-sintered-products
18. Verma, P.K., Vasudev, H., Singh, G. and Singh, S., 2024. A review on the microstructural and biomedical properties of implants manufactured using additive manufacturing. *Journal of The Institution of Engineers (India): Series D*, *106*, pp.1587–1601.
19. Akbarnozari, A., Ben-Ettouil, F., Amiri, S., Bamber, O., Grenon, J.D., Choquet, M., Pouliot, L. and Moreau, C., 2020. Online diagnostic system to monitor temperature of in-flight particles in suspension plasma spray. *Journal of Thermal Spray Technology*, *29*(5), pp.908–920.
20. Khan, M.N. and Shamim, T., 2014. Effect of operating parameters on a dual-stage high velocity oxygen fuel thermal spray system. *Journal of Thermal Spray Technology*, *23*(6), pp.910–918.
21. Lv, C., Zhou, X., Zhong, L., Yan, C., Srinivasan, M., Seh, Z.W., Liu, C., Pan, H., Li, S., Wen, Y. and Yan, Q., 2022. Machine learning: An advanced platform for materials development and state prediction in lithium-ion batteries. *Advanced Materials*, *34*(25), p.2101474.
22. Fadlullah, Z.M., Tang, F., Mao, B., Kato, N., Akashi, O., Inoue, T. and Mizutani, K., 2017. State-of-the-art deep learning: Evolving machine intelligence toward tomorrow's intelligent network traffic control systems. *IEEE Communications Surveys & Tutorials*, *19*(4), pp.2432–2455.
23. DeMille, K.J. and Spear, A.D., 2022. Convolutional neural networks for expediting the determination of minimum volume requirements for studies of microstructurally small cracks, Part I: Model implementation and predictions. *Computational Materials Science*, *207*, p.111290.
24. Wang, Y., Zhang, M., Lin, A., Iyer, A., Prasad, A.S., Li, X., Zhang, Y., Schadler, L.S., Chen, W. and Brinson, L.C., 2020. Mining structure–property relationships in polymer nanocomposites using data driven finite element analysis and multi-task convolutional neural networks. *Molecular Systems Design & Engineering*, *5*(5), pp.962–975.
25. Benayad, A., Diddens, D., Heuer, A., Krishnamoorthy, A.N., Maiti, M., Cras, F.L., Legallais, M., Rahmanian, F., Shin, Y., Stein, H. and Winter, M., 2022. High-throughput experimentation and computational freeway lanes for accelerated battery electrolyte and interface development research. *Advanced Energy Materials*, *12*(17), p.2102678.
26. Mitropoulos, T., Avrampos, P. and Vosniakos, G.C., 2022. Determining favourable process parameters in computer numerically controlled polishing of metal surfaces. *International Journal of Manufacturing Research*, *17*(3), pp.310–325.
27. Chen, L., Zhang, W., Nie, Z., Li, S. and Pan, F., 2021. Generative models for inverse design of inorganic solid materials. *Journal of Materials Informatics*, *1*(4), pp. 310–325.
28. Ward, L., O'Keeffe, S.C., Stevick, J., Jelbert, G.R., Aykol, M. and Wolverton, C., 2018. A machine learning approach for engineering bulk metallic glass alloys. *Acta Materialia*, *159*, pp.102–111.
29. Liu, Y., Wu, J., Wang, Z., Lu, X.G., Avdeev, M., Shi, S., Wang, C. and Yu, T., 2020. Predicting creep rupture life of Ni-based single crystal superalloys using divide-and-conquer approach-based machine learning. *Acta Materialia*, *195*, pp.454–467.

30. Liu, Y., Ding, L., Yang, Z., Ge, X., Liu, D., Liu, W., Yu, T., Avdeev, M. and Shi, S., 2023. Domain knowledge discovery from abstracts of scientific literature on Nickel-based single crystal superalloys. *Science China Technological Sciences*, *66*(6), pp.1815–1830.
31. Gu, B., Sheng, V.S., Wang, Z., Ho, D., Osman, S. and Li, S., 2015. Incremental learning for v-support vector regression. *Neural Networks*, *67*, pp.140–150.
32. Ward, L., Dunn, A., Faghaninia, A., Zimmermann, N.E., Bajaj, S., Wang, Q., Montoya, J., Chen, J., Bystrom, K., Dylla, M. and Chard, K., 2018. Matminer: An open-source toolkit for materials data mining. *Computational Materials Science*, *152*, pp.60–69.
33. Zhang, Q., Chang, D., Zhai, X. and Lu, W., 2018. OCPMDM: Online computation platform for materials data mining. *Chemometrics and Intelligent Laboratory Systems*, *177*, pp.26–34.
34. Zhang, Y., Wen, C., Wang, C., Antonov, S., Xue, D., Bai, Y. and Su, Y., 2020. Phase prediction in high entropy alloys with a rational selection of materials descriptors and machine learning models. *Acta Materialia*, *185*, pp.528–539.
35. Kaufmann, K. and Vecchio, K.S., 2020. Searching for high entropy alloys: A machine learning approach. *Acta Materialia*, *198*, pp.178–222.
36. Zhou, Z., Zhou, Y., He, Q., Ding, Z., Li, F. and Yang, Y., 2019. Machine learning guided appraisal and exploration of phase design for high entropy alloys. *npj Computational Materials*, *5*(1), p.128.
37. Huang, W., Martin, P. and Zhuang, H.L., 2019. Machine-learning phase prediction of high-entropy alloys. *Acta Materialia*, *169*, pp.225–236.
38. Hamrani, A., Medarametla, A., John, D. and Agarwal, A., 2024. Machine-learning-driven optimization of cold spray process parameters: Robust inverse analysis for higher deposition efficiency. *Coatings*, *15*(1), p.12.
39. Perna, A.S., Auriemma Citarella, A., De Marco, F., Di Biasi, L., Viscusi, A., Tortora, G. and Durante, M., 2025. Optimizing cold spray deposition on thermoplastics: A machine learning approach focused on powder properties. *Journal of Materials Engineering and Performance*, *34*(8), pp.6527–6538.
40. Ren, J., Zhou, T., Rong, Y., Ma, Y. and Ahmad, R., 2022. Feature-based modeling for industrial processes in the context of digital twins: A case study of HVOF process. *Advanced Engineering Informatics*, *51*, p.101486.
41. Eskandari, A., Gupta, M. and Joshi, S., 2025. Hybrid thermal spray: A pathway to realize novel coating microstructures and properties. *Journal of Thermal Spray Technology*, *51*, pp.1–28.
42. Dutta, V., Thakur, L., Singh, B. and Vasudev, H., 2022. A study of erosion–corrosion behaviour of friction stir-processed chromium-reinforced NiAl bronze composite. *Materials*, *15*(15), p.5401.
43. Singh, M., Vasudev, H. and Kumar, R., 2021. Corrosion and tribological behaviour of BN thin films deposited using magnetron sputtering. *International Journal of Surface Engineering and Interdisciplinary Materials Science (IJSEIMS)*, *9*(2), pp.24–39.
44. Sharanabasava, H., Raviprakash, M., Prasad, C.D., Ramesh, M.R., Phanibhushana, M.V., Vasudev, H. and Kumar, S., 2023. Microstructure, mechanical and wear properties of SiC and Mo reinforced NiCr microwave cladding. *Advanced Materials and Processing Technologies*, *9*(2), pp.24–39.
45. Singh, M., Vasudev, H. and Singh, M., 2022. Surface protection of SS-316L with boron nitride based thin films using radio frequency magnetron sputtering technique. *Journal of Electrochemical Science and Engineering*, *12*(5), pp.851–863.
46. Prashar, G., Vasudev, H., Thakur, L. and Bansal, A., 2023. Performance of thermally sprayed hydroxyapatite coatings for biomedical implants: A comprehensive review. *Surface Review and Letters*, *30*(1), p.2241001.

47. Vasudev, H., 2020. Wear characteristics of Ni-WC powder deposited by using a microwave route on mild steel: Microwave cladding of Ni-WC. *International Journal of Surface Engineering and Interdisciplinary Materials Science (IJSEIMS), 8*(1), pp.44–54.
48. Parkash, J., Saggu, H.S. and Vasudev, H., 2022. A short review on the performance of high velocity oxy-fuel coatings in boiler steel applications. *Materials Today: Proceedings, 50*, pp.1442–1446.
49. Singh, P., Vasudev, H. and Bansal, A., 2024. Effect of post-heat treatment on the microstructural, mechanical, and bioactivity behavior of the microwave-assisted alumina-reinforced hydroxyapatite cladding. *Proceedings of the Institution of Mechanical Engineers, Part E: Journal of Process Mechanical Engineering, 238*(3), pp.1060–1071.
50. Singh, S.P., Verma, P.K., Singh, S., Kapoor, M. and Balu, R., 2024. Microstructural characterization and electrochemical behaviour of HVOF sprayed WC-Co-Cr coatings on 316L boiler steel stainless steel. *Results in Surfaces and Interfaces, 16*, p.100248.
51. Verma, P.K., Minhas, A.S., Singh, P. and Kumar, S., 2024. Slurry erosion behaviour of thermal sprayed Al2O3 and Cr2O3 coatings for turbine steels. *AIP Conference Proceedings, 2986*(1), p.020012.
52. Verma, P.K., Kumar, V. and Vasudev, H., 2024. Bioactivity and corrosion analysis of thermally sprayed hydroxyapatite based coatings. *Journal of Electrochemical Science and Engineering, 14*(4), pp.419–439.
53. Verma, P.K., Vasudev, H. and Kumar, V., 2025. Biocompatibility and corrosion performance of HA-based LVOF sprayed and post-annealed coatings on SS 254 surgical-grade bio-implant material. *Ceramics International, 51*(10), pp.13207–13225.
54. Verma, P.K., Vasudev, H. and Singh, G., 2025. Improving the corrosion performance of SS 304 stainless steel using HVOF-sprayed WC-Co-Cr coatings in chloride environments. *Journal of Bio-and Tribo-Corrosion, 11*(4), pp.1–13.

Index

D

E

F

I

J

L

M

S

T

U

V

W

X

Y

Z

For Product Safety Concerns and Information please contact our EU representative GPSR@taylorandfrancis.com
Taylor & Francis Verlag GmbH, Kaufingerstraße 24, 80331 München, Germany

www.ingramcontent.com/pod-product-compliance
Ingram Content Group UK Ltd.
Pitfield, Milton Keynes, MK11 3LW, UK
UKHW020156260926
475702UK00047B/474
9781041220985